Re-Thinking the Cogito

AF693440

Continuum Studies in Philosophy
Series Editor: James Fieser, University of Tennessee at Martin, USA

Continuum Studies in Philosophy is a major monograph series from Continuum. The series features first-class scholarly research monographs across the whole field of philosophy. Each work makes a major contribution to the field of philosophical research.

Aesthetic in Kant, James Kirwan
Analytic Philosophy: The History of an Illusion, Aaron Preston
Aquinas and the Ship of Theseus, Christopher Brown
Augustine and Roman Virtue, Brian Harding
The Challenge of Relativism, Patrick Phillips
Demands of Taste in Kant's Aesthetics, Brent Kalar
Descartes and the Metaphysics of Human Nature, Justin Skirry
Descartes' Theory of Ideas, David Clemenson
Dialectic of Romanticism, Peter Murphy and David Roberts
Duns Scotus and the Problem of Universals, Todd Bates
Hegel's Philosophy of Language, Jim Vernon
Hegel's Philosophy of Right, David James
Hegel's Theory of Recognition, Sybol S.C. Anderson
The History of Intentionality, Ryan Hickerson
Kantian Deeds, Henrik Jøker Bjerre
Kierkegaard, Metaphysics and Political Theory, Alison Assiter
Kierkegaard's Analysis of Radical Evil, David A. Roberts
Leibniz Re-interpreted, Lloyd Strickland
Metaphysics and the End of Philosophy, HO Mounce
Nietzsche and the Greeks, Dale Wilkerson
Origins of Analytic Philosophy, Delbert Reed
Philosophy of Miracles, David Corner
Platonism, Music and the Listener's Share, Christopher Norris
Popper's Theory of Science, Carlos Garcia
Postanalytic and Metacontinental, edited by Jack Reynolds, James Chase, James Williams, and Ed Mares
Rationality and Feminist Philosophy, Deborah K. Heikes
Role of God in Spinoza's Metaphysics, Sherry Deveaux
Rousseau and Radical Democracy, Kevin Inston
Rousseau and the Ethics of Virtue, James Delaney
Rousseau's Theory of Freedom, Matthew Simpson
Spinoza and the Stoics, Firmin DeBrabander
Spinoza's Radical Cartesian Mind, Tammy Nyden-Bullock
St. Augustine and the Theory of Just War, John Mark Mattox
St. Augustine of Hippo, R.W. Dyson
Thomas Aquinas & John Duns Scotus, Alex Hall
Tolerance and the Ethical Life, Andrew Fiala

Re-Thinking the Cogito

Naturalism, Reason and the Venture of Thought

Christopher Norris

continuum

Continuum International Publishing Group
The Tower Building
11 York Road
London SE1 7NX

80 Maiden Lane
Suite 704
New York NY 10038

www.continuumbooks.com

© Christopher Norris, 2010

First published 2010
Paperback edition first published 2012

All rights reserved. No part of this publication may be reproduced or transmitted in any form or by any means, electronic or mechanical, including photocopying, recording, or any information storage or retrieval system, without prior permission in writing from the publishers.

British Library Cataloguing-in-Publication Data
A catalogue record for this book is available from the British Library.

ISBN: HB: 978-1-4411-7154-2
PB: 978-1-4411-1821-9

Library of Congress Cataloging-in-Publication Data
A catalog record for this book is available from the Library of Congress

Typeset by Newgen Imaging Systems Pvt Ltd, Chennai, India
Printed and bound in Great Britain

For Shelley Campbell

Contents

Acknowledgements

I am especially grateful for all sorts of reasons to Tim Andrews, Robin Attfield, Reg Coates, Ray Davies, Dave Hilldrup, Douglas Houston, Dave Hume, Dick James, Wendy Lewis, Vesna Main, Gordon May, John Mealing, Rhian Rattray, Alison Scott-Baumann, David Skilton, Beatty Smith, Rob and Helen Stradling, Alison Venables, and Barry Wilkins; to Laurence Peddle, Helen and Alice (not to mention Tilly) for their great kindness in a difficult time; and once again to my colleagues, friends and postgraduate students (overlapping categories, needless to say) in the Philosophy Section at Cardiff University. Jonathan Webber gave me reason to re-think a couple of crucial points through his several clear-headed interventions in seminar debate at just the right stage of my work in progress on this book. Aside from regular points of contact such as our weekly staff and postgraduate discussion groups there were several conferences, seminars and other invitations farther afield where I had a chance to air some relevant ideas and benefit from various well-informed comments and criticisms. For these opportunities I should like to thank (among others) Gary Cape, Harry Cowan, Steven Craig, Clive Cazeaux, Alan Norrie, Manuel Barbeito Varela, Mario von der Ruhr, and Friedel Weinert.

Let me mention – so as to pre-empt the charge of covert revisionism – that this book marks a fairly sizable change of mind with respect to some of the arguments advanced in previous (including several quite recent) works of mine such as *Language, Logic and Epistemology: A Modal-realist Approach* (London: Macmillan, 2004) and *On Truth and Meaning: Language, Logic and the Grounds of Belief* (London: Continuum, 2006). In particular, I now take a more robust line on the need for a thoroughly naturalistic treatment of issues in epistemology, ethics, and philosophy of mind, along with a more sanguine view of the prospects for reconciling full-strength naturalism with an adequate normative, i.e., reason-based or rational-evaluative approach to those same issues. Where I once thought that the normativity requirement could not be met by any of the full-strength naturalisms currently on offer and thus counted strongly against them I am now persuaded – with a few reservations flagged up along the way – that they offer our last, best hope of escaping the dead-end epistemological predicament for which present-day toilers in both main camps, 'analytic' and 'continental' alike, have Descartes and Kant chiefly to thank. We had much better do what philosophers (most philosophers at any rate) seem to find exceptionally difficult, that is, set aside all the lingering qualms that result from an attachment to some more-or-less qualified version of that same dualist paradigm and push right through with the naturalizing project that is well under way in some areas of their own discipline and, more consistently, in other branches of the social and human sciences. Of the various people who helped to move my thinking in this direction – whether through discussion or by offering me space to develop my ideas in print – let me thank in particular Matthew Beaumont, Richard Corrigan, Filip Grgic, Ruth Groff, Mervyn Hartwig, Dan Latimer, Joseph Margolis, Jeffrey Perl, Tom Rockmore, Nick Turnbull, Dmitri Vardoulakis, Jeremy Wisniewski, and Pete Wolfendale.

This will be the sixth book of mine that Sarah Campbell has had a large hand in commissioning and steering through to publication with Continuum, so my thanks go to her yet again and also to David Avital for their constant supply of encouragement and guidance. As ever it is to Alison, Clare and Jenny that I am most deeply and lastingly indebted for their love, patience, kindness and (I very much hope) forgiveness during the period of its writing. Never were those qualities placed under greater strain and never did human beings respond with greater generosity. Let me finally express my gratitude to the editors and publishers of *The Polish Journal of Philosophy, Prolegomena, Revue Internationale de Philosophie,* and *Southern Humanities Review* for giving my arguments a first airing in their pages and also for permission to reproduce some of that material here in revised and expanded form.

Cardiff, November 2009

Introduction

Naturalism and/or Rationalism: Contexts of the Current Debate

I

In this book, I shall argue that philosophy of mind in the mainstream-analytic, i.e., the dominant Anglo-American line of descent has very largely run out of steam or arrived at the point of having pretty much exhausted its always rather exiguous stock of proprietary concepts and categories. This state of exhaustion is everywhere apparent in the relentless narrowing of focus, the cult of technical (or pseudo-technical) expertise in fields where this is simply not called for, the obsessive dwelling on a handful of well-worn topics for debate, and the onward march of an academic culture given over to the almost exclusive pursuit of a 'research' agenda set entirely by its own self-absorbed interests and priorities. In fact one might fairly conclude from a trawl through the certified analytic house journals that this is a 'degenerating research-programme', in Imre Lakatos's sense of the term.[1] That is to say, it is an enterprise that has long outlasted its creative or productive phase, that has accrued a great number of unresolved problems, and has by now put up such a thick 'protective belt' of auxiliary hypotheses designed to ward off any threats to its own waning hegemony that it is hard to know what could properly count as refuting (or indeed corroborating) evidence.

More specifically, I shall put the case that this state of exhaustion has much to do with the resistance to philosophical naturalism in any robust or unqualified form. It has resulted at least in part from the sheer amount of intellectual energy and resourcefulness devoted either to arguing against naturalistic approaches in epistemology, ethics, cognitive psychology, and philosophy of mind or else to the search for some viable middle-ground position that would take the naturalist challenge partially on board while all the same retaining some elbow room for more traditional (crypto-Cartesian) conceptions of the mind/body 'relationship'. Of course there are exceptions to the general rule – work of a decidedly analytic character that none the less breaks new conceptual ground or at least opens up some novel perspective – but these are both rare and achieved very much despite and against the widespread pressure to conform.

What counts as a serious or constructive intervention, along with what counts as a more-or-less acceptable departure from the norm, is decided according to a range of criteria that no doubt vary in their detailed specification from one expert domain to another but all of which have a certain deep-laid (in fact discipline-constitutive) feature in common. Thus they typically suppose the proper role of philosophy to involve the analysis of pre-existent concepts, or the attempt to get clear about received ways of thinking and talking. In this view it is the business of philosophers to describe, contrast, elucidate, and (sometimes) refine or criticize those modes of thought without so far mistaking that role as to set themselves up in the different business of inventing or creating new concepts.

One main line of argument throughout this book will be that philosophy has sold itself short by accepting such a drastically scaled-down view of what it might and ought to achieve, that is, once it manages to throw off a burden of inherited but still very active beliefs or presuppositions concerning its proper role. Another line of argument – closely connected with that – has to do with the continued prominence of Cartesian or crypto-Cartesian commitments in the thought of philosophers who would most likely reject or profess themselves baffled by any such imputation. Where these two lines come together is in making the case that if only these thinkers could overcome their deep resistance to full-strength naturalism and recognize that in truth there is no deep discrepancy or ultimate conflict of aims between the naturalist and rationalist outlooks in philosophy of mind, epistemology and ethics, then there would be no need for the kinds of often highly subtle and resourceful but always inconclusive since endlessly problematical 'solutions' currently on offer. My own proposal – modest enough because I take it to be just a somewhat more forthright statement of the way things are going in some of the more fruitful regions of present-day debate – is that they henceforth switch their main focus of interest from an issue that has for too long exerted a well-nigh mesmeric power of fascination over thinkers in Descartes' wake to an issue, or rather a range of closely interrelated issues, that can promise far more in the way of productive and open-minded debate. Quite simply it is wrong – a false though perhaps understandable delusion – to suppose that all real and (especially) all major scientific or other notable advances in thought must ipso facto occur in the spotlight of conscious awareness. For if that were the case, then it is hard to conceive how thinking could ever break through those barriers of received, inherited, or taken-for-granted belief that tend to define what should count – at least for the mind in such a state of intently self-monitoring conscious activity – as duly accordant with this or that item of accredited truth.

Most likely what has obscured this fact about the workings of human intelligence is the idea that consciousness (indeed a heightened mode of consciousness) *must* be the precondition for any exercise of thought at its fullest creative or inventive stretch since, contrary to the claim I have just made, it is the restraining effect of unconscious, subliminal or unexamined beliefs that puts up the greatest resistance to change across the whole gamut of human intellectual endeavour. However, as I argue in Chapter 3, such notions go against the large amount of evidence offered by thinkers – mathematicians and physicists among them – who profess to have achieved some decisive breakthrough while undergoing mental states that ranged from profound sleep to dreaming, verbal/visual free association, or semi-conscious reverie.[2] Besides, there is good warrant from some of Freud's more discerning and philosophically acute readers to suppose that the unconscious, at least when conceived in adequate, i.e., non-'vulgar' psychoanalytic terms, can itself lay claim to powers of inventive or creative-exploratory thought that must sometimes be acknowledged to exceed the compass of conscious (let alone self-conscious) accessibility.[3] Without perhaps endorsing some of Lacan's more tortuous anti-Cartesian sallies – such as his quip that 'where I think "I think, therefore I am", that is where I do not think to think' – one can none the less take his authentically Freudian point that the conscious mind is not always (if ever) entirely master in its own house.[4] One can then go on to derive the implication that attends any such scaling-down of conscious vis-à-vis unconscious or preconscious capacities of mind. That is, it will then become evident that the latter have a far greater role than is normally supposed in the kinds of heterodox or innovative thinking that by their very nature surpass the explanatory scope of a purely apodictic or first-person-privileged phenomenological conception.

This in turn means distinguishing clearly between, on the one hand, *thought* as that which alone makes possible every kind of major scientific, philosophical, creative, ethical, or socio-political advance and, on the other hand, *consciousness* (locus of the so-called 'hard problem') as that which alone, so its adepts maintain, holds an answer to the deepest questions posed by our distinctively human nature.[5] In my view, the failure to make this distinction has led to a great deal of wasted or at any rate largely misdirected philosophical effort. Moreover, it has had the unfortunate effect – as I said earlier – of obscuring the extent to which thought can run ahead of conscious awareness or whatever falls within the present range of consciously accessible ideas and thereby achieve just the sorts of advance that must otherwise remain either deeply mysterious or downright inconceivable. Hence the oddity, to an analytic ear, of phrases such as 'conceptual creativity' or 'conceptual inventiveness', suggesting as they do that this might be a process whereby the conscious, self-conscious and reflective *cogito* somehow manages *both* to discover new regions of a given, objectively existent ideational terrain (as implied by the very term 'concept') *and* – in a radically constructivist or anti-realist sense – to produce new modes of conceptual envisioning out of its own creative-inventive resources.[6] That is to say, those phrases would appear to constitute a near oxymoron or at least the kind of abusive coupling that indicates the presence of a category-mistake and the need for some more careful policing of the relevant philosophic bounds.

Of course the contrast just mentioned, or something very like it, will be familiar enough to anyone who has followed the debate between realists and anti-realists in mathematics, philosophy of science, ethics, and other fields, especially as that debate has shaped up over the past three decades in response to Michael Dummett's large-scale project for re-casting it in strongly metaphysical but also – primarily – linguistic or logico-semantic terms.[7] However, I would argue that it is really just another of those pseudo-dilemmas or misconceived topics of dispute, along with the supposed conflict between naturalism and rationalism, which we can best get over by coming to see – to accept without qualm or qualification – that it does not and indeed cannot confront us with any kind of ultimate choice. Rather there is a basic sense in which rational procedures even at a relatively high level of abstraction *cannot but* be rooted in our various, more or less inventive ways of exploring a mind-independent reality that none the less responds to those investigative efforts by bearing out our theories, conjectures, and hypotheses to a greater or lesser extent.[8] It is here that the two main themes of my book come most closely into contact. For it seems to me that any promising solution to the one pseudo-dilemma, that of naturalism versus rationalism, will by the same token have a strong claim to resolve the other, i.e., the issue – implicitly a chief bone of contention between the 'analytic' and 'continental' schools – as to how philosophy can be both rigorous (or conceptually disciplined) and creative (able to produce fresh-minted, striking, or truly innovative concepts that point a way beyond existing cognitive or intellectual limits). More than that, any solution along such lines will help to discredit the idea put about by those on the more extreme wing of anti-analytic – as distinct from the milder brands of post-analytic – thought: namely, that the sole alternative to continuing in that old, scholastic or hidebound analytical way is to view philosophy as just another 'kind of writing' and one that does best in the way of open-ended metaphorical or narrative recreation of the self if it seeks to emulate the poets, novelists, or critical 'strong' misreaders such as Harold Bloom.[9]

II

Perhaps we can best come to see what is wrong with proposals like this by the simple expedient of replacing 'creative' by 'inventive', and then – perhaps keeping in mind its Latin root *invenire*, 'to hit upon, come across, discover, find out, aptly (inventively) devise' – taking the point that such advances very often first occur at a preconscious level or at any rate outside the supposedly spot-lit zone of Cartesian 'clear and distinct ideas'. That is, they come about not through some incomprehensible leap in the power of conscious or reflective understanding but rather through the capacity of thought to engage resourcefully with various kinds of challenge that tax its present powers of comprehension to the utmost and beyond, and which possess such a (literally) thought-provoking character precisely in virtue of their doing just that. Philosophers have lately made a strong start in this direction by casting doubt on the whole notion of privileged first-person epistemic access, along with the equally tenacious belief – none the less so for its having caused so much philosophic trouble – that consciousness should or could play a leading role in our most significant moments of advance across the entire range of human scientific and cultural achievement. Among them are the naturalizing thinkers whose work receives a good deal of commentary in the following pages and those, like Stephen Stich, who have drawn attention to the crucial role in our cognitive activity of the various 'subdoxastic' (i.e., dispositionally effective but consciously inaccessible) states, processes, or events which can be shown to enter into a great many of our everyday or specialized choices and judgements.[10]

Most striking about such arguments is the extent of their break not only with old-style or unreconstructed Cartesianism but also with the sundry compromise forms – among them those anatomized here – that can be shown to smuggle back all manner of Cartesian contraband under cover of a quasi-naturalist rhetoric. Indeed, it is a part of my case in this book that analytic philosophy has very often been marked by just such residual yet still potently mischief-creating traces of mind-body dualism despite its revisionist (or downright rejectionist) stance with regard to the entire heritage of thought within which Descartes routinely figures as the source of all our epistemological woes.[11] Moreover, that desire for a compromise settlement typically goes along with a kindred resistance to the very idea that philosophy might both discover and create – 'invent' in just the way that the word's etymology so nicely suggests – possibilities of thought that exceed the compass of established or pre-existent concepts. That is, it fails to recognize the wide variety of preconscious, epistemically opaque, or (in the nowadays fashionable idiom) subdoxastic routes by which thinking can arrive at a breakthrough stage in some given project of enquiry. I have not made regular use of this latter notion – despite its fitting rather well, in some respects, with my own line of argument – because another chief aim of Stich and others who share his 'experimental' approach to the problems of philosophy is drastically to curtail what they see as the absurdly inflated pretensions of 'armchair' or a priori philosophizing. That is, they have no time for any project that purports to come up with substantive (i.e., non-trivial or informative) results through an exercise of reason that admits no empirical or documentary (e.g., psychological) checks by way of corroborating evidence.[12] I shall argue, on the contrary, that there is and always must be room within any such naturalized approach for that integral component of right reason, whatever the precise specification, in the absence of which it would suffer from a chronic normative deficit.[13]

What is required in order to remedy this otherwise irremediable lack is a due acceptance of the role played in our processes of knowledge-acquisition not only by the kinds of logically regimented, e.g., hypothetico-deductive thinking that rationalists typically place centre-stage but also by the kinds of creative-intellectual advance that alone have the capacity to provide such thinking with its vital subject-matter. In addition, as Philip Kitcher very usefully insists, there is little hope of devising a plausible epistemology for the formal and physical sciences unless we build in an adequate account of the way that certain salient features of the topic-domain can be seen to permit, invite, or solicit certain answering modes of human cognitive response.[14] Here again, one can see how impoverished (or downright unworkable) is a rationalist outlook unanchored in the modes of our being in the natural world, but also how crucially defective is any naturalistic outlook that fails to recognize the full extent of the mind's creative powers, that is to say, its indispensably active role in the diverse procedures of conceptual exploration and inventive hypothesis-testing whereby significant progress most often comes about, whether in the formal or the natural sciences. Hence, in Chapters 4 and 5, my extended discussion of two representative thinkers – Spinoza and Alain Badiou – who must certainly be counted arch-rationalists as regards their explicit philosophical aims, commitments and priorities. Yet at the same time – or so I shall argue – their thinking bears witness to the curious but highly revealing fact that a rationalism pressed to the limits (and beyond) of consistent application will always at some point encounter the need to make sense of its worldly or historical placement as well as acknowledging, at least implicitly, the intimate tie between reason and the various enabling as well as limiting conditions of our physical or natural embodiment.[15]

This in turn leads on to a fuller recognition than most philosophers (as opposed to literary critics) would happily endorse of the extent to which the interpretation of canonical texts likewise involves a stretch of the intelligence – inventive, creative, yet none the less disciplined or rigorous for that – without which reading would amount to no more than an exercise in passively compliant thematic or logico-grammatical parsing.[16] No doubt there have been plenty of analytic types who went a good bit further in the critical-revisionist direction than would ever be countenanced by those of a more conservative or orthodox mind. Thus Bertrand Russell, in a founding text of this tradition, sees nothing wrong or intellectually presumptuous about proposing that ordinary (natural) language falls short in certain crucial respects, that it lacks sufficient clarity of sense or logico-semantic structure for truly philosophical purposes, and therefore that it needs to have that structure teased out through a process of conceptual analysis that renders its meanings perspicuous and hence amenable to reformulation in more precise, less ambiguous or misleading terms.[17] Then again, there are 'moderate revisionist' projects that seek to update old thinkers in the light of supposedly more adequate modern concepts and methods, along with those others – e.g., reconstructive readings in the mode of Strawsonian 'descriptive' (rather than prescriptive) metaphysics – that overtly reject any such agenda but which none the less manage to smuggle in some large shifts of emphasis and orientation.[18] However, as I have said, these departures from the norm with regard to one canon of orthodoxy – that which enjoins a due respect for authorial intent as arrived at through a likewise due respect for matters of historical-intellectual context – are often thought to pass muster in so far as they respect another, more important since distinctively philosophical (or analytic) norm, namely that of showing how those 'same' old ideas and arguments can be given an unexpected new lease of life through being systematically recast in this or that present-day conceptual idiom.[19] Where these approaches all converge – and where they find common ground with the otherwise very different outlook of Wittgensteinian or Austinian 'ordinary language' philosophy – is by taking it for granted that philosophers have their work cut out in providing a more perspicuous grasp of concepts that are already there for clarifying treatment, whether in the critical-reconstructive mode after Russell or, following Wittgenstein, with a view to therapeutically leading us back to the common sense wisdom enshrined in everyday talk.[20]

The fact of this convergence despite large differences of view in other respects is, I think, the main reason why most analytic philosophers would (and do) register a sense of unease, intellectual discomfort, or downright incredulity when confronted with claims like that of Gilles Deleuze that philosophy should indeed create or invent rather than merely describe and analyse concepts.[21] If the legacy of logical empiricism still lies heavy on the analytic enterprise despite all the vigorous post-Quinean efforts to shake it off then perhaps its most lasting (and damaging) effect has been to perpetuate the notion, at whatever preconscious level, that all philosophy should properly aspire to the condition of the analytic statement, or – failing that – to the kind of straightforwardly empirical warrant that was once thought to characterize scientific observation sentences.[22] Of course such ideas are now widely considered to have gone the way of old-style logical positivism, having suffered a long series of concerted assaults at the hands of objectors who have stressed (among other things) the self-undermining character of the verification-principle, the theory-laden nature of observation sentences, the under-determination of theory by evidence, and – a main plank in all these arguments – the hopelessness of any epistemic appeal that rests on our somehow having direct sensory-perceptual access to the world and must therefore be seen as still subscribing to the now much-derided 'myth of the given'.[23] Yet, although that way of thinking nowadays has few defenders in its overt or programmatic form it has none the less left so deep an imprint on the modus operandi of most philosophy in the analytic mainstream as to constitute something like a depth-grammar of that enterprise throughout its evolution over the past five or six decades and more. Thus there is still the dominant assumption that philosophy is (or at any rate should be) primarily devoted to the analysis of concepts, and moreover that well-formed concepts suitable for philosophic treatment are such as must always already have their place marked out or their role sufficiently defined in some existing language or context of enquiry, whether of an everyday ('ordinary') type or belonging to a relatively specialist, maybe scientific mode of discourse.

What unites the two main lines of analytical descent from Frege/Russell and Wittgenstein/Austin is their reluctance to conceive that it might be not only a proper or legitimate but a vital part of philosophy's vocation to create new concepts or invent new possibilities of conceptual exploration that take thinking beyond anything available to it in the way of extant or ready-made topics for debate. The chapters in this book are all intended to indicate paths beyond the chronic impasse (as I see it) of so much present-day mainstream discussion, an impasse whose origins go back far beyond the travails of logical positivism and can be seen to lie chiefly – leaving Descartes out just for once – in Kant's recourse to a representationalist theory of mind and knowledge. This eventually led, via the 'linguistic turn', to just about every current variation on the idea that truth must be epistemically constrained, i.e., cannot be conceived as surpassing the limits of human cognitive or intellectual grasp and therefore – since language itself must be thought of as marking the horizon of knowledge – that talk of truth had better be replaced by talk of the specific conditions under which some given utterance properly counts as making sense according to the prevalent norms or criteria of some given, whether everyday or specialized language-community. Hence, as I argue in Chapter 6, the convergence between certain prominent strains of 'continental' (mainly French-influenced) post-modernist thought and certain likewise prominent strains of 'post-analytic' philosophy which themselves have grown out of a deep-laid scepticism as regards any claim to knowledge or truth that would affect to transcend the formative conditions of its own cognitive, conceptual, linguistic, cultural, or other such jointly enabling and limiting contexts.

My own view, developed at length in what follows, is that only by adopting a radically different approach – one that establishes a firm distance from that whole chapter of developments – can philosophy hope to overcome the impasse and strike out in some truly promising new directions. There are two such alternative lines of thought that diverge very sharply as regards their leading premises and deepest philosophical commitments but which do have one major characteristic in common. That is, they both reject the representationalist paradigm, whether in its Kantian or present-day 'linguistified' form, and both try to break its hold by redirecting attention to conceptual resources that have been largely ignored on account of that long detour into regions that have now turned out to be inherently under-provided with the kind of sustenance that philosophy requires if it is not to become endlessly lost in disputes of its own often pointless engendering. One is the naturalized outlook that has emerged across various topic-domains – in particular epistemology, philosophy of science, philosophy of mind, and cognitive psychology – partly by way of a marked reaction against the sorts of limiting or dead-end development sketched in the preceding four paragraphs and partly by way of a positive conviction that philosophy can best get itself through and beyond them by aligning its interests more closely with those of the natural sciences.[24] Here I put the case for full-strength naturalism rather than the various halfway or compromise versions that have lately entered the field, a case that finds support (I maintain) in the problems and dilemmas to which these latter fall prey when their implications are consistently followed through. The other main alternative is a likewise fully fledged rationalism which doesn't accept the kind of limiting condition – the restriction to judgments arrived at by bringing sensuous intuitions under adequate concepts – that Kant imposed on what he saw as the unfortunate tendency of pure (speculative) reason to create all manner of puzzles and perplexities through its striving to soar into regions of abstract thought beyond any such proper and needful restraint.[25]

My two chief exemplars of the high rationalist tradition are Spinoza and Alain Badiou, a contemporary thinker whose work I discuss in Chapter 5 and also, more briefly, toward the end of this introduction. Badiou is notable above all for his view of mathematics as the paradigm case of inventive-creative as well as rigorously logical thought, along with his insistence that 'mathematics is ontology' and also – following directly from that – his pursuit of this insight and its far-reaching consequences across a range of disciplines and subject-areas that would strike most analytic philosophers as ambitious to the point of intellectual abandon.[26] Although rationalism might seem the polar opposite of naturalism – at least the sort of uncompromising naturalism in question here – that appearance is deceptive in so far as it takes for granted the dualist outlook that is firmly rejected by naturalizers and rationalists alike, i.e., the idea that there is a gap to be filled or an epistemic divide to be got over between subject and object, mind and world, or representation and reality. That 'the order of thought is the order of things' is a Spinozist dictum which still strikes a radical, even scandalous note when set against all the concerted efforts of philosophers – from Descartes down – to fill that gap by some epistemological means even while effectively keeping it open or widening it yet further. This is one reason for Spinoza's prominent role here, especially in Chapter 4 where I examine various aspects of his modern reception-history, among them his current high standing as a thoroughgoing ontological monist and opponent of Cartesian dualism who pressed far toward some of the more scientifically-informed positions adopted by present-day cognitive psychologists and philosophers of mind.[27] Reviled during his lifetime and for two centuries thereafter as an atheist, materialist, heretic, and corrupter of morals he was yet held up – especially by the German and English Romantic philosopher-poets – as a 'god-intoxicated' pantheist mystic who sought to transcend all the bad antinomies bequeathed by thinkers like Descartes and Kant. Devoted philosophically to an order of truth *sub specie aeternitatis* – 'under the aspect of eternity', 'from a standpoint transcending all temporal or historical restrictions' – Spinoza was none the less very actively engaged in the turbulent political events of his time, adopting what was (by contemporary standards) an ultra-progressive or radical-left stance on issues like democracy, freedom of speech, and (above all) liberty of religious conscience.

It is on these grounds chiefly, that Spinoza has come in for such extremes of obloquy or vilification on the one hand and admiration or eulogy on the other. Both responses have tended to be reinforced by his heterodox views concerning the indissoluble union of mind and body, the failure of Descartes' attempt to establish their mutually exclusive character, and – following directly from that – the impossibility of conceiving the soul as enjoying some privileged (God-given) mode of existence outside and beyond the confines of physical, bodily, or temporal life. Hence Spinoza's remarkably modern conception of the mind as in some sense an 'idea of the body' and of the body as likewise an 'idea of the mind', neither of which can exist in isolation from the other. More than that, one is liable to mis-state the case even when arguing for a version of the mind/body or mind/brain identity thesis since the mere use of those terms – implying as it does that they have some distinct and well-defined sense – carries with it the risk of falling back into that same deep-laid illusion. Such was the error that Descartes expressed in an uncompromising, drastically dualist and thus highly problematical form. However – as Spinoza was quick to acknowledge – it was also one that most people (then as now) routinely accept in a common sense, intuitive, quasi-self-evident way. This is why he laid maximum stress on the claim that such intuitive or common sense 'knowledge' was very often no such thing but rather the result of our naively accepting the supposed self-evidence of the senses or the various received beliefs and ideologies that happened to prevail at some given time.

Spinoza's model here was basically that of the advancement of knowledge in the natural sciences, especially as these had developed in the wake of Copernican-Galilean astronomy. Such progress involved a willingness to cease relying on the witness of 'obvious', un-self-critical or unreflective sensory experience and instead give credit to the often highly counter-intuitive results of detailed empirical observation combined with the most advanced current techniques of mathematical analysis. This was the crucial stage of transition that occurred with the passage from Spinoza's 'first' to his 'second kind of knowledge', the former having to do with the realm of 'confused' or 'imaginary' (i.e., sense-based) ideas while the latter marked the advent of 'adequate ideas' that had undergone a process of conceptual review, rectification and critique. Philosophic thought should emulate the achievements of physical science and renounce any lingering attachment to a common sense or vulgar-empiricist outlook of passive trust in the deliverance of mere sensory-perceptual warrant. It could then bid fair to have achieved its own equivalent of the great revolution in physical science whereby post-Copernican astronomers no longer trusted in the sheer self-evidence that the Sun revolved around the earth since, after all, one could see it every day rising at dawn and sinking earthward in the evening.

III

Spinoza thus belongs squarely to the rationalist tradition in his confident belief that reason was inherently equipped – at least when exercised with due care and in the absence of powerful opposing forces – to arrive at non-self-evident truths through its own capacity for lucid, rigorous and self-critical thought. However he was not at all prone to under-estimate the strength or persistence of such opposing forces, and indeed devoted a good deal of his intellectual energies to explaining their nature and effects. For the most part they resulted either from the mind's natural tendency to fall back upon received ideas or common sense illusions as a buffer against anything that threatened the socio-cultural-intellectual status quo, or else from various well-entrenched systems of belief – chief among them religious dogma and political ideologies – which likewise worked to suppress any threat of a challenge to orthodox habits of thought. What Spinoza found most depressing or alarming about the situation in the Dutch Free Republic of his time was the fact that its hard-won liberties of thought and conscience – those that had once made it a (relatively) safe haven for free-thinking Jewish intellectuals like himself – were in serious danger of being undermined by the forces of resurgent religious fanaticism and sectarian political strife. Hence his pre-eminent place among the thinkers of a 'radical enlightenment' that was in many ways far ahead of its time and even – as celebrants would nowadays claim – abreast of current scientific and cultural developments that philosophy is still struggling to take on board.[28]

For some, like the neurologist Antonio Damasio, Spinoza was a great pioneer in recognizing the extent to which human emotions – whether the 'joyful', life-affirming and positive or the 'sad', depressive and negative affects – are inseparably bound up with both our physical-bodily and intellectual-cognitive modes of being.[29] For others, like Donald Davidson, he pointed the way toward a form of 'anomalous monism' whereby it is possible to assert that every mental event has its physical correlate (i.e., a corresponding brain-event) and yet that there exist no strict or exceptionless psycho-physical laws that would simply reduce the mental to the physical and hence entail a determinist doctrine which leaves no room for notions such as free will or moral responsibility.[30] This was one of the charges brought against Spinoza by his religious and political detractors, a charge often couched in the all-purpose defamatory language of 'atheism' or 'heresy'. Indeed it is the same kind of worry that is still very much in the air when philosophers nowadays strive to maintain some middle-ground position, like Davidson's, that would pay due regard to the findings of modern science – especially in areas such as neurophysiology and cognitive psychology – whilst yet leaving room, in principle at least, for the claims of free will or autonomous rational agency. Other versions of this currently popular stance include the idea that thoughts or items of phenomenal (sensory-perceptual) experience should be conceived as 'supervening on' the physical brain-states to which they correspond, or as somehow 'emergent from' those states in such a way that the physical is prerequisite to the mental while the latter is not wholly reducible to the former as hard-line physicalists or central-state materialists would hold.[31] Or again, the case is put in terms of 'token-token' rather than 'type-type' correlation so as to stress the open-ended variety of possible linkages or fits between physical and mental events, and thus – like Davidson – avoid any outright determinist or reductionist outcome.

Spinoza is often invoked in this context of debate although it is far from clear that he would have gone along with such halfway solutions to the various dilemmas bequeathed by Cartesian dualism. In present-day terms, he might also have looked very much askance at the various attempts to strike some moderate or middle-ground stance between a thoroughgoing naturalism premised on the causal closure of the natural world (brains and therefore minds included) under the laws of physics and a strong autonomist outlook premised on the fact – the supposed self-evidence – of human free will as a manifest exception to those laws and a sphere wherein reasons, rather than causes, are the ultimate court of appeal. In Chapter 1 I seek to diagnose the sources of the chronically becalmed, not to say stagnant condition of much current work in analytic epistemology and philosophy of mind. More specifically, I argue that the widespread turn toward various forms of conceptual and linguistic analysis has for the most part given rise to a somewhat reworked or semi-naturalized version of themes that go back via Kant to Descartes. Despite often claiming to have marked a decisive break with such 'metaphysical', i.e., misguided and self-deluding modes of thought the analytic project can in truth be seen to have inherited all the same problems and merely recast them in such a way as to disguise that deeper continuity. Those problems are posed with greatest force with regard to subject-areas where philosophy is – or at any rate feels itself – most acutely under threat from developments in other, especially scientific quarters. Most sensitive on this score are thinkers engaged with topics in epistemology and philosophy of mind since it is here above all that the perennial themes of philosophical reflection tend to feed back into the discipline's elective self-image as the place where consciousness comes most closely into contact with its own operative scope and limits as determined through an exercise of self-critique in the Kantian or phenomenological mode.

However – so I argue – that conception of philosophy's proper role has now run aground not only on various dilemmas intrinsic to the nature of any such project but also on the kinds of challenge put up by those other, 'non-philosophical' disciplines that might lay claim to possessing stronger credentials in this area. I put the case that philosophy will have nothing very useful to say about the 'hard problem' of consciousness or the distinct though closely related issue of free will versus determinism unless it adopts a consistent and thoroughgoing naturalist outlook fully responsive to developments in neuroscience and cognitive psychology. On the negative or diagnostic side I offer an account of the various means by which analytic thinkers have tried – and ultimately failed – to escape or circumvent the restrictions of that old Cartesian-Kantian dualist paradigm. These include, as I have said, a whole range of compromise notions such as 'supervenience', 'emergent properties', or 'anomalous monism', along with the distinction between humanly intelligible *reasons for* and physically efficacious *causes of* human action. In more constructive vein I argue that the best way forward is through a naturalistic yet normatively adequate approach that takes its bearings – if not its full stock of methodological commitments – from advances in neuroscience and cognitive psychology. As I have said, that approach might also find philosophical inspiration in

Spinoza's strikingly prescient working-out of a non-reductive yet uncompromising monism applied to the mind/body and free-will/determinism issues. Only by achieving a decisive break with the dualist paradigm can philosophy hope to move beyond this residual attachment to ideas that have survived mainly on account of their protean character but also on account of their undoubted appeal to certain deep-laid philosophic as well as everyday or common-sense intuitions. Yet the strength of that appeal is no less open to challenge from a different, scientifically-informed philosophical quarter than other such items of intuitive self-evidence – like the belief in a geocentric cosmos – that have proved a temporary (even if tenacious) obstacle to progress in various fields of enquiry. What is required is another 'Copernican revolution', one that goes beyond and directly against the exclusive focus on human epistemic capacities and limits that Kant connoted by his use of that phrase.

Chapter 2 develops the same line of argument by looking at some of the issues, problems and self-imposed dilemmas that emerge from Harry Frankfurt's well-known essay 'Freedom of the Will and the Concept of a Person'.[32] That essay has exerted a widespread influence on subsequent thinking in ethics and philosophy of mind, especially through its central idea of 'second-order' desires and volitions. These are taken to involve the capacity of human agents – at any rate those who have achieved a certain level of reflective and autonomous moral judgement – to endorse or to strengthen certain of their first-order wants, appetites, or inclinations and to treat others as falling below their own best standards of acceptability. In the latter case, so Frankfurt argues, the criterion of personhood – as distinct from mere human species membership – is precisely the possession of a moral will that enables such discriminations to be made and, moreover, to play a decisive role in the agent's process of ethical and intellectual self-formation. Thus persons, properly so called, are defined by contrast to that class of human beings whom he rather quaintly describes as 'wantons', that is to say, those who fall short of fully achieved humanity in the relevant respects since – for whatever reason – they are unable to assume responsibility for certain crucial or definitive life-choices, such as the choice to break or continue with a drug-taking habit. Frankfurt's approach thus promises a third-way solution to certain long-standing issues – chiefly those of free will versus determinism and the mind/body problem – that have up to now resisted the best efforts of philosophical deliverance or therapy. It looks very much like the kind of answer that would avoid the 'high priori road' of any Kantian or suchlike 'metaphysical' approach by adopting a broadly naturalized conception of human moral agency while not going so far down the path toward wholesale ethical naturalism as to lose the benefits (of personhood, choice, self-knowledge, and at any rate relative autonomy) that come with the Kantian conception.

However, I suggest that this appearance is deceptive and that Frankfurt's way of addressing these issues – especially his leading idea of second-order desires and volitions – lies open to a long-familiar range of objections from both a naturalist (anti-Kantian) and a strong autonomist (anti-naturalist) quarter. More specifically, I show that his notion of moral will as possessing a multiplex structure whereby higher-order volitions can reject or countermand the promptings of unregenerate first-order desire is one that must inherently give rise to various problems of a logical, metaphysical, and – most importantly in this context – ethical character. They are the same sorts of problem that typically arose with earlier attempts to carve out a space for freedom of will against the perceived encroachments of scientific determinism, while also prefiguring the difficulties faced by subsequent thinkers (such as John McDowell) who have likewise sought to resolve that issue through a quasi-Kantian and Wittgensteinian appeal to the 'second nature' embodied in our various acculturated practices and modes of thought.[33] Such compromise solutions will always end up by leaning (or collapsing) in one or the other direction, i.e., toward a form of cultural-linguistic determinism that dare not quite speak its name or an autonomist doctrine hedged around with so many de rigueur naturalizing caveats and qualifications as to render it largely nugatory. I conclude that a thoroughgoing naturalism is the only response that can meet the kinds of challenge increasingly mounted from various scientific quarters, notably those of neurophysiology and cognitive psychology. Frankfurt's essay will most likely go down as a striking but failed attempt to shore up the residue of those Kantian-autonomist defences against the encroachments of a naturalism assumed – wrongly, I believe – to offer no remotely adequate resources for conceiving of human beings in humanly acceptable terms. What it brings out with particular force is the regressive or circular character of all such rearguard strategies for heading off the perceived threat of a fully fledged (as distinct from 'anomalous' or qualified) monist outlook with regard to the mind/body or, as it is more often configured nowadays, the mind/brain relationship.

Indeed, this is not so much a 'relationship' in any philosophically defensible sense of that term but rather a matter of strict identity between the physical and mental domains, as opposed to the basically Cartesian dualist picture that has proved so tenacious – in a variety of scaled-down compromise forms – no doubt because it goes along with some of our deepest-laid intuitive or folk-psychological habits of thought. In Chapter 3 I take a basically deflationary rather than dismissive, reductionist, or downright eliminativist line as regards the 'hard problem' of consciousness. On this account the problem has acquired such special-case status as a result of two erroneous premises that have between them worked to skew the debate in various dead-end directions. The first is the residual Cartesian idea – still very active albeit rarely acknowledged as such – that thought finds its highest, most advanced or fully achieved and articulate form only at the level of conscious (perhaps self-conscious or reflective) awareness. The second, closely related to that, is the idea that disciplines like philosophy of mind, cognitive psychology, and epistemology must also be seen to meet this challenge since otherwise they will lack the kind of apodictic or self-validating warrant that is afforded by a lucid (i.e., fully conscious) grasp of what knowledge amounts to in those various fields of enquiry. Although many would nowadays reject such claims – not least as a result of concerted attacks on the notions of 'private language' or first-person privileged epistemic access – they are still clearly detectable in many quarters of debate.

Here I argue, conversely, that thought may very often range far beyond the limits of conscious or reflective understanding. Moreover, this *must* be the case wherever it is a question of some major new discovery or conceptual breakthrough – whether in the natural, the formal or the human sciences – which requires that thinking should achieve a decisive advance beyond previous, at the outset likewise conceptually hard-won even if thereafter deeply embedded ideas, suppositions or standing beliefs. In this respect my argument stakes its distance from a number of widely canvassed alternative positions, among them the 'new mysterian' claim that human beings are simply not intelligent enough – that they lack the conceptual wherewithal or sufficiently developed powers of thought – to resolve the 'hard problem' of consciousness in other than simplistic or hopelessly question-begging terms.[34] In my view, such defeatist doctrines merely turn a problem into a dead-end dilemma by assuming that if thought cannot achieve the wished-for state of transparent, self-intimating, conscious *rapport-à-soi* then it cannot have anything of value to say and had best remain silent on the topic. This can seem a sensible option only on the widespread though mistaken view – one that the sciences left behind as a precondition of achieving any progress beyond the naïve self-evidence of the senses – that the human perspective or 'view from inside' must have a privileged status. When reinforced by the commonplace appeal to qualia, phenomenal appearances or 'what it's like' to undergo some given mode of sensory-perceptual experience this amounts to a form of anthropocentrism which (at least from the standpoint adopted here) has no more rightful place in debates about thought and consciousness than in other, up to now more successful fields of investigation.[35]

IV

At this point – for reasons explained in the previous paragraph – I turn from naturalism to rationalism as the second of those two alternative paths that offer a promising route of escape from the dead-end predicament of much mainstream philosophic thought. Here my chief exemplar is Alain Badiou, a thinker whose deployment of mathematics – in particular, of concepts and procedures drawn from post-Cantorian set theory – is aimed very squarely against the sorts of restriction that Kant imposed by requiring that thought (more precisely: that 'pure' or speculative reason) be subject, at any rate for cognitive purposes, to the rule that sensuous intuitions be brought under adequate or correspondent concepts of understanding.[36] As I have said, that demand was at the very heart of Kant's epistemological turn and played a central role in his project to demote both ontology and purebred rationalism from their erstwhile high standing as disciplines or branches of the philosophic enterprise. Badiou re-asserts the capacity of thought in its jointly analytic and creative-exploratory role to achieve various kinds of epochal advance not only as regards mathematics, logic and the formal sciences but also in politics, ethics (despite his deep suspicion of any project bearing that name), and other regions of investigation where such thinking succeeds – albeit very rarely – in breaking through the limits hitherto imposed by some deep-laid consensus of judgement or belief. At the same time, and for the same reason, he reaffirms the primacy of ontological enquiry (here conceived *more mathematico*) as against both the widespread post-Kantian idea of epistemology as first philosophy and the subsequent turn toward language or discourse as the ultimate horizon of meaning, knowledge, and truth. For him, these purported revolutions in thought – whether Kant's or that proclaimed by proponents of the linguistic turn – are 'revolutions' only in the older, decidedly conservative sense best captured by the saying 'what goes around, comes around'. That is, they amount to an implicit endorsement of whatever *antecedently counts* as meaningful, correct, valid, warranted, knowledge-conducive, or ethico-politically justified by the lights of some existing belief-community or prior consensus of suitably qualified judgement.

Hence Badiou's claim that this whole long chapter of post-Kantian developments should in truth best be seen as a retrograde movement of thought, one which – especially in its Wittgensteinian or linguistic-communitarian guise – represents something like the sophists' revenge for the even longer history of snubs and put-downs that they have suffered at the hands of philosophers from Plato to the present. Yet with just a few exceptions his work has been ignored or routinely dismissed by Anglophone philosophers and discussed at any length – or with an adequate knowledge of his large and demanding oeuvre – only by cultural and critical theorists. This is unfortunate in various ways, not least because Badiou is himself a philosopher by training and avocation, and also because his thinking is some of the most resourceful, inventive and potentially fecund (as well as the most technically and conceptually demanding) to be found in present-day philosophy. Here I offer an account of that thinking with particular reference to Badiou's highly original work in the philosophy of mathematics and, more specifically, his exploration of those far-reaching ontological issues raised by developments in post-Cantorian set theory. His approach is more ambitious and adventurous than the kinds of discussion mostly carried on by analytically trained philosophers although – I should stress – none the less rigorous or mathematically accomplished for that. Indeed, when compared with most work in that 'other' (analytic) tradition, Badiou's shows a much higher degree of intellectual creativity as well as a far greater depth of engagement with the 'truth-procedures' (that is, the heuristics of problem-solving and paradox-resolution) that have typified the progress of set-theoretical methods and concepts.

I then go on to explain – again with a view to allaying suspicions in the analytic camp – how Badiou can make the seemingly unwarranted leap from philosophy of mathematics, via a set-theoretically grounded ontology, to questions (some of them urgently topical) in the socio-political and ethical domains. Most significant here is the cardinal distinction – as in the title of his major book *Being and Event* – between the realm of ontology as that which belongs to some existing or pre-constituted order of things and the realm of events or whatever transpires in such a way as radically to disrupt, transform, or revolutionize that existing order. It is by way of this distinction that Badiou is able to argue his case for mathematics as the basis for a critical ontology not only of the formal and physical sciences but also of those projects in the socio-political sphere where progressive or emancipatory thinking is likewise bound up with certain crucially enabling or formative as well as restrictive or constraining material factors. Thus, thinking on occasion runs up against some unforeseen (and strictly unforeseeable) occurrence that stretches its conceptual resources to the limit and beyond, or which finds no place in an extant ontology (i.e., a presumptively complete and definitive account of being in its various modes and aspects) based on the pre-established methods and procedures of some current stage in the history of some particular discipline or practice. Such is the 'event', as Badiou defines it: a term that should properly be reserved for just those kinds of properly epochal change – whether landmark discoveries in the formal or natural sciences, political revolutions, or radical transformations of ethical or artistic practice – which establish new terms for what henceforth counts as an instance of truth-apt, progressive, or good-faith commitment. Indeed, his own thinking is remarkable chiefly for taking so strong and principled a stand against just about every major direction of the present-day philosophic tide. Thus he has affirmed the absolute priority of ontology over epistemology, truth over knowledge, knowledge over the means or mode of its linguistic representation, and the claim of mathematics as *prima philosophia* over those of various currently favoured rival candidates.

In this connection – and so as to head off likely sceptical responses from the analytic quarter – I emphasize Badiou's distinctly heterodox but deeply informed and authoritative treatment of set theory and issues in the philosophy of mathematics. Here he strikes out in a direction far removed from the kinds of often typecast and deadlocked dispute (such as that between hard-line realists and anti-realists) that have typified recent analytical discussion.[37] Thus he sees the constant extension and refinement of set-theoretical concepts from Cantor to the present as an object-lesson in the way that knowledge typically accrues through successive encounters with a limit on its present-best powers of rational comprehension. That sense of falling-short, or presumptive anomaly, in turn provides the impetus that 'forces' thought (through a formal operation that Badiou lays out in precisely specified set-theoretical terms) to move beyond its current, restricted stage of development and achieve what was hitherto a strictly inconceivable advance. It is in this respect chiefly that his work provides such a striking exemplification of my central point in this book regarding the non-opposition (indeed the mutually supportive relationship) between naturalism and rationalism as philosophic outlooks whose presumed antagonism most likely comes of their making large and hence, it might seem, conflicting claims to our intellectual allegiance. On the widely-held incompatibilist view, rationalism involves a firm commitment to certain guiding or founding precepts – chief among them the priority of reasons over causes in matters of human understanding, self-knowledge, and moral conscience – which must entail a flat rejection of naturalism, given its (supposed) reductionist/determinist drive to preclude all notions of autonomous personhood or rational agency. Thus naturalism – at least in its full-strength versions – is taken to involve a more or less overt denial of the human capacity to engage in certain kinds of rationally oriented enquiry (including the various 'truth-procedures' that Badiou lays open to view) whose exercise requires the willed deployment of just such autonomous cognitive and intellectual powers. However, as I have said, this presumption has its source not so much in some deep-laid (maybe ineradicable) conflict between rationalism and naturalism but rather in the culture-specific history of thought from Descartes, via Kant to the present which has raised that issue – along with its close relative, the putative dilemma about free will and determinism – to a high point of philosophic puzzlement.

Of course any way of resolving, dissolving or dismissing these problems will itself be sure to manifest a certain preference in one or the other direction. Thus rationalists will say that the path to salvation lies (where else?) through a due acknowledgement of the extent to which rational or normative criteria always enter into our assessment of naturalized or causal-explanatory hypotheses, while naturalists will say that those criteria are grounded – at least to the extent that they possess good warrant – in properties concerning the natural domain (from the macro to the micro, brain-states included) over which those hypotheses range. Then again, some compatibilists of an ecumenical – not to say fudgy – disposition will venture to suggest how we can indeed have the best of both worlds without serious compromise by adopting a double-aspect theory – like Donald Davidson's 'anomalous monism' – according to which we can perfectly well find room for both reason-based (justificatory) grounds and causal-explanatory accounts just so long as we keep them safely apart and don't allow either to extend its claims beyond their own proper remit.[38] That is to say, there is a language characterized by talk of reasons, decisions, choices, commitments, justifications, etc., and another language whose currency is that of causes, dispositions, ambient stimuli, motivating factors, etc., such that the two may be thought to co-exist without any of the problems routinely brought up by incompatibilists on condition that they are used in different contexts accordant with their own distinctive (non-conflicting since wholly incommensurable) standards of valid application.

In truth this 'solution' is no such thing but more of an argumentative shuffle – or a handy means of evading the issue – with its source in the Wittgensteinian idea of valid judgements as having their truth-conditions fixed (more precisely:

their conditions of warranted assertibility set) by their agreed-upon role within this or that presently existing language-game, communal discourse, or cultural 'form of life'.[39] Thus it falls far short of anything that could satisfy a rationalist like Badiou, for whom it would figure (along with other variants of the linguistic turn) as just another instance of the modern retreat to an age-old sophistical ruse. Nor would it cut much ice with anyone of a strong naturalistic persuasion for whom it would typify the error of supposing that the task of explanation could ever be fulfilled – or, as Wittgenstein would have it, rendered otiose – by the mere appeal to what people normally incline to say in this or that acculturated context of utterance. That rationalists and naturalists have this much in common, i.e., their shared opposition to one prominent strand in present-day philosophic thought is of course a wholly negative ground of comparison and hence no sufficient justification for my claim that they can and should be viewed as mutually supportive – indeed reciprocally co-dependent – philosophical positions. What that claim requires if it is to stand up despite the dominant belief that they are somehow radically at odds is the positive argument pursued at greater length in the following chapters, i.e., that naturalism, so far from undermining the main tenets of a rationalist outlook, in fact provides the only adequate means by which rationalism can actually get a hold on the world. That is, it offers the sole point of purchase whereby rationalism can make good its claim to offer genuine, real-world applicable rather than merely notional or speculative knowledge.

No doubt it will be said that this is exactly what Kant aimed to achieve through his self-professed Copernican revolution in the history of philosophic thought, that is, his intent to set philosophy back on its epistemological feet by rejecting both the purebred rationalist way of ideas and the scepticism-inducing Humean empiricist premise that sensory experience is our sole source of knowledge in the absence of anything plausibly ascribable to innate or a priori sources.[40] However, the alliance of naturalism and rationalism envisaged here is one that very definitely doesn't go by way of that epistemological turn that Kant did so much to promote, and even less by way of the subsequent turn toward language, discourse, or representation which resulted from the scaling-down of Kant's more metaphysically loaded claims. Rather, it involves a naturalist outlook concerning the capacities of human reason which takes them to be firmly rooted in various aspects of our sentient dealings with the world that involve an irreducibly physical dimension, and which cannot be explained without some reference to the fact of our embodiment as creatures with a certain range of highly evolved perceptual and cognitive capacities including those that equip us for thinking at a high level of abstract conceptualization. By the same token, it involves a rationalist outlook according to which this naturalized approach cannot exclude the normative dimension – the appeal to certain likewise irreducible standards of right reason, logic, consistency, consilience, argumentative warrant, and so forth – in the absence of which we should be left with a species of raw naturalism wholly unable to justify claims with regard to our knowledge of the growth of knowledge or our power to exercise discriminative judgement in assessing truth-claims of various sorts.[41]

V

That these are very far from being merely 'technical' issues in epistemology and philosophy of mind should be evident from even the briefest reflection on those major changes in prevalent patterns of belief that have characterized our social, ethical, legal, and other institutional ways of life over the past (say) two centuries. Among the more striking recent advances in the history of moral thought is the idea that it is wrong – and in some way philosophically confused – to persecute people for holding heterodox or widely unacceptable beliefs.[42]

This advance may be claimed to have come about through the progressive secularization of moral values and the growing awareness that mere strength of personal conviction, especially in matters of religious faith, is no substitute for reasoned argument or open participant debate. Of course there are those – true believers or 'fundamentalists' of whatever persuasion – who reject this (to them) decadent way of thinking and maintain that anyone not sharing their faith is intellectually and morally at fault. They hold that beliefs should be arrived at through a deliberative exercise of will that one can either choose to make (thus gaining salvation) or fail to make (thus damnably rejecting the benefits on offer). From this point of view – call it 'doxastic voluntarism' – we are fully responsible for whatever we believe and can therefore be held fully to account for any errors, confusions, or failures of doctrinal commitment.[43] If beliefs are indeed volitional or subject to our powers of deliberative choice then there is no avoiding such ultimate responsibility and no taking the liberal line that people very often believe what they do through various kinds of cultural influence, formative background, ideological conditioning, and so forth. Still less can one argue – like many liberal reformers and opponents of religious persecution during the past four centuries – that it is precisely our right to this free exercise of moral and intellectual conscience that orthodox types would drastically curtail through their insistence on our all cleaving to the one true faith. For, as they see it, the only freedom that counts is the freedom either to think correctly and hence be confirmed in that faith or else to misapply one's belief-forming capacities and hence be confirmed in error.

Nor is this attitude by any means confined to the evangelical types who tend to arrive on one's doorstep at the most awkward times and promptly engage in a detailed exegesis of some blood-curdling passage from the Bible which proves that you had better believe what they say if you don't want to end up like one of those tormented characters in a Bosch painting. Peter Geach – a famously acute logician and philosopher of language – wrote an essay called 'On Worshipping the Right God' which went some more complicated ways around in reaching much the same conclusion.[44] This was basically a piece of Christian (Catholic) apologetics which drew on certain fairly technical arguments in philosophy of language in order to make its point, but which also pressed the point home by use of a neat (if tendentious) topical analogy. Thus Geach offered the example of a working-class voter canvassed during the run-up to a 1950s election who, when asked which party he would support, replied that he intended to vote Conservative since Harold Macmillan was a 'good union man'. However, what he hadn't grasped was that the word 'Unionist' in 'Conservative and Unionist Party' referred to the Tory commitment that Northern Ireland should remain a part of the United Kingdom, rather than to any trade-union links. So this poor deluded chap was destined to vote against his own perceived best interests through a deficiency of knowledge – or a misattribution of party-political creeds – that led him into grievous error. In the same way, Geach suggests, it is no use taking the kind of good-willed, vaguely ecumenical approach that would count it sufficient justification that somebody professed a sincere belief in God, no matter what their particular conception of the deity's nature and attributes. For it has to be the 'right', the one-and-only true God that they believe in, which for Geach means a deity defined precisely according to the tenets of Catholic faith. Otherwise the issue regarding their sincerity is quite beside the point since their object of belief is a pseudo-object – a false god – and their professions of faith altogether lacking in veridical content. Like the deluded voter they have failed to determine the correct range of defining criteria for the object-term in question, and have thus quite simply not referred to 'God' in any proper (theologically-warranted) sense of the word.

Geach argues his case partly on logico-linguistic grounds, that is, by appealing to Frege's cardinal distinction between 'sense' and 'reference'.[45] According to Frege, the 'sense' of an expression is its meaning as given by the various imputed properties, attributes, or distinctive features that we associate with it, while its 'reference' is the real-world object, person, or event that the expression serves to pick out. Thus 'sense determines reference' in so far as this picking-out can reliably or properly occur only on the basis of our having some prior knowledge – some grasp of the relevant identity-criteria – by which correctly to distinguish and denominate the object concerned. Also the Fregean theory claims to offer a solution to certain long-standing conceptual problems in philosophy of language. Among them is the puzzle as to how statements of identity between co-referring expressions – such as 'The Morning Star is identical with the Evening Star' – may not be merely tautological, i.e., empirically vacuous truths-of-definition but rather communicate genuine discoveries or items of information. What distinguishes cases like this ('A=B') from straightforward tautologies ('A=A') is the fact that they assert a referential equivalence between two expressions each of which has a different sense or range of descriptive criteria. Thus 'the Morning Star' and 'the Evening Star' both refer to the planet Venus even though this fact was once an astronomical discovery and might still come as news to any number of people deceived by celestial appearances and by the commonplace assumption that different names always and infallibly denote different objects. Hence Geach's argument with respect to the decisive, all-important matter of believing in 'the right God'. For it is just his contention – following Frege – that 'God' is not a proper name that directly picks out its intended referent but should rather be thought of as a definite description that entails certain highly specific identifying attributes. Thus it either succeeds in that purpose by getting the attributes right or fails altogether by getting them wrong and not managing to pick out anything at all. In which case, people who claim to believe in 'God' but who conceive him in terms other than those laid down by the dictates of orthodox faith are ex hypothesi sadly deluded and – in Geach's inquisitorial view – intellectually and morally so much at fault as to be headed straight for the eternal bonfire.

My point here is partly to illustrate the kind of dogmatism that results from the conjunction of a rigid stance on matters theological with a willingness to push right through with certain theses in philosophy of logic and language. However, it is also to emphasize the clash between any such hard-line conception of doxastic responsibility and that other, typically more 'liberal' or secularizing view which maintains that it is wrong – philosophically confused as well as morally objectionable – to blame other people for holding what we take to be false, misguided, infidel, heterodox, or downright obnoxious beliefs. On this view, doxastic commitment is by its very nature something that eludes the apportioning of praise or blame in so far as it results from a process of belief-formation over which we have little (some would say no) control since its sources very often go so far back and exert so strong a grip on our minds as to leave us no genuine choice in the matter. Such has been the argument of those – from the ancient Greek materialists to Spinoza, Shelley, John Stuart Mill, and present-day advocates of philosophical determinism like Ted Honderich – who reject the idea that anyone should be accounted morally (let alone criminally) at fault for holding some belief that conflicts with the dominant orthodoxy or local state of authorized best opinion.[46]

At this point, as we have seen, opponents often respond by invoking the distinction between *reasons* and *causes*, one that has been central to another tradition of philosophic thought whose chief representatives – whatever their otherwise large differences of view – include Plato, Descartes, Kant, and Wittgenstein. Thus, if we take beliefs to be a product of causal, cultural, or other kinds of antecedent conditioning rather than an outcome of responsibly exercised rational or moral choice then we shall treat their holders as mere creatures of circumstance lacking the most basic attributes of human free will and autonomy. To which they frequently add the claim that it is very much easier to adopt a determinist view of other people's actions, principles, and beliefs than it is to adopt the same attitude from a first-person standpoint. That is, we just *do* have the powerful conviction – whatever our philosophic views to the contrary – that in reaching some deeply-held belief or deciding on some principled course of action in consequence of that belief we are doing so not through causal compulsion but precisely as the upshot of rational reflection and autonomous moral choice.

Besides, it may be thought that determinism self-deconstructs in so far as any argument in its favour must be committed to a pair of downright contradictory theses, namely (1) that our beliefs are beyond our control since causally determined, and (2) that we should opt for determinism rather than free will since the former (unlike the latter) follows from an un-self-deluded exercise of rational intelligence. This point may seem to have considerable force if one looks at the kinds of argument put forward by philosophers who have opposed religious persecution or other forms of doctrinal intolerance by denying that beliefs are volitional and hence that anyone deserves either praise or blame on account of their doxastic commitments. It emerges most strikingly in Spinoza's case since he combined a thoroughgoing determinist creed whereby everything could ultimately be explained through the concatenations of cause and effect with the idea that true wisdom and a certain kind of freedom (i.e., the willing recognition of necessity) were attainable by those – and only those – who managed to achieve this hard-won knowledge.[47] In Marx also there is the well-known problem of explaining just how to reconcile a doctrine of economic determinism – even if, as Engels famously put it, one that might apply only 'in the last instance' – with a belief that human beings are capable of bettering their material and social condition through modes of consciously willed practical agency.[48]

Perhaps more telling is the point that thinkers like Mill and the poet Shelley advanced their case for freedom of conscience, especially in the religious sphere, on the ground that any penalties attached to non-compliance with the dictates of received belief were themselves morally, intellectually, and socially pernicious.[49] On the one hand, such penalties were grossly unjust in so far as they assumed a standpoint of outright doxastic voluntarism and hence failed to make due allowance for the extent to which beliefs are *not* volitional – not subject to change or revision at will – but rather a matter of what people think (what in some sense they are obliged or compelled to think) through a process of rational and moral deliberation. On the other hand, these thinkers maintain, it is wrong to penalize heterodox beliefs since this represents a drastic and unwarranted encroachment on the basic liberty of each individual to engage in that process to the best of their ability and thereby arrive at conclusions in keeping with their sense of intellectual and ethical justice. Yet there is a paradox here, as opponents are again quick to point out: namely, that the second line of argument presupposes just the kind and degree of doxastic freedom – of unconstrained choice as regards the exercise of those same rational capacities – which the first line of argument seems to close off.

Thus one finds Mill, in some of his most eloquent passages, attacking what he sees as the iniquities of Christian doctrine (especially the ideas of eternal damnation and of Christ's vicarious atonement for sin) on grounds of both their morally repugnant character and their being imposed upon credulous minds by force of orthodox fiat. Where his argument might appear distinctly strained is in charging every person with the responsibility to think for herself and not fall victim to such forms of institutional thought-control while also rejecting any punitive measures against non-believers or religious dissidents on the grounds that belief is not volitional and hence not properly subject to praise or blame. The locus classicus here is a famous passage in Mill's *Examination of Sir William Hamilton's Philosophy* where he asserts that 'I will call no being good who is not what I mean when I apply that epithet to my fellow creatures; and if such a being can sentence me to hell for not so calling him, to hell I will go'.[50] And again: 'a being who can create a race of men devoid of real freedom and inevitably foredoomed to be sinners, and then punish them for being what he has made them, may be omnipotent and various other things, but he is not what the English language has always intended by the adjective holy'.[51] This is all (in my opinion) extremely well said and goes to the heart of what's morally as well as intellectually wrong about orthodox Christian doctrine. Still it does bring out a deep-laid tension in the thinking of those – like Spinoza and Mill – who belong squarely to the great line of proto- or post-Enlightenment moral and socio-political thought. It is the tension between that liberal precept according to which beliefs are (at least to some extent) involuntary and should therefore not be subject to blame or punitive sanction and another, just as typically liberal precept which follows Kant's Horatian dictum *Sapere aude* – 'dare to know', 'think for yourself', 'have the courage of your own convictions' – and thus requires that people should indeed take responsibility for the full and free exercise of their own best judgement.

This is not to say (in the paradox-mongering fashion of some postmodernists) that the whole project of Enlightenment thought self-deconstructs on the impossibility of both thinking for yourself and obeying someone else's (e.g., Kant's) downright injunction to do just that.[52] Rather, it concerns the problems engendered by Kant's attempt to wrest a domain of ethical autonomy – of rationally exercised freedom of will under the dictates of moral law – from the otherwise implacably determinist order of physical cause and effect. First there is the problem that has become more acute with subsequent (post-Kantian) advances in scientific fields like neurophysiology and cognitive psychology. This is the question as to why – on what rational as opposed to fideist grounds – we should assume the existence of any such 'internal' or first-person privileged 'mental' realm that lies intrinsically beyond reach of causal explanation. Philosophers may argue that this is to miss the point since what counts as an adequate 'explanation' for the purposes of physical science, i.e., one couched in terms that reduce to some detailed specification of brain-states, neuronal firings, increased activity in this or that cerebral region, etc., cannot possibly count when it come to explaining our experiences, motives, meanings, beliefs, and other such altogether different goings-on.[53] Kant went a long and highly complicated way around in making this distinction whereas nowadays – after

Wittgenstein – it is more often made through a kind of cultural-linguistic policing operation. That is to say, it involves the much simpler expedient of declaring that causal explanations belong to one particular 'language-game' (that of the physical sciences) which has no place in the alternative language-games of ethics, theology, psychology, aesthetics, or any area of discourse where reasons – rather than causes – are the furthest one can get by way of justifying grounds.

However the argument then runs smack into a second problem, namely, that to have (or to think one has) some reason for adopting this or that belief or for acting on a certain principle is itself a decisive causal factor in determining the content of one's thoughts and beliefs or the nature of one's ethical acts and commitments. This may have to do with a wide range of predisposing attitudes or values, from those imbibed as a result of one's formative cultural background to those that have their source in some philosophic doctrine (such as Kantian deontological ethics) or some reactive tendency to challenge or reject any doctrine that claims one's allegiance. Nevertheless it can be argued – even (or especially) in the latter sorts of case – that despite the strong preference that most persons feel for regarding their beliefs as rationally justified rather than causally determined, still there is an ultimate, rock-bottom sense in which the reasons for so believing or acting are themselves capable of explanation in causal as distinct from purely rational or ethical-justificatory terms. For if the 'reasons' adduced are to possess any kind of genuine motivating force and not come down to mere products of whim or passing inclination then they will possess it in virtue of their playing a role in our deepest, most settled and firmly-held modes of belief. Yet to that extent precisely – so the argument runs – they will exert both a determined and a determining influence on everything we think and do, 'determined' (that is) in so far as they result from a range of anterior causal factors and 'determining' in so far as they then function as the springs and motives of our every last thought and deed.

Thus the vaunted distinction between reasons and causes – whether drawn in Kantian ('metaphysical') terms or, after Wittgenstein, with reference to the disparate 'language-games' of science and ethics – is one that can easily be turned right around so as to treat any talk of 'reasons' as just a shifty device for evading the unpalatable truth that our beliefs cannot be other than causally determined. Nor is there much comfort to be had for defenders of doxastic voluntarism in the appeal to certain currently modish ideas from the scientific domain such as quantum uncertainty and the limits of determinism as regards our knowledge of objects and events on the microphysical scale.[54] For arguments of this sort incur the double liability of (1) ignoring the simple fact that quantum phenomena are restricted to just that scale, and (2) vainly trying to vindicate the precepts of free will and rational autonomy by adducing 'evidence' which at most yields a notion of sheer unmotivated randomness. That is to say, any comfort gained from the idea that determinism doesn't go all the way down is more than offset by the unfortunate result that the conception of ourselves as rational agents with significant scope for ethical choice becomes just another product of wishful thinking. This tendency for arguments to bite back, so to speak, or flip straight over into some counter-argument with dire effects on their intended gist is evident in the various ambiguities or semantic slippages that often infect their language. Thus, according to context, 'determine' means either 'exert an exceptionless or law-like causal influence such that our actions and beliefs are necessarily beyond our willed or conscious control', or 'resolve to act always on just those beliefs – or for just those reasons – that we have arrived at through our own intellectual or moral volition'. However, as we have seen, these two apparently opposite or mutually exclusive meanings are in fact much harder to disentangle and indeed exemplify much of the confusion that has characterized this whole debate. For the quantum-based argument is just one example of the way that thinkers are apt to swing across from one to the other horn of this dilemma, i.e., from the bugbear of a thoroughgoing determinism which leaves no room for the exercise of autonomous choice to a reactive (wholesale indeterminist) doctrine which likewise empties that notion of any substantive or meaningful content.

VI

I have suggested that there is no answer to be had from Wittgensteinian or other such purported 'solutions' that would treat this as merely a pseudo-dilemma brought about by our failing to recognize the existence of different language-games for different contexts, those of causal explanation on the one hand and of rational belief-ascription on the other. This has been the prevalent approach among Anglophone thinkers since analytic philosophy embraced the idea that such problems can mostly be cleared up by examining our forms of linguistic usage, whether in order to diagnose confusion or else to assert (like Wittgenstein) that everything is alright with the way we ordinarily talk and that it is only philosophers who get into muddles by wrenching expressions out of their everyday linguistic and cultural contexts.[55] Still, it is an approach that will strike many people as leaving the problems very squarely in place since no amount of patient therapeutic coaxing-down from the supposed heights of 'metaphysical' abstraction can prevent our being tugged both ways on issues such as free will versus determinism, or again – closely related to that – doxastic voluntarism versus the claim that beliefs are not volitional just in so far as they are held on adequate (rational or principled) grounds.

Nor are those issues a source of disturbance only to ethical theorists who typically take a somewhat abstract, hence problem-inducing view of matters that might otherwise be of little concern. For they are likely to arise in a great range of contexts, among them debates about responsibility under law, where the issue often falls out between conservative upholders of autonomy and free will who reject all ideas of 'extenuating circumstance' or of causal (e.g., socio-cultural) factors beyond the individual's control, and advocates of a more-or-less qualified determinist outlook who regard such factors as crucial to any apportionment of criminal or moral blame.[56] They are also involved in the kinds of everyday dilemma that occur when people are confronted with some instance of morally repugnant belief for which they would unreservedly condemn the holder, were it not for certain facts about his or her upbringing, social situation, or formative experience that are felt to constitute at any rate a partially exonerating causal explanation. Thus, one may well deplore various forms of racial, ethnic, cultural, or religious prejudice while none the less seeing – perhaps by a difficult moral-imaginative stretch – how those convictions came to exert such a hold and license such bad behaviour. Less dramatically, it is the sort of allowance we make for people with irrational but fairly innocuous beliefs (like creationists, UFO-watchers, or doorstep bible-bashers) whose delusions very likely have their ultimate source in some causally relevant aspect or episode of their background history. Thus we shall often do best – adopt the most humane but also most ethically responsible approach – if we acknowledge that certain beliefs may not be within the holder's power to abandon or to modify at will, even through the utmost exercise of rational thought or reflective self-criticism.

Such is at any rate the argument of those who endorse some (however qualified) version of the social or psycho-physical determinist case. For them it is a definite sign of moral progress – rather than a symptom of our 'liberal', for which read 'degenerate' or 'decadent' times – that in matters of law as in matters of ethical judgement we have become more willing to accept the complexities of human motivation and the extent to which causal factors play a role in explaining (if not justifying) modes of conduct or belief that we find morally repugnant. These thinkers put it down to a refinement in our sense of natural justice, along with a greater depth of psychological insight, that the law should nowadays make increased allowance for 'diminished responsibility' or 'mitigating circumstance' as the kinds of consideration that properly apply when assessing the gravity of certain crimes. And conversely, they will view it as a sign of large-scale social and moral regression when judicial systems see fit to impose the very harshest penalties with minimal regard for such extenuating factors. Indeed, it is nothing short of barbaric, from this point of view, that society should ever sanction the practice of capital punishment, given what we have lately come to understand concerning the non-voluntary character of much human behaviour and the extent to which actions, like beliefs, are less at our conscious or deliberate command than is supposed by conservative policy-makers for whom such talk is itself just a symptom of latter-day moral decline. That this regression has gone so far in certain US states as to condone the execution – some would say judicial murder – of adolescents, mentally retarded offenders, and victims of massive social deprivation or sexual/emotional abuse is further reason (they might well argue) to count it a grievous affront to any civilized moral conscience.

Yet at this point the question surely arises: *by what standard* can we make such judgements and trust that others will agree or come around to a similar way of thinking? That is, how can we advance such claims unless through the appeal to an exercise of critical, enlightened, reflective, and ethically responsible thought which invokes precisely those values that are questioned – or assigned a more limited role – by the proponents of determinism in however nuanced or qualified a form? After all, is there is not something wrong with a moral or ethico-political outlook which makes such an effort to see all around other people's situations and allow for the various causal factors that should enter our moral reckoning while none the less implicitly claiming for itself – and requiring of other like-minded persons – a capacity for autonomous thought that admits no such limits on its power of rational justification? Hence the argument voiced by Kant and taken up by a good many present-day (mostly conservative) thinkers: that we offend against others by not holding them fully responsible for their actions, and thus not respecting their autonomy as rational agents in the way that we ourselves should wish to be respected.[57] For Kant, this basic precept of 'enlightened' thought was one which required not only that everyone be subject to the same, strictly non-negotiable moral imperatives but also that every miscreant should suffer a condign measure of retributive justice – including capital punishment – regardless of whatever extenuating factors might be instanced on their behalf. In so thinking, Kant was simply following out the logic of his own twofold requirement in matters of practical reason or ethical judgement. These were (1) the categorical imperative according to which we should always act on a principle or maxim such that we can will its universal adoption without running into moral dilemmas or downright logical contradictions, and (2) the precept that we should always treat other persons as autonomous agents or 'ends' in themselves rather than as offering a means toward the fulfilment of our own, self-interested aims or desires.

This is not the place for a detailed rehearsal of the problems with Kant's argument, among them – most famously – the fact that his absolute veto on lying (since if everyone lied then nobody would trust anyone and lying would become quite pointless) is a precept that could not be acted on without producing morally repugnant consequences in certain situations. My point is rather that Kantian ethical rigorism, combined with his absolute insistence on treating persons as autonomous moral agents quite apart from any merely 'pathological' (i.e., in Kant's sense heteronomous) desires, inclinations, or causal promptings, is an approach that leads to some dubious ethical as well as legal and socio-political conclusions. On the other hand, it cannot well be denied that determinism is apt to strike most people as an even more offensive doctrine when extended beyond its home domain of natural-scientific enquiry to the realm of moral agency and choice. After all, we very often do find reason to judge that certain beliefs are intrinsically bad and that their holders are at fault – morally to blame – on that account, whether through their passive, uncritical acceptance of pernicious (e.g., racist) doctrines or through what might be called the active false logic of xenophobia, sexism, religious prejudice and other such irrational (ideologically driven) habits of thought. Thus, for instance, we may be tempted to explain the morally repulsive mindset and violent behaviour of British National Party supporters in terms of their educational shortcomings, their socially blighted or underprivileged lives, their exposure to intensive racist propaganda, the extent of unemployment in their home region, or whatever. Yet beyond a certain point – the point where moral judgement comes into play – we are apt to set aside those mitigating factors, along with the outlook of psycho-social determinism which lends them a genuine if limited weight, and switch to the alternative, less charitable view that such beliefs and behaviour have to be condemned on consequentialist as well as on deontological (i.e., principled) grounds.

To this extent we are then adopting a version, albeit perhaps a qualified version, of the cardinal Kantian premise: that in order to respect other people's dignity as autonomous and thinking persons we must also hold them responsible for any failure to exercise those rational-deliberative powers to best, morally optimal effect. Where the qualifications typically come in – and where most ethicists would nowadays depart from Kant – is with regard to the various kinds of injustice or the forms of narrowly prescriptive, dogmatic, or context-insensitive judgement that can easily result from any too rigid application of that precept. Yet the basic idea of such moral autonomy, and hence of doxastic voluntarism, is one that we implicitly subscribe to whenever we feel ourselves justified in considering some belief (whether expressed in word or deed) to be ethically culpable, and whenever that judgement itself presupposes an appeal to certain shared (i.e., normative and inter-subjective) standards of evaluative warrant. So there is a sense in which Kant may be thought to have got it right concerning the ethical 'kingdom of ends' and the impossibility of claiming such warrant except on condition that we hold ourselves and others accountable to those same standards. Besides, if we once allow determinism to get a grip on our moral thinking then the way will be open to an attitude of *tout comprendre, c'est tout pardonner*, that is, the idea that the more we understand about other people's complications of motive, temperamental oddities, pressures of circumstance, etc., the more we shall incline to excuse or exonerate their various faults. In which case – so this argument runs – we will probably end up through straightforward parity of interest by granting ourselves the same kind of moral licence. Such was Kant's chief objection to what he saw as the inadequate (again, 'pathological') character of any ethical theory, like Hume's, based on nothing more than those shared inclinations of sympathy and fellow feeling that supposedly formed the basis of all our other-regarding sentiments. If this were sole ground of morality – one that entailed no respect for moral laws transcending the nature of our sociable instincts or mere creaturely needs – then, according to Kant, ethics must find itself in the same hopeless predicament as epistemology when subject to the rigours of Humean sceptical doubt. Only by acknowledging the legislative power of practical reason in the moral sphere and of conceptual understanding as concerns our grasp of objects and events in the empirical domain could philosophy break out of this sceptical dilemma, answer the 'scandal' of Hume's challenge, and make good its otherwise forfeited claim as a discourse of reason and truth.

As I have said, this led Kant to adopt an outlook of extreme moral rigorism which went so far toward discounting those Humean considerations – among them the interest in mutual tolerance and a due regard for our shared humanity, including its more instinctual, less rationally elevated aspects – as to find little resonance in present-day ethical thought. Thus one is tempted to echo Quine's pithy remark (in a somewhat different context) that 'the Humean condition is the human condition', and suggest that it bears just as much on issues of moral judgement as on epistemological matters.[58] That is to say, any reckoning of the former kind which fails to make due allowance for those various causal, instinctual, and socio-cultural factors that find no place on the Kantian account is sure to produce wrong judgements by drastically underrating the complexity of human situations and motives. Yet there is still the very basic, morally inescapable sense in which we *do* hold people worthy of praise or blame for what they believe and – even more – for those beliefs that they are willing to avouch or make manifest through various ethically committed deeds or actions. At the time of writing there is a group of highly organized, US-based and well-funded 'Christian Zionists' who are apt to stop you on the street and put forward a variety of arguments – always with detailed biblical warrant – for what I take to be some pretty obnoxious claims. Briefly put, their belief is that the Old Testament prophecies can only be fulfilled if the State of Israel extends its dominion over all those territories (including the post-1948 Palestinian homeland) to which it lays 'legitimate', i.e., biblically warranted claim, even if this involves the mass deportation or – as they will say if pressed – the mass slaughter of any who resist that claim. Such is the consummation devoutly to be wished by anyone who has read the true, i.e., the prophetic significance of recent Middle Eastern political events. For when this comes about, as infallibly foretold, then the Jews will undergo an instantaneous conversion to Christian faith, thereby fulfilling the message of the Gospels and marking the advent of a new dispensation when the elect will forthwith be 'ravished' to Heaven, leaving all the rest – doubters and infidels – to eternal perdition.

Here one has to choose between treating these people as basically harmless though gullible victims of intensive religious (or politico-theological) propaganda, and treating them as rational persons who could and should have subjected such beliefs to a process of rational scrutiny that revealed their thoroughly absurd, inhumane, and morally depraved character. Where the issue becomes more urgent is in the case of repugnant and irrational doctrines that produce actions of a similar kind, from terrorist atrocities such as we have witnessed with appalling regularity in recent years to state-sponsored wars of a strongly ideological and (very often) religiously motivated character. After all, as Aristotle was the first to recognize, there are certain forms of 'practical syllogism' which involve a conjunction of premises such that the most appropriate, indeed only adequate conclusion is one that issues in a consequential act or deed rather than another, logically entailed proposition or statement. That is to say – in the more up-to-date parlance of cognitive psychology – there is a tight link between the realm of propositional contents, attitudes, desires, etc., and the realm of purposive, real-world oriented action where those 'inner' goings-on are more-or-less effectively carried into practice. Therefore we can best, most reliably interpret other people's behaviour as the practical outcome of various beliefs and motivating interests whose content is no longer closed off in some shadowy realm of first-person privileged access but is open to a range of explanatory hypotheses with substantive causal as well as rational grounds.

Such cases may range from straightforward ascriptions of meaning and motive where the action concerned (whether deed or speech-act) leaves little room for doubt in this regard to more complicated instances of mixed or confused motivation where we are hard put to say whether – or just how far – the agent/speaker may be subject to causes or subliminal promptings beyond their conscious or rational grasp. Of course we are well practised at this sort of thing since it makes up a large part of our everyday communicative dealing with those whose intentions towards us and others we need to understand pretty much off the bat and without going through a lengthy process of conscious analytic or deliberative thought. What this approach helps to bring out is the complex interplay of reason-based and causal-explanatory modes of understanding that allows us to make these kinds of judgement. It does so basically by adopting the default strategy of counting other people rational and right in most matters – likelier than not to have some practical end or expressive purpose in view – and apt to set about achieving it by the best, most effective means to hand. Otherwise, if that strategy fails, then we often fall back on a predominantly causal rather than reason-based mode of explanation, i.e., one that interprets their beliefs and behaviour in terms of cultural conditioning, ingrained prejudice or various psycho-pathological factors.

I find this analysis persuasive in so far as it offers at any rate the promise of an answer to various difficult issues in (among other fields) philosophy of mind, language and action. On the other hand, it leaves us no nearer a solution – even, one might think, with more problems on our hands – if applied to the issue of doxaxtic voluntarism versus the claim that beliefs are not volitional, i.e., that in some sense we are compelled to believe what we do, whether through causal influences beyond our rational control or through our coming to acknowledge the force of certain (to us) wholly convincing arguments, reasons, or principles. This is where the determinist detects a crucial flaw in the case put up by defenders of free will who combine that thesis with a doctrine of human rational and moral autonomy. For, if that case is to carry much weight, then it will need to show not only that our thoughts, beliefs, and actions are 'free' in the basic non-determinist sense 'causally unconstrained' but also that such freedom is compatible with – indeed inseparable from – the willingness to acquiesce in those beliefs that strike us as possessing rational warrant or superior moral authority. However, so the hard-line determinist will argue, this means that we are just exchanging one kind of unfreedom for another, or escaping the trap of causal determinism only to run straight into the other, less obvious but (from the autonomist's viewpoint) even more constrictive and insidious trap of having our beliefs and actions dictated by force of rational conviction. Certainly there would seem little scope for the exercise of a genuine, as opposed to merely notional kind of freedom if the choice falls out (as this argument would have it) between, on the hand, a random, rationally unmotivated way of believing or acting and, on the other, an acceptance of reasons and principles that are thought to exert a compelling power over one's modes of belief-formation and practical-deliberative thought.

VII

As I have said, this is one of the main problems with arguments for free will that derive comfort and (supposed) philosophical support from those elements of indeterminacy that are taken to characterize 'objects' and 'events' on the microphysical scale. However it is also a problem for those who acknowledge that any freedom devoid of rational-justificatory warrant is a 'freedom' not worth having but who recognize the force of that dilemma wished upon them by opponents who adopt the above-mentioned line of argument. My point in all this is not to take the determinist's side and maintain that free will is just an illusion – perhaps a necessary illusion – by which we manage to shield ourselves from knowledge, i.e., the truth of determinism that would otherwise have morally, psychologically, and socially corrosive effects. Rather determinism is one of those doctrines, like scepticism about the 'external world' or the existence of other minds, which can gain a strong hold on our thinking in certain, distinctly philosophical frames of mind yet will always very soon run up against a host of opposing considerations. Among them – not least – are our strong belief to the contrary based on the experience of making up our minds in some matter of intellectual or moral conscience and, largely in consequence of this, our natural inclination – unless we still subscribe to some form of long-discredited behaviourist doctrine – to think of other people in kindred terms. All the same the old worry won't go away, especially for those who have kept abreast of the latest developments in neurophysiology and cognitive science. For their findings must surely give pause to anyone – die-hard Cartesians apart – who would wish to hold a firm, non-negotiable line against causal (as opposed to reason-based) explanations in philosophy of mind and action.

Also, to repeat, there is a strong case from the ethical standpoint for taking more account of causal factors in the process of belief-formation than is allowed by those staunch upholders of free will and moral autonomy who follow Kant in their conviction that any compromise on this point is an insult to human dignity.

To put it bluntly: such thinking leads straight on to iniquitous judicial practices like that which has filled up hundreds of US death-row cells with prisoners who, by any halfway civilized moral reckoning, would be counted unfit to plead or as victims of a massively unjust social order. Extreme cases, like hard cases in law, are not always the best source of guidance in such matters but they do help to focus the issue with uncommon clarity. Five years ago as I write, there was news of an atrocity which, it may be thought, called for nothing less than outright and unqualified moral condemnation, namely the mass-murder of hundreds of Russian school children along with their teachers, parents and grandparents by a group of largely Chechyen Islamist militants. In this context it might seem morally obtuse to suggest that 'explanations' of any kind – historical, ideological, religious, psycho-biographical, or whatever – might begin to get a hold on the overriding question as to how human beings could commit such crimes unless through some ultimate depravity of nature that defies our utmost powers of rational and ethical grasp. Yet it cannot be entirely beside the point to instance such factors as the brutally repressive war that had been waged by Russian forces against the population of Chechnya, the extent to which that war was perceived as an attack on Islamic values and traditions, the influence of a powerful national-separatist ideology that thrived on this religious doctrine, and – not least – the fact that the movement was led by a man whose entire family and many of whose closest relatives had been killed as a result of Russian attacks.

With a few notable exceptions – among them Jonathan Glover's fine book *Humanity: A Moral History of the Twentieth Century* – philosophers have tended to fight shy of testing their arguments against real-world events that pose the moral issue in such stark and philosophically intractable terms.[59] They have been inclined to keep 'metaphysical' questions such as that of free will versus determinism in a separate compartment from the kinds of debate that concern themselves directly with the rights and wrongs of particular historical, religious, or socio-political conflicts. To be sure, recent years have witnessed a marked reaction against the sorts of meta-ethical discussion that were dominant in Anglophone moral philosophy just a few decades back, and which have now given way to a growing interest in applied or practical ethics. Yet this work still tends to go on in relative isolation from, on the one hand, those broader metaphysical issues and, on the other, meta-ethical debates about the language, logic, or conceptual status of moral discourse in general. Nor perhaps is it surprising that this should be the case, given what we have seen of the problems that result when the issue is raised as to whether and just how far human agents can be held responsible for the nature and consequences of their beliefs. Still they are problems of a kind that philosophy must face up to – and without the common recourse to various modes of evasive or face-saving talk – if it is to play any useful and distinctive role in the wider debate on these issues. It seems to me that the supposed conflict between rationalism and naturalism is one major contributory factor here, and that it has done a good deal not only to block any really fresh thinking about the free-will/determinism issue but also to perpetuate and reinforce the above-mentioned division of labour among philosophers of ethics. All the more important, I would suggest, that we come to see it as just another product – albeit an exceptionally hard-wearing product – of that same old Cartesian legacy which has given us (among other vexing dualisms) the mind-body problem, the subject/object distinction, and the realist versus anti-realist dispute in its manifold present-day forms.

Moreover, these issues cannot be resolved in any remotely satisfactory way by adopting some midway stance – whether a response-dispositional approach in epistemology, a framework-relative conception of knowledge and truth, or a moderate naturalism designed to head off charges of rampant reductionism – that yields so much vital ground on both sides.[60] Thus there is little to be hoped from any partial naturalization of reason – or, for that matter, any kindred attempt to infuse naturalism with elements of a scaled-down rationalist philosophy of mind – if what results is a rationalism shorn of its most distinctive (indeed constitutive) features and a naturalism likewise deprived of its chief virtues, among them its anti-dualist outlook and its promise of an end to the misconceived dispute between causal and reason-based modes of accounting for thought and action. That we can have full-strength naturalism and full-strength rationalism without the least need for such compromise deals is a fact about human beings and their place in the world that follows directly from their nature as thinking animals, or – as the classical adage more precisely has it – from their peculiar status as creatures constitutionally 'capable of reason' rather than creatures somehow guaranteed to be 'rational' simply in virtue of that very nature.

This possibility could have been lost to view only through the curious effect of bifurcated vision brought about by the grip of Cartesian notions on so much of our everyday common-sense as well as our more specialized philosophical thinking. Indeed, one likely response to that last sentence of mine will be for readers to reflect 'How can he presume to make routine and standard use of the word "thinking", coming as it does with such a large burden of associated crypto-Cartesian ideas about the mind as a realm of private or inner goings-on?'. And of course they could make the same point about my usage of 'reflect' in the subsequent sentence where I imagine them raising that objection, as likewise to the verb 'imagine' which I used to very similar, as it might seem plainly self-refuting effect just now. However it is wrong – and another clear sign of the extent to which Cartesian assumptions have colonized so many quarters of the current debate on this topic – to suppose that such talk of 'thinking', 'reflecting', 'imagining', or any of their manifold analogues, even when deployed in the context of an overtly anti-Cartesian argument, must all the same involve some residual dualist commitment. After all, it has been among the chief lessons imparted by anti-dualists from Ryle to the present that we can carry on using such language – properly 'using' rather than merely 'mentioning' or decking out with queasy quote-marks – just so long as we accept that it denotes nothing like Descartes' disembodied *res cogitans* or that notion of the solitary, self-absorbed revolver of private thoughts famously captured in Rodin's sculpture *Le Penseur.* Even if we are less than fully persuaded by Ryle's specific recommendation, i.e., that we substitute adverbial qualifiers for substantive nouns – 'doing things thoughtfully, carefully, intelligently, imaginatively', and so forth, in place of their reified Cartesian nominal forms – still the general point holds good.[61] In short, there is nothing contradictory or inconsistent about taking a robustly anti-dualist and naturalistic line while continuing (since really we have little choice) to deploy a whole range of words, phrases, and idioms that possess an irreducibly intentional content and hence an inescapable reference to certain distinctive aspects of human experience that cannot be adequately described or conveyed in purely physicalist terms.

So the objector is off-beam in supposing that any use of a 'mentalist' or intentional vocabulary is enough to convict the user of harbouring residual or unavowed Cartesian beliefs. Of course I should myself be careful here and not too quick in pressing charges like this since a good proportion of this book is given over to showing in detail how that old Cartesian (or Cartesian-Kantian) legacy is still managing to set the agenda and influence – almost surreptitiously dictate – the terms on which these matters are discussed even by philosophers of an expressly naturalistic leaning. Where the difference lies, or so I would claim, is in the way their lingering attachment comes out as a refusal or at any rate a strong disinclination to press the naturalist case right through to its various, in their view philosophically disturbing conclusions. My point, on the contrary, is that such worries are wholly misplaced since in truth there is no reason – inverted Cartesian qualms apart – why the thoroughgoing naturalist should not be just as thoroughly committed to an intentionalist language of thinking, reflecting, meaning, intending, and so forth. Indeed it is the besetting vice of many debates in this area to set the terms or raise the stakes in such a starkly uncompromising way that any talk of that kind must be seen either as a striking vindication of the 'old', supposedly discredited dualist paradigm or else as an unfortunate backsliding into antiquated 'folk-psychological' beliefs. Yet of course the latter case cannot be made – would quite simply lack the most basic vocabulary including the word 'belief' and all its near-synonyms – unless by falling back on precisely such talk, or by going some tortuous (e.g., behaviourist) ways around in order to avoid it while often smuggling it back through the use of proxy or surrogate terms. However there is no need for all these credibility-stretching ploys and stratagems if one accepts the central thesis of my book: that naturalism and rationalism are perfectly compatible just so long as the naturalism finds room (as surely it must) for an adequate range of cognitive capacities or rational-discursive powers while the rationalism is duly purged of any Cartesian, that is to say, first-person-privileged, apodictic, or consciousness-based grounds of appeal. For if consciousness and thought are clearly distinguished, as they can be with reference to Descartes' writings other than the *Meditations*, then it becomes quite possible to take on board even a hard-line physicalist or central-state-materialist position like that adopted by the Churchlands – thinkers who would advise that we dump every last remnant of folk-psychology – and yet to suffer absolutely no embarrassment in using all manner of intentionalist or thought-related locutions.[62]

All the same, after a somewhat bullish introduction, I should perhaps do more to acknowledge not only the depth of resistance to naturalism of this sort among a great many philosophers but also how deeply that resistance goes into the self-image nurtured and defended by philosophy's advocates from Plato down, even where they took issue between themselves on just about every other count. Hence no doubt the corresponding degree of animosity shown toward those who challenged that image, from the ancient materialists to Hobbes, Spinoza, and nowadays (albeit in a less strident form) toward radical mind-brain reductionists such as Paul Churchland.[63] Some of those earlier thinkers – along with others in the same, radically heterodox line – were picked out by Louis Althusser in his late writing as figures who had stood squarely athwart the mainstream of Western philosophy through their adoption of an 'aleatory materialist' outlook that rejected all the standard conceptual props of idealism and (metaphysical) materialism alike.[64] What unites them – again despite some otherwise large differences of philosophic aim – is that their thinking represents a sizable challenge or a downright threat to that widespread consensus regarding the need for philosophy to carve out a certain elective domain within which its own practitioners can enjoy, perhaps to a higher since more self-conscious degree, the kind of intellectual and ethical autonomy that properly distinguishes human beings (or persons) from other sentient creatures and everything else besides. As we shall see in various contexts, most tellingly in Harry Frankfurt's attempt to reformulate this distinction as a matter of degrees or relative placements on a qualitative scale, the encounter with naturalism is one that imposes considerable strain on that particular, until very lately predominant idea of philosophy's accredited status or legitimate role. Accepting it at anything like full-strength, as opposed to the various scaled-down versions currently on offer, can feel very much like just giving up on philosophy and switching to one or other of those nowadays more naturalistically-oriented disciplines (from sociology to cognitive psychology) that have shown greater willingness to place due emphasis on the second term of their contested designation as social or human sciences.

That philosophers none the less can and should do so – *can* without any such affront to their intellectual dignity and *should* as a matter of retaining that dignity on respectable, scientifically conversant and humanly responsible terms – is the main point I seek to convey in this book and the message that I hope will be taken away by readers (most likely a majority) as yet undecided on the question.

Chapter 1

Living with Naturalism, Full-Strength: Why Philosophers Find It Hard

I

Since 'naturalism' is a term used in various contexts and in various context-related senses I had better start off by getting clear about what I shall mean by it here and throughout this book. The sort of naturalism that I have in mind is the sort that philosophers and thinkers in other disciplines refer to when they reject the idea that there is something very special, something exceptional or downright unique about human beings that cannot be accounted for in terms provided by any branch of the natural sciences. That is to say, it is the view that there is nothing about subject-areas such as epistemology, ethics, psychology, historiography, anthropology, political science, aesthetics, or literary criticism that places them intrinsically (rather than provisionally and temporarily) beyond the descriptive-explanatory compass of disciplines such as physics, chemistry, or biology.[1] Very often this goes along with the further, 'hard-core' naturalist or reductionist claim that in the end it will be physics that provides – or could in principle provide – all the answers worth having since physics has to do with the fundamental objects, properties, structures, dispositions, causal powers, and so forth, that must be thought to underlie and explain whatever belongs to those other, ipso facto less 'fundamental' regions of enquiry.

Such was the bottom-up 'unity of science' thesis which achieved its widest currency from the 1920s through the 1950s owing largely to the doctrines of logical positivism and its various, e.g., logical-empiricist successor, movements.[2] In this view the disciplines (or pseudo-disciplines) that occupied the upper reaches of that scale – from anthropology, sociology and psychology to ethics, aesthetics, literary criticism, and so forth – were redeemable only in so far as they sought to emulate the methods of the natural sciences and were else best treated as so many means for the conveyance of emotive, subjective, expressive or likewise non-cognitive (hence scientifically vacuous) meanings. In which case, clearly philosophy would have to give up its pretensions of epistemological, ethical, normative or rational-evaluative grandeur and henceforth accept something more like a Lockean 'under-labourer' role vis-à-vis the natural or physical sciences. This doctrine just as clearly had wider implications beyond the academic or specialist philosophical sphere, that is say, implications that bore decisively on certain very basic (though now strongly challenged) components of the hitherto prevailing human self-image with regard to such matters as rational and moral autonomy. For it made a full-scale programme of rejecting any principles, values, or reason-based as opposed to causal-explanatory criteria that would presume to lift human actions and beliefs above the inherently determinist realm of physical processes and events. Such was, at any rate, the strong or uncompromising naturalist programme espoused by the proponents of a robustly science-led conception of philosophy's present and future role. If it has more recently come under pressure from thinkers within the analytical tradition, then this has not been by way of a full-scale challenge to the science-led pecking order endorsed by those erstwhile confident proclaimers of the doctrine. Rather, as in W. V. Quine's hugely influential 'Two Dogmas of Empiricism', the attack has most often been directed toward a certain philosophic aspect of the positivist/empiricist case – namely its conceptual (logico-semantic) and epistemological framework – while continuing to honour physical science as its methodological lodestar, albeit a conception of physical science lacking those basic normative standards whose absence might well be thought to render that appeal highly problematic.[3]

It is fair to say that this 'post-analytic' movement of thought has kept up its naturalist subscriptions in so far as it holds out against those other, more linguistically or hermeneutically oriented approaches, whether influenced by Wittgenstein or Heidegger, that press much further in a strong-descriptivist direction.[4] On the other hand, Quine's express commitment to a science-led physicalist outlook is oddly compromised – even (as I have argued elsewhere) effectively undermined – through his adopting an outlook of radical meaning-holism and conceptual-scheme-relativism which, if taken at full strength, would remove any grounds for assessing scientific statements with respect to their empirical, theoretical, or causal-explanatory warrant.[5] According to him, theories are always underdetermined by the best empirical evidence to hand while that evidence is itself always theory-laden and hence incapable of deciding the issue except in accordance with some foregone doctrinal commitment or theoretical parti pris. I shall not here dwell on the various objections raised against Quine's arguments, most of which have to do with their drastic normative deficit and failure to explain how progress in knowledge could ever come about or – more to the point – ever be known to have come about.[6] What I do want to stress is the fact of his continuing to grant pride of place to the natural sciences with physics very much primus inter pares when it is a matter of laying down guidelines for the conduct of enquiry in other disciplines, philosophy included. Thus Quine remains adamant that philosophers had much better look to physics as the best, most advanced and reliable source of instruction for such purposes even though his doctrine of wholesale ontological relativism would, if acted on consistently, leave science (or at any rate philosophy of science) bereft of any principled basis for its claim to pre-eminence in that regard. Moreover, this provides the best – perhaps only – convenient handle with which to get a hold on the otherwise highly exception-prone distinction between 'analytic' and 'continental' thought as those terms figure in current philosophical usage. Where they have followed markedly different paths is with respect to the former's more pronounced (if technically diverse and far from universal) attachment to a naturalistic approach in matters of ontology, epistemology, and philosophy of mind.

This applies even to thinkers who have pushed pretty far with the 'linguistic turn', that is to say, with a language-first outlook that would seem to bring them out in broad agreement with the continental drift toward hermeneutically inspired notions of truth as a notion applicable only within certain culture-specific or language-relative contexts of understanding.[7] On the face of it, this claim is about as remote as possible from the naturalist premise that truth has primarily to do with certain salient features of physical reality, itself ex hypothesi taken to include all the relevant aspects of human conceptual, psychological, mental, social, and cultural experience. And indeed there is a continuing lively debate between those – the reductionists or hard-line physicalists – who would insist that such distinctions are merely a remnant of old, 'folk-psychological' habits of talk even when expressed in more refined philosophical terms and those who deny that any such programme could possibly be carried through while still purporting to explain or describe what goes on at the subjective, affective, perceptual, phenomenal, conscious, or self-conscious level.[8] Then again, there are philosophers who strive to articulate some form of compromise doctrine – such as Donald Davidson's 'anomalous monism' – that would seek to have it both ways and acknowledge the claims of the natural sciences, neurophysiology in particular, while rejecting any claim that mental/psychological events might reduce without remainder to the kind of explanation supposedly available from that quarter.[9] Still anyone who has followed these debates with an eye to developments on both sides of the 'analytic'/'continental' rift will most likely have registered this difference between, on the one hand, a prevalent mode of thought for which naturalism is always a contender (whether accepted or rejected) across the whole range of philosophical subject-areas and, on the other, a context where there is still the dominant presumption that the natural and the human or social sciences had better get along through a more-or-less amicable parcelling-out of their respective disciplinary domains.

This claim holds good even when – as so often – analytic philosophers point out various problems with philosophical naturalism, such as the regular failure (as they see it) to provide adequate criteria for distinguishing rational from irrational modes of thought, or warranted from unwarranted instances of inductive inference, or valid from invalid chains of deductive reasoning.[10] Above all this objection tends to come up in the sphere of ethical or socio-political debate where again it typically takes the form of an argument that naturalism cannot (so to speak) in its nature provide the kind of justificatory warrant that such arguments surely require if they are to rise above the status of value-neutral observations about human needs, desires, inclinations, aptitudes, proclivities, and so forth.[11] Still, there is a widely prevailing analytic assumption that any statement of the case against 'bald' naturalism will need to accommodate other, more moderate versions in so far as they point a way beyond the dilemmas of Cartesian dualism and its various likewise problematical latter-day descendants. All the more so, since this might offer the prospect of a reconciliation – at any rate some form of entente cordiale – with the physical sciences, especially with neurophysiology via the mediating offices of cognitive psychology. Hence the curious fact that philosophers from Quine to Davidson and Rorty have pressed their more or less extreme variants of the linguistic turn – the claim that language is in some sense the ultimate horizon of intelligibility – while at the same time managing to square that claim with the commitment to at any rate a nominal or face-value version of naturalist thinking.[12] What this shows is that a good many analytic (or post-analytic) philosophers feel a need to make common cause with the high-prestige discourse of the natural sciences no matter how strained the alliance might be when examined more closely or how acute the tension between scientific naturalism and any form of the language-first, strong-descriptivist, or discourse-constructivist doctrine.

My purpose here is not only to pursue these issues in recent philosophical debate but also to address the more general kinds of concern that have resulted from the deepening impact of a naturalistic or scientifically-informed world view on thinking about moral, social, and political issues. Not that these latter should be conceived as belonging to a realm quite apart from debates within epistemology, metaphysics, philosophy of language, or ethical theory such as tend to preoccupy many academic philosophers, especially (though not exclusively) within the mainstream-analytic line of descent. In fact, it seems to me that such apparently specialized or even (it might be thought) hyper-cultivated philosophic issues are very often within reach of worries and concerns that have a much wider present-day resonance. No doubt the philosophers' way of expressing those concerns will typically involve a technical idiom or mode of address that appears to inhabit a realm quite apart from the sorts of question about knowledge, experience, selfhood, freedom, agency, and moral responsibility that are apt to strike any reasonably well-informed person when absorbing the latest front-line reports from neurophysiology, cognitive science, or any such field where the claims of naturalism tend to conflict with traditional ideas. Thus, it is not hard to see what wider anxieties are finding a voice and, supposedly, a means of resolution when Davidson comes up with his account of 'anomalous monism' as a theory that occupies the sensible, non-dilemma-prone middle ground between hard-line physicalist reductionism on the one hand and, on the other, some toned-down variant of Cartesian dualism that presumes to disregard those scientific claims or at any rate discount them for its own distinctive purposes.[13] Indeed I would suggest that even the most technically complex and specialist debates within philosophy of mind, epistemology, or ethics are apt to touch such a sensitive nerve – or to prompt such wider and perhaps more urgently topical reflections – as soon as the naturalism issue comes into view.

This raises the further question, much discussed in the current philosophic literature, as to where we had best turn for guidance in such matters. If the answer here is 'best turn to the scientists' then of course it remains to be decided which precise branch or region of the natural sciences might offer the most hopeful prospect of illuminating insights. After all, if it is a matter of carving out some space for the experiential life-world as against the (supposedly) implacable laws of physical determinism then a lot will depend on the assigned order of priority between (say) physics and biology, or neurophysiology and cognitive science, or those rival sub-disciplines within a fiercely contested field such as sociobiology that accord different weightings to natural-scientific or causal-explanatory and rational-reconstructive or hermeneutic modes of enquiry. Then again, there is the scale that runs from a science-led (physicalist or central-state-materialist) position in philosophy of mind, to a qualified dualist or 'anomalous monist' theory of the Davidsonian sort, and thence – at the opposite extreme – to a 'folk-psychological' doctrine which maintains that all such revisionist theories are wholly beside the point since manifestly at odds with the wisdom enshrined in everyday or common-sense modes of thought.[14] Thus it is no longer a question of the old antagonism that ranged typecast science-and-technology champions against likewise typecast spokespersons for the humanities, mostly literary critics, and took them to represent well-defined positions on either side of the 'two cultures' controversy.[15] Rather it is a question of different, more or less sharply conflicting emphases within an intellectual culture that has come a long way since those debates of the 1960s and 70s (very often *dialogues des sourdes*) that presupposed not only the fact of that drastic rift but also the idea that one could plump straightforwardly for one or the other option.

II

What has changed since then is partly a matter of the extent to which naturalistic ideas have penetrated the other, i.e., the humanistic or 'literary' milieu and thereby created a situation where adopting anything like an overt anti-naturalist position is tantamount to rejecting science tout court, or running the risk of association with various present-day irrationalist creeds. Quite simply, it is felt, there is so much evidence piling up in various branches of the natural sciences – especially in disciplines like genetics, molecular biology, neurophysiology, and cognitive science but even in fields with a claim to authority on matters that were once the humanist's exclusive preserve – that one can only be displaying ignorance, dogmatism, or judgment-distorting ideological adherence if one presumes to doubt or challenge that authority on non-naturalistic grounds. Thus philosophical defences of concepts such as free will, autonomy, rationality, or creativity have three basic options with regard to this perceived invasion of their home turf by a dominant scientific world view. They can always adopt the line of least resistance, back away from any such outright confrontation, and avoid the whole issue by taking for granted its manifest irrelevance to their particular topic-domain or line of enquiry. Then again, they can opt to face the challenge – or face it down – by asserting (in Wittgensteinian fashion) that it is plainly misconceived and a product of the presently widespread confusion that mixes up the language-game of natural-scientific or causal explanation with that of assigning reasons, motives, or justificatory grounds.[16] Otherwise – presently the most favoured option – they can make terms with the naturalist world view, albeit in a nuanced or qualified form, while seeking to head off its more troublesome (i.e., reductionist, strong physicalist, or determinist) implications. This they claim to achieve by espousing some version of the double-aspect theory that allows them to preserve a conceptual space for such supposedly distinctive human attributes as those listed above without thereby coming into conflict, or at any rate flat-out conflict, with the claims of natural science.[17]

Among the latter, broadly compatibilist or conciliatory approaches are Davidson's idea of 'anomalous monism' and the various theories of emergence or supervenience that consider those attributes – along with others of a physically-based but (so it is argued) non-physically-reducible type – as belonging to a sui generis class subject to no such sharply dichotomous parcelling out of domains. However, the trouble with all these appeals to an alternative, supra-dualist mode of explanation is the fact that most often they are either couched in highly elusive, even evasive terms which resist any kind of substantive explication or else lean over in one or the other (physicalist or dualist) direction despite their claim to have emerged on the far side of any such false dilemma.[18] Thus it has to be asked of emergence/supervenience advocates, as likewise of the Davidsonian anomalous monist, just how their positions are supposed to avoid this endemic toppling tendency, or in just what respect their favoured solutions can be held to mark a decisive advance on more traditional, i.e., outright dualist or (less commonly) downright monist ways of framing the issue. After all, it is very far from obvious that merely devising a term or phrase that suggests some conceivable or earnestly wished-for means of transcending the dilemma is enough to actually break its hold, or transcend it as a matter of achieved philosophical grasp.

Of course this is by no means the first time that philosophers have claimed to overcome those vexing dualisms of subject and object, mind and nature, or free will versus determinism through a process of conceptual *Aufhebung* that subsumes them into some higher-order synthesis or moment of achieved transcendence. However, the advocates of a third-way alternative in analytic philosophy of mind or epistemology are not, it is fair to say, well disposed toward any such Hegelian deliverance from all our inherited problems. Nor indeed are they inclined to endorse any version of the Kantian attempt to represent those problems as merely the upshot of an overweening will-to-know on the part of pure (metaphysical-speculative) reason that can best be held in check by an exercise of disciplined critical thought designed to beat the bounds of legitimate, i.e., jointly empirical and conceptual sense.[19] Or rather: where thinkers of an analytic mind such as John McDowell or Robert Brandom recommend a return to Kant or Hegel in order to think through and beyond these dilemmas they do so – as I have argued elsewhere – through a Wittgenstein-influenced or broadly pragmatist conception of what properly should count as getting clear or thinking straight about philosophic problems and dilemmas.[20] The result of this, predictably enough, is very like the result of those attempts to resolve the issue between physicalism and mind/body dualism by invoking some idea like supervenience, emergence, or anomalous monism as a putative escape route from all such presumptively false or misconceived dichotomies. That is to say, it begs all the relevant questions as to how these notional 'solutions' are meant to go beyond the level of linguistic-conceptual reform to that of substantive philosophical problem-solving, or whether such basically redescriptive or therapeutic strategies can do the kind of work that is required of them if they are actually to meet the naturalist challenge, rather than adroitly sidestep the issue through an act of stipulative redefinition.

Nor could one expect such proposals to carry much weight with those for whom this is a real and urgent as opposed to merely technical or academic concern. They include not only philosophers strongly committed to a naturalist or anti-naturalist standpoint in epistemology, ethics, metaphysics, or philosophy of mind but also scientifically-informed laypersons for whom it may have a crucial bearing on their sense of personal identity, selfhood, or moral autonomy. Thus there exists at any rate a rough equivalence between the range of attitudes adopted by reflective but non-academically-trained individuals with regard to this sort of question and the various positions espoused by thinkers across the whole gamut of presently recognized philosophical positions. At the one extreme are those hard-line physicalist, central-state materialist, or eliminativist doctrines that confidently look forward to replacing 'folk-psychological' talk about beliefs, desires, meanings, intentions, and so forth with scientifically-informed talk about brain-states, neuronal firings, or observable links between various regions of synaptic activity.[21] At the opposite extreme are those, nowadays mainly reactive, movements of thought which incline to treat such ideas as just another instance of science (or scientism) pushing its claims far beyond their legitimate sphere. For them, quite simply, there is no making sense of the thesis that a physical specification of the type thus envisaged could possibly come close to describing, explaining or capturing the sorts of subjective or phenomenological experience that are captured so well – understandably enough, given their long-term adaptive pre-history – by 'folk-psychological' (intuitive or common-sense) modes of expression.[22] In between are those various compatibilist approaches that see no need for such a drastic polarization of views since, after all, human beings are physically embodied creatures as well as creatures for whom the self-evidence of conscious, subjective, or qualitative mind-states – states describable only in terms of 'what it's like' for sentient beings of just that kind – is something that has to be reckoned with on any adequate, i.e., humanly meaningful or intelligible account.[23] However, these ecumenical options often share the weakness of 'anomalous monism' and the kindred positions outlined in the preceeding paragraphs, i.e., that when one asks what precisely they amount to in specific philosophical (let alone scientific) terms one is left with the answer 'nothing very much save a verbal device for finessing the problem or a handy technique for temporarily pushing it out of sight'.

My point – and it bears repeating – is that these are not merely the sorts of issue that preoccupy certain present-day cognitive psychologists and philosophers of mind but, on the contrary, issues that are apt to strike anyone of a moderately reflective (call it 'philosophical') cast of mind who has made some effort to keep up with recent scientific developments. Moreover, they are questions that tend to arise with particular force and insistence when it is a matter of trying to get straight about the basics of what it is to be human, to take responsibility, to choose, to decide, to succeed (or fail) in one's projects or commitments, and so forth. It is the same with our thinking about other people's motives and actions when, as often happens, we are hard-pressed to judge whether they should rightly be credited (or blamed) for some particular course of conduct or thought of as deserving neither praise nor blame since they seem to have had little or no choice in the matter. Here it is worth remarking that we are generally more prone to extend allowances of this sort to other people than to ourselves, that is to say, more inclined to adopt a broadly determinist outlook when it comes to their virtues or vices than when faced with some cheering or dispiriting instance of ethically evaluable first-person conduct. That is to say, the 'view from inside' is one that makes it feel very much as though we are fully responsible for those various acts, good or bad, that issue from us as autonomous or freely willing agents, whereas with other people we are a lot more likely to be struck by the range of material, circumstantial, social, environmental, cultural, psychological and other such (supposedly) causal factors that may have conspired to make them behave that way.

Of course this contrast doesn't always hold good since there will always be those who find a means to excuse any amount of bad behaviour on their own part by overtly or tacitly invoking just such a range of presumed extenuating factors. Or again, there are those – on the whole more amiable – sorts of person who refuse to take credit where credit is due by reason of having enjoyed the right (i.e., virtue-enhancing or good-conduct-facilitating) kind of formative culture. To which one should add, for symmetry's sake, that the former self-exculpating types may very well be sticklers for the idea of absolute moral accountability when it comes to apportioning blame to other people, while the latter stern self-deniers of merit on their own ethical score may just as well lean over backwards to accord other people full credit for this or that instance of praiseworthy conduct. However, despite these odd temperamental exceptions, it is generally safe to say – on introspective as well as on hetero-phenomenological grounds – that individuals are prone to regard (or experience) their first-person doings and sayings as the upshot of autonomous, freely-willed acts of choice while more disposed to cast around for circumstantial or causal-explanatory factors when confronted with the doings and sayings of others. No doubt it may be claimed – and has been by a good many behaviourists, central-state materialists, debunkers of 'folk-psychological' talk, and determinists of various strength – that this is just the old Cartesian idea of privileged first-person epistemic access, or the deluded (though oddly tenacious) belief that knowledge of one's own presently occurring thoughts or mental states is a special since strictly indubitable, incorrigible, or intrinsically self-validating kind of knowledge. On the other hand that idea clearly goes so deep into our everyday or common sense as well as our more philosophical notions of the self and what constitutes its unique *rapport-à-soi* that no amount of such attempted re-education is likely to have much lasting effect. And if the neuroscientists fail to convince with almost daily announcements of some striking new advance in their detailed mapping of the brain and its micro-structures then philosophers have still less chance of succeeding through their efforts to correct or reform our commonsensically ingrained habits of talk. Thus their arguments are apt to create no more than a relatively specialist debating-point whether launched from a hard-line physicalist quarter or in the name of some other philosophical persuasion – nowadays most often with its proximate source in Wittgenstein – that adopts a somewhat milder tone yet also takes for granted the demise of Cartesian dualism.[24]

That compound adjective 'hetero-phenomenological', which I slipped in craftily without comment a few sentences back, was first introduced by Daniel Dennett in a related context of enquiry and is one which I think earns its philosophic keep despite being such a mouthful.[25] What he meant by it, briefly, was the sort of approach to issues in cognitive psychology and philosophy of mind that would accord well enough with his distinctive combination of a basically physicalist (i.e., science-led or neuro-physiologically informed) with a non-reductionist (i.e., subject-oriented, report-based or phenomenological) way of addressing those issues. Thus the reference to phenomenology signifies a willingness in principle to take the testimony of human subjects on board – to consult them and respect their intuitions where these can reasonably claim some degree of first-hand epistemic warrant – while the 'hetero'-prefix indicates Dennett's resolve to place firms limits on the scope of that privilege and not allow it to become anything like a Cartesian appeal to the *cogito* as solitary master in its own house. This phenomenology will be other-directed, that is to say, will take its investigative bearings not from the first-person thinking subject turned in upon its own processes of thought and delivering its certitudes solely on the strength of that presumed self-authorizing warrant but rather from a notion of the first-person subject as one among many, all of them potential sources of insight as regards the nature, structure and modalities of human experience and consciousness. In short, what Dennett is out to achieve here is an approach that would break with that mythic idea of the subject as sole and indisputable authority on its own inner states – that image of consciousness as a 'Cartesian theatre' where mental events pass before the mind's eye in constant lucid review – which has managed to exert such a powerful grip on the Western philosophical imagination from Descartes to Husserl and beyond. What he wishes to substitute for it is an alternative, still distinctly metaphoric but (he believes) phenomenologically as well as scientifically more accurate concept of the mind as joint compiler, editor and reader of 'multiple drafts' which undergo a constant process of revision with various topics periodically coming to the fore – inhabiting the spotlight of conscious awareness – and different formulations perpetually competing for pride of cognitive or epistemic place.

Thus his project has a strong claim to have appreciably shifted the terms of debate with regard to those long-standing issues that other philosophers had sought to resolve by the application of elusive or evasive notions like 'emergence' and 'supervenience'. Or again, Dennett's proposals have a far more substantive feel about them – and a sense of correspondingly greater problem-solving power – than the coining of a slogan such as Davidson's 'anomalous monism' whose suggestion of somehow having managed to resolve or transcend the paradox contained in their own phrasing is never cashed out in more specific philosophical terms. Were Dennett's approach indeed capable of matching these high hopes, then his idea of 'hetero-phenomenology' would have taken us a long way toward making better sense of those classic antinomies (the mind/body problem and free will versus determinism) and perhaps toward reducing, by helping to explain, those asymmetries between first-person and third-person perspectives on experience, choice and responsibility discussed earlier. Yet there is a curious sense – one familiar to anyone who has spent much time reading around in the rapidly proliferating literature devoted to the 'hard problem' of consciousness – that for all Dennett's undeniable powers as a wonderfully gifted hetero-phenomenological describer and theorist of his own and other people's manifold modes of conscious being-in-the-world he has somehow taken us no further toward bridging that troublesome gap. That is to say, his descriptions do an excellent job at both levels, the descriptive-evocative and (albeit with some disclaimers) the physical or neuro-scientific, but cannot – any more than previous treatments of the issue – provide even the remotest glimpse of how these two dimensions might be reconciled. What we are left with at the end of all Dennett's brilliantly ingenious, scientifically well-informed and philosophically refined (not to mention witty and conceptually inventive) variations on this theme is a range of undeniably acute insights on both sides of the Cartesian split that can make it seem even deeper and wider since so much intelligence has gone into his arguments for counting it a pseudo-problem.

III

Up to now I have been talking about two main issues – the mind/body (or mind/brain) question and the antinomy of free will and determinism – which philosophers for the most part tend to treat separately. That is, discussions of them tend to occur in different forums (conferences, seminars, books, journals, internet sites, and so forth) since the former is construed as belonging to such specialized subject-areas as epistemology, philosophy of mind, cognitive psychology, and maybe at a pinch neuroscience while the latter is taken to fall squarely within the field of ethics or – at even more of a pinch – neuroscientifically informed sociobiology. Nor is it by any means a hopeful sign of imminent or longer-term rapprochement that some would see the best prospect for both parties in accepting the pinch and ceasing to repine at the steady advancement of scientific knowledge into regions that were once thought immune from any such naturalistic (for which read reductionist or downright determinist) threat. On the contrary, it is just that prospect of a grand alliance between these currently high-profile scientific disciplines that has tended to generate a sharpened degree of alertness and a mustering of ever more resourceful counter-arguments among philosophers bent upon turning back the tide of a physicalist world view with – so it is assumed – deep-laid designs on our claims to autonomy, free will, and genuine selfhood. This desire to avoid a conflation of issues that might then present a more convenient target for wholesale reduction along naturalistic lines is perhaps one reason why philosophers have tended to keep a certain distance between the mind/brain and the free-will/determinism spheres of debate. All the same their connectedness is hard deny, at least as regards the role they have played in the post-Cartesian line of descent, whether among its direct inheritors on the continental-rationalist side or among those in the British empiricist camp who, as we are often reminded nowadays, shared a lot more with their rationalist counterparts – including the 'way of ideas' as a starting-point for epistemological reflection – than they might have wished to acknowledge.[26] At least it is pretty clear that if someone came up with a major advance (since full-scale 'solutions' are scarcely to be thought of) in the treatment of either issue then this would surely have large implications for our thinking about the other.

All of which leaves the question wide open as to whether philosophy holds out the promise of achieving any such advance, or – as it might more caustically be phrased – whether philosophers have anything of the least interest or value to contribute. At this stage it won't cut much ice for philosophers to say, as they might once have said with some conviction, that theirs after all is the discipline which takes thinking – and thinking about thinking – as its home-ground or chief topic domain, and hence that they possess at least good prima facie warrant for claiming such special expertise. Nor would that argument have been restricted to those in the Cartesian-rationalist tradition or believers in the mind's capacity to bear apodictic, self-evident, or a priori witness to its own nature and existence in and through the very act of thought. Empiricists might not have the philosophic means, or indeed the inclination, to assert anything so crucially dependent on that rationalist idea of privileged first-person epistemic access, or of consciousness as a perfectly transparent medium wherein the *cogito* could be assured of discovering clear and distinct ideas. Yet they could plausibly claim – again with a view to securing philosophy's distinctive entitlement to pronounce on such matters – that their epistemology was better equipped in that regard since it avoided any dubiously self-validating appeal to the subject as ultimate source and guarantor of knowledge, and replaced it with a far more sensible (since experience-based and strictly non-aprioristic) way of proceeding.

This might seem to give empiricists a clear argumentative edge over their rationalist opponents when it comes to issues – like determinism versus free will or physicalist monism versus its dualist or variously qualified (e.g., anomalous monist) variants – where empiricist claims are likelier to count for something in a largely science-dominated context of debate since they start out from a position having more in common with the scientific world view. After all, they can point to the historical fact that the natural sciences have often struck up a close and mutually supportive relationship with various forms of empiricist epistemology, from the ancient Greek atomists to those who found in Locke (as he likewise found in them) a sensible adherence to the scope and limits of empirical observation, and thence to various recent, philosophically updated but recognizably kindred modes of thought. To that extent they would appear better placed than the rationalists – at any rate purebred Cartesians – in so far as the basic empiricist conception of the mind and its relation to the various modalities of human sensory-perceptual experience is one that already incorporates a sizable measure of adjustment to natural-scientific methods and assumptions. However this claim has a drawback on account of its already conceding some crucial ground to those in the hard-line physicalist or central-state-materialist camp who would argue that philosophy no longer has anything of value to offer concerning such questions since they now belong squarely to the province of disciplines like neurophysiology and the most advanced (i.e., more 'scientific' and less 'philosophical') branches of cognitive psychology. For it is no great distance – as appears very plainly in Quine's landmark essay 'Two Dogmas of Empiricism' – from a naturalist stance that still allows some room for the exercise of human rational-evaluative powers as a source of normative warrant, to a wholesale physicalist stance whereby such notions are treated merely as symptoms of an obsolete folk-psychological or (worse still) a lingering philosophic-metaphysical attachment to outworn 'mentalist' notions.[27]

Hence the drastic tension in Quine's thinking between a normative conception of science implicitly based on the principle of inference to the best, most adequate rational or causal-explanatory account and a drastically reductive empiricist creed which, if carried right through, would leave science in pride of place but devoid of any such normative standards. Hence also the ease with which naturalism of this sort translates – as with Quine – into a fixed antipathy toward any talk of meaning, intention, belief, or other such (in his view) hopelessly ill-defined since irredeemably mentalist or psychologistic terms of address.[28] It is the same slide by which moderate, that is, compatibilist or avowedly mind-and-free-will-accommodating versions of the naturalist position tend to lean over – or collapse – into just the sorts of hard-line or intransigent physicalist doctrines that they had striven to avoid. So empiricist approaches may seem to offer the best prospect of rapprochement with the natural sciences while in fact yielding ground at precisely the point where that opposing, e.g., strong-reductionist or central-state-materialist thesis is apt to stake its most confident claims. This is nowhere more vividly exemplified than in Quine's falling back from a naturalized epistemology that could in principle find room for normative standards of epistemic warrant to a starkly reductive physicalist stance that allows nothing more in the way of such warrant than a crudely doctrinaire version of Skinnerian behaviourist or stimulus-response psychology.[29] By now there is a fairly wide agreement that Quine's kind of radical-empiricist approach suffers from a marked normative deficit, or a failure to explain the crucial difference between progressive, knowledge-conducive or truth-oriented episodes of scientific theory-change and those that turned out – perhaps in the fullness of Popperian falsificationist time – to have possessed no such virtues. Moreover that deficit on the epistemological side can be seen to go along with a likewise inadequate grasp of what distinguishes the kinds of thought-process that can typically be relied upon to produce such results from other ways of thinking that typically lead to flawed, partial, erroneous, or downright false states of belief.

Thus, it is far from clear that the problems in question have been brought any closer to solution by this widespread revolt against Cartesian aprioristic ideas in favour of a broadly empiricist outlook that would seem, on the face of it, much better suited to achieving a workable modus vivendi with the methods of a naturalized or science-led approach. To be sure, there have been various tactical alliances struck up and one could reasonably claim – from a sampling of the relevant literature – that 'empiricist' is a fair if somewhat rough and ready description of the stance in matters epistemological adopted by most present-day subscribers to a physicalist, central-state materialist, or eliminativist (anti-'folk'-psychological) conception of mind/brain identity. Nor is this at all surprising, given their shared genealogy in a movement of thought – more precisely, a recurrent and often reactive impulse – directed against the perceived liability of purebred philosophic reason (whether in its Platonist, Cartesian, Kantian, or Husserlian-phenomenological modes) to conjure up a realm of forms, essences, clear and distinct ideas, transcendentally derived a priori concepts, or eidetically grounded primordial intuitions wherein it would enjoy the privilege of guaranteed first-person epistemic access. It is largely through the will to have done with all such – as the critics maintain – delusory sources of apodictic support that philosophy of mind and cognitive psychology have lately taken so marked a turn toward empiricist, or at any rate strongly anti-rationalist, conceptions of the best way forward. On the other hand there is a growing sense – not only among those of a converse rationalist or mind/brain dualist persuasion but also among some adherents to a broadly naturalist standpoint – that empiricism suffers from certain inbuilt problems, including that gross normativity deficit, which call into question both its claims as a properly science-led epistemology and its warrant as an adequate basis for thinking about issues in philosophy of mind.

At its most direct, this would be the question as to how such thinking could itself make a start or achieve any worthwhile results if it were truly subject to the kind of constraint – the hard-line behaviourist reduction of thought to a matter of observable or reflex stimulus-response – that is so prominent and problematical a feature of Quine's empiricist outlook. Such is also the main objection to Quinean epistemology and philosophy of science, i.e., that he presses so far in a jointly physicalist and wholesale framework-relativist direction as to leave no room for those operative standards (of evidential warrant, descriptive yield, explanatory scope and depth, predictive power, and so forth) without which there would seem no rationally assignable grounds for the choice of one over another scientific theory. Thus the lack of any non-framework-relative ontology (or adequate conception thereof) goes along with a lack of causal-explanatory resources and this in turn – as a result of Quine's primitive behaviourist psychology – with what comes across as a strikingly (indeed self-disablingly) low estimation of human cognitive and intellectual powers. Hence the tendency among more recent naturalistically-inclined philosophers to move beyond the kind of depthless and inert physicalism that Quine inherited from his logical-empiricist mentors and which he then bequeathed, albeit in drastically relativized form, to thinkers across a wide range of (e.g.) 'post-analytic', social-constructivist, and Rortian neo-pragmatist camps. On this later strong-naturalistic account, typified by Wesley Salmon's pithy injunction to 'put the "cause" back into "because"', what's needed is a much sturdier realist outlook grounded in a firm metaphysical commitment to the existence of natural kinds and of the various causal powers, properties, dispositions, and underlying (microstructural or depth-ontological) attributes that constitute the kinds in questions.[30]

One source for this return to a basically Aristotelian conception is the emergence of a closely analogous approach to issues in philosophical semantics – first developed by Saul Kripke and Hilary Putnam – which proposes a theory of reference-fixing based on kind-specific attributes or essences, and a similarly grounded account of how names hold firm across even the most dramatic or large-scale episodes of scientific paradigm-change.[31] Thus the 'new' Kripke-Putnam theory of naming, necessity and natural kinds was one that promised a solution to those various problems of meaning-variance or incommensurability between different paradigms, frameworks, or conceptual schemes which had lately arisen as a direct consequence of arguments such as those of Quine and Kuhn, but which could seen to have their ultimate source in the earlier 'descriptivist' theory of sense and reference developed by Frege and Russell. More than that – and of greater relevance in the present context – it allowed for a stronger normative component in the process of explaining how knowledge might accrue through successive, sometimes paradigm-transformative advances in the state of scientific knowledge while none the less preserving sufficient continuity of reference to ensure that scientists (and scientifically-literate laypersons) could be reasonably certain that they were in fact referring to the same thing. This normative component entered through the claim that such advances betokened an increasing depth-explanatory grasp of just what it was that characterized the kinds in question along with their properties, attributes, structures, causal dispositions, etc., and also – inseparably from that – an ever more refined and sophisticated power of empirically grounded and theoretically informed reasoning on the best evidence to hand.

Thus the 'new' causal theory in its various forms – linguistic, logico-semantic, metaphysical, ontological, and natural-scientific – didn't work out as a hard-line physicalist or determinist creed with the consequence (as might be expected) of excluding or at least very strongly playing down that element of rationally motivated theory-choice. Rather it signalled the way toward a standpoint beyond those vexing antinomies of mind and nature, subject and object, the rational and the causal, or even (conceivably) free will and determinism that had so hobbled the course of philosophical debate on these topics from Kant to the present. This it did by, in effect, overleaping that history and returning to an earlier (though suitably updated) conception of the role of human thought and agency vis-à-vis the realm of natural causes, processes, and events. What results is a radically different perspective on the problems that first took hold with Kant's heroic but failed attempt to reconcile the twin claims of transcendental idealism and empirical realism, and which then underwent a succession of shifting conceptual or terminological guises right down to their logical-empiricist and subsequent post-analytic or linguistically-oriented forms. The new theory makes a key point of treating the causal-explanatory and the rational-normative as strictly inseparable aspects of human cognitive activity which cannot give rise to the familiar range of stubbornly persistent post-Cartesian or post-Kantian dualisms except through a basic misconception of their jointly operative nature. Thus one could put a strong case for viewing those difficulties with the old descriptivist (Frege/Russell) theory pointed out by Kripke and Putnam as a logico-semantic variant of the same root dilemma, one that has by now become so deep-laid and so much a feature of the philosophic scene as to constitute almost an inbuilt condition of reflective enquiry in epistemology and philosophy of language. That is to say, just as Kant strove to resolve but if anything further deepened and complicated the problems bequeathed by Descartes – chief among them the mind/body issue and the antinomy of free will and determinism – so analytic philosophers failed to discover the wished-for escape route from similar problems through their turn toward language (or various conceptions of language) as a substitute for the old-style empiricist or rationalist 'way of ideas'.

It seems to me that this kinship between the two chapters of philosophic thought comes out most strikingly if one considers a particular (perhaps the most famous) instance of Kant's difficulty in making good his claim to reconcile transcendental idealism with empirical realism. This is the flat contradiction involved in asserting (1) that ultimate or noumenal reality lies altogether outside and beyond the realm of sensory-perceptual or cognitive concepts and categories, but also (2) that it somehow has an impact on our receptive faculty through what can only be construed under the category of causal (or quasi-causal) agency. It is not hard to see in this epistemological quandary a displaced version of the free-will/determinism issue, one that prefigures the kindred sorts of problem that arise in Kant's Second *Critique* when he strives to connect the categorical imperative and other such high-level edicts or abstract maxims by which the autonomous moral subject supposedly lays down laws for its own proper conduct with the kinds of particular or situated ethical predicament in which human agents often find themselves.[32] It is then more than likely that any guidance offered by Kant's universalist maxims will at best provide only the vaguest, most generalized or all-purpose ethical guidelines, and at worst the kind of inflexibly absolute and context-insensitive rules that might very well produce bad, even wicked behaviour on the part of moral agents drilled in such ways of thought. However my point here is not to belabour these well-known problems with Kant's doctrine of practical reason. Rather it is to point up their connection with that curious muddle in the First *Critique* about noumenal reality and whether or not it can be accessed by way of categories – such as causality – which, as he insists, belong firmly on the side of phenomenal cognition, or the realm of humanly achievable knowledge wherein sensuous intuitions are 'brought under' concepts of understanding. For it is here, I suggest, that one can find a main source of those sundry dualisms that have plagued the discourses of epistemology and ethics from Kant right down to the present, and all of which involve some form of the divorce between causal-explanatory and rational-normative modes of thought. Moreover, they can now be seen to have carried across into that version of the latter-day linguistic turn – the Frege-Russell descriptivist theory – that made reference ultimately dependent on sense and sense in turn dependent on some given state of knowledge with regard to that referent whose nature and salient (defining) features were thereby relativized to this or that language, discourse, framework, or conceptual scheme.

From here it was no great distance to those other, more extreme or 'strong'-descriptivist notions – such as Rorty's – which pushed this way of thinking to its ultimate conclusion, i.e., the idea that since reality is always under some description or other, therefore we should give up the delusory quest for an order of truth that might somehow (impossibly) transcend all such partial or limiting perspectives.[33] Instead, Rorty counsels, we should henceforth accept that interpretation goes all the way down, that our descriptive languages are utterly contingent or interest-relative, and therefore that the best thing philosophers can do – in company with novelists, poets, literary critics, cultural theorists, ethnographers, et al. – is try out new and inventive (especially metaphorical) styles of talk whose aim is not correctly or truthfully to represent but rather to re-describe that notional 'reality' in a way that catches on since it happens to chime with a wider sense of communally shared values and beliefs. All the same, he is keen to reject any imputation of cultural or social-linguistic relativism, asserting that he is as realist and objectivist as anyone about the 'brute reality' of (say) the photons that impacted on Galileo's eyeball, as likewise on the eyeballs of his opponents the orthodox Catholic astronomers of Padua, when Galileo took himself to be viewing the moons of Jupiter and they took themselves to be viewing no such thing.[34] Where Rorty's relativism kicks in is through his claim that this baseline level of reality is irrelevant to the scientific issue between them, as likewise to the present-day philosophic issue between realists and anti-realists, constructivists, or relativists since it is only at the stage when it comes under one or another rival description or theory that debate starts up and those differences of interpretation enter the field.

IV

What this amounts to is a full-strength or ultra-polarized version of the same dichotomy that I have traced back to Kant and forward to various forms of analytic and post-analytic (e.g., Quinean) thought. It involves a yet more extreme version of that deeply problematical and self-disabling split between a domain of physical objects and events conceived in inertly empiricist terms and a sensory-perceptual-cognitive domain where theories, hypotheses, or 'interpretations' seem to be envisaged as floating altogether free of any other than notional anchorage in the physical world. More precisely, as shown by Rorty's example of Galileo and the Paduan astronomers, it involves an epistemic displacement or reduplication of that split in the way that causal 'impacts' are taken to occur through a process of purely passive or blankly receptive sensory intake (rather than uptake) that neither informs nor is informed by any higher-level modes of perceptual, cognitive, or rational-evaluative thought. Indeed it is among the great ironies of post-1960 epistemological debate that the Quine-Kuhn doctrines (not to say dogmas) of the 'underdetermination of theory by evidence' and the 'theory-laden character of observation-statements' have overwhelmingly been pushed in this paradigm-relativist direction rather than deployed to just the opposite, i.e., realist and rationalist effect. On the latter view they would count strongly in support of the claim that if even our most basic or common sense perceptions are in some degree theoretically informed then we can take it that the physical sciences have by now reached an advanced stage of jointly empirical, theoretical and causal-explanatory progress. In which case there is no reason to adopt a sceptical or relativist position – rather every reason to adopt just the opposite stance – in this idea of the reciprocal entwinement of mind and world as a presupposition of any philosophic or historical approach that would account for our knowledge of the growth of scientific knowledge.

Thus Rorty is getting things exactly back-to-front when he argues – very much on the strength of that thesis and others of a similar relativist, constructivist, or framework-internalist import – that following the linguistic (or hermeneutic) turn across every field of philosophical enquiry it is simply self-evident that bottom-line terms of endorsement like 'true', 'real', or 'rational' are nothing more (though nothing less) than signs of approval according to our present cultural-linguistic-descriptive-theoretical-evaluative lights. On the contrary, what drops out of this account – what has to drop out if it is to reach the desired ultra-relativist conclusion – is the extent to which enquiry is enabled, provoked, guided, elicited, constrained, or sometimes blocked and forced to pursue some alternative path by encountering theoretically informed observational data that may either corroborate or disconfirm its standing beliefs and commitments. In short, we are not obliged to accept the idea that those Quine-Kuhn theses (as one is tempted to call them, the two last dogmas of post-empiricism) necessarily lead to a paradigm-relativist upshot or a language-first, strong-descriptivist outlook which leaves no room for talk of truth or reality except by way of compliments routinely paid to those beliefs that we find culturally appealing or conducive to the wider communal good. That idea is the final outcome of a way of thinking which, as I have said, constantly reveals its Cartesian and Kantian lineage, and the effects of which are everywhere apparent in the kinds of self-imposed dilemma that have continued to vex the discourse of analytic philosophy long after it took the turn toward language as (supposedly) a means of resolving or escaping them. Where they still strike home with most power to disturb is in relation to philosophy of mind or to issues in other areas – including epistemology, cognitive science, and philosophical semantics – that have lately come to bear much of that inherited conceptual strain. It was left for Rorty to push this whole line of thought to the limit (or beyond) by concluding from the jointly epistemological and logico-semantic thesis that everything known to us is known or cognized under some description or other to the wholesale scheme-relativist claim that such descriptions do indeed go 'all the way down', and that therefore we had best give up thinking in those old metaphysical-realist terms.

That the resultant problems have much to do with the free-will/determinism issue should be clear from what Rorty himself takes as the wholly desirable or 'edifying' prospect of a culture that has learned to live without philosophy – at least without philosophy as hitherto and currently practised – and also learned, partly through shaking off the dead hand of philosophical realism, that the world is wide open to creative redescription in accord with our best social, cultural, and individual interests. Thus Rorty's updated, post-linguistic-turn version of the Jamesian 'will to believe' is one that starts out by completely decoupling the questions 'What is true?' and 'What can I know?' from the question 'What can I reasonably hope for?', but then goes on to sink the difference or invert the order of priority between them in such a way as to make 'truth' and 'knowledge' dependent on – or relative to – whatever hopes and desires happen to prevail within some given community of beliefs.[35] In the process, moreover, there is a shift of emphasis regarding the further (on the face of it crucial) question as to just what constitutes a 'reasonable' hope in the light of those various material, physical, scientific, social, psychological, and other factors that may set limits as well as aims or goals for any such transformative enterprise. Bertrand Russell famously objected to James's pragmatist idea that belief could be volitional – a matter of willed or elective assent to ideas, propositions or doctrines that may be thought psychologically or socially desirable – on the twofold grounds that it involved the possibility (indeed the strong likelihood) of self-deception and was also a ready-made excuse for various forms of morally delinquent conduct through failure or refusal to confront their likely consequences.[36] In Rorty's case likewise that faith in our capacity to change the world by changing our language-games, descriptions, metaphors, or 'final vocabularies' is close kin to the strain of wishful thinking – of trust in the performative efficacy of hoping it might be so – which forms such a prominent aspect of American intellectual as well as popular culture.

What gives his case its particular diagnostic force in the present context is the fact that Rorty's extreme version of the Jamesian-pragmatist creed can so clearly be seen to result from his situation as a thinker who expressly disowns any lingering attachment to old-style epistemological or mainstream-analytic modes of thought yet who remains very much (despite all his strenuous disavowals) in thrall to their agenda. This is not merely a question of his having rather undermined his official stance by continuing to engage in often fairly technical kinds of debate on topics in epistemology and analytic philosophy of language, albeit just as often starting out or ending up with some remark that cautions against according them too much earnest attention. More than that, it is a matter of his having pushed the inherited stock of epistemo-linguistic dilemmas to a point of maximal strain, as for instance by positing that drastic dichotomy between 'brute causal impacts' on Galileo's retina and the 'interpretation' that he (unlike the orthodox astronomers) placed upon them in keeping with his heliocentric world view. What this yields is a purely notional freedom or space for creative 'redescription' that is effectively quarantined from contact with physical, material or natural reality since it has renounced the very idea that descriptions might achieve a greater or lesser degree of verisimilitude – or approximative truth – with respect to that which they purport to describe.

Nor is it at all evident how, having pressed so far with the antirealist-constructivist argument, the process can then be halted and turned around so as to give the redescriptive strategy some purchase on realistically informed rather than merely wishful conceptions of our future social, political and cultural good. For if one thing emerges with unmistakable force from Rorty's work it is the fact that once this rift has appeared in its various post-Kantian forms – the gap between mind and world, subject and object, concept and intuition, apriorism and naturalism, rational-normative and causal-explanatory modes of thought – then the way is open to an relativist creed which fully exploits the opportunities thus created for an interpretative freedom wholly unconstrained by standards of veridical or evidential warrant. Moreover it points up the close kinship between this chapter in the history of recent epistemo-logico-linguistic enquiry and those longstanding metaphysical debates concerning the mind/body or free-will/determinism issues that have likewise entered a modern phase distinguished by its focus on the question as to whether – and if so, then precisely how – human meanings, intentions, judgments and beliefs can be reconciled with a naturalized epistemology and ethics. In fact, I would go so far as to say that metaphysical quandaries of that sort are the unspoken agenda or hidden motivation of a good many present-day philosophic wrangles, like those recounted above, that present themselves under a different (e.g., logico-semantic or cognitive-psychological) aspect so as not to become too plainly embroiled in such old and, as they might seem, philosophically insoluble problems. At any rate they have scarcely been laid to rest despite the varied solutions proposed by philosophers from Kant on down, among them thinkers like Quine and Rorty whose assurance of having emerged on the far side of all those vexing (especially Kantian) antinomies can be seen to have resulted from a failure to perceive the problems with their own radically empiricist, holistic, and framework-relativist positions.

The main lesson here, I suggest, is that philosophy is and will surely remain ill-equipped to meet such challenges so long as it adheres to the agenda set by those two major episodes in its modern history – the Kantian epistemological turn and the analytic turn toward language as (supposedly) the ultimate horizon of intelligibility – which can now be seen to have led it into this longstanding and, as it seems, philosophically terminal impasse. What is required to begin with is a clearer, historically informed awareness that these are issues of comparatively recent provenance and moreover issues that would not loom anything like so large if we could strike up a closer, more live and productive relationship with certain thinkers or traditions of thought that either predated that entire (two-centuries-long) chapter of intellectual history or have somehow managed to avoid its full impact. Again it is not just an academic matter of finding some alternative scholarly or philosophic furrow to plough, although this might itself be welcome enough to anyone whose patience is sometimes strained by their professional obligation to keep up as best they can with the organs of mainstream-analytical debate. More than that, it is a matter of trying to articulate a whole different way of approaching the mind/brain and free-will/determinism issues that would no longer seek – what is by now pretty much a forlorn hope – to resolve those issues on some more-or-less qualified or nuanced version of their own inherently dualist and hence inevitably problem-creating terms.

This goes back to my opening point about the need to clarify these various, often less than evident or straightforward links between the sorts of debate that preoccupy analytic philosophers and the sorts of debate (or the kinds of anxiety) that typify much 'popular' discussion of such topics. By critically reviewing the reasons for their having become so powerfully gripped by a range of (perhaps) misconceived or badly formulated questions philosophers might well have something of value to say in the context of a wider social concern with the place of human meanings, beliefs, and intentions in a culture increasingly given over to science-led or naturalistic ideas of what counts as genuine knowledge. For instance, Spinoza was one pre-Kantian thinker who advanced a radically heterodox conception of the mind-body nexus and of the means by which people might be brought to relinquish their 'folk-psychological', i.e., illusory though deeply-held notions of human free will or autonomy.[37] Only thus, he believed, might they also be brought to accept what amounted to a doctrine of thoroughgoing psycho-physical monism that in many ways anticipated the arguments of present-day central-state materialists or advocates of an out-and-out naturalist position according to which we can now look forward to a time when the advent of a mature neurophysiology will enable us finally to dispense with all forms of folk-psychological talk. Hence Spinoza's conception of the mental and the physical as two 'attributes' of the self-same substance, with the former conceived – in likewise thoroughly monist terms – as an 'idea' of the latter, or a direct correlate of bodily states that may themselves be affected (for better or worse) by mental activities, processes, and events.

It is not surprising that Donald Davidson picked up on these strikingly prefigurative aspects of Spinoza's thought in an essay that enlisted him as, in effect, a proponent of anomalous monism – Davidson's preferred solution to the mind/body or mind/brain dualism – *avant la lettre*.[38] Yet anyone who reads the two of them will most likely be struck by the contrast between the clarity, assurance, and perfect confidence in his own (however heterodox) proposals that are such notable features of Spinoza's *Ethics* and the distinct sense that Davidson is here finessing the issue, or that anomalous monism is really just a formula for having it both ways and evading (rather than resolving or confronting) the main problems. That is, the latter is then apt to seem more like an instance of fully-fledged naturalism or full-strength reductionist physicalism that dare not quite speak its name, and which thus falls back on the appeal to anomaly – to the idea of an exceptionless but non-law-governed correlation between mind-states and brain-states – in a move that effectively places the issue in quarantine. That this is conceived strictly as a matter of token-token rather than type-type correlation, and therefore as entailing no commitment to a strong physicalist/ reductionist stance, seems to me just another handy means of avoiding the main philosophical issue. For that issue has nothing to do with questions of just how and where certain human thoughts, beliefs, feelings, desires, and so forth have their corresponding sites, centers, networks, pathways, or modes of neuro-physical instantiation in the human brain. Rather – as so often in philosophy – it has to do with the question whether these sorts of question make any kind of sense or whether, as the dualist would have it, they should be seen as yet more cautionary instances of the gross category-mistake that thinks to collapse mind-talk into brain-talk without significant remainder. To which the physicalist will respond that in truth it is the unreconstructed dualists as well as those other, semi-reconstructed types with an attachment to fuzzy middle-ground ideas like supervenience, emergent properties, or anomalous monism who have themselves fallen into the category-mistake of supposing mind-talk to be anything more than just that, i.e., an already creaky folk-psychological idiom which will at length be rendered obsolete by the rapid advance of neuro-scientific knowledge.

Of course one has to make due allowance for Davidson's awareness of the complex and often highly-charged debate that has surrounded this topic since Kant and even more so as a result of recent developments in philosophy of mind and neuroscience. Despite the general analytic taste for tackling issues as if from scratch and without overmuch (if any) deference to the views of earlier thinkers – an attitude reinforced by Davidson's famous nonchalance in that regard – still this matter of inheritance, of just when philosophers come on the scene and what sorts of issue currently define the philosophical agenda, is one that needs taking into account when seeking to understand why a thinker like Davidson should need to hedge his bets so carefully. In fact, the answer fairly leaps off the page, as with a good many other treatments by analytic philosophers whose seeming indifference to the issues bequeathed by those earlier thinkers doesn't prevent such work from being strongly affected – provoked, stimulated, sometimes inhibited, or (as in this case) induced to adopt a highly qualified or hedged-about stance – by what they had to say. Thus with regard to Davidson's take on Spinoza, it is clear that he stops well short of any full commitment to the Spinozist conception of an all-embracing causal-determinist order wherein human creatures would take their place as more or less powerful, more or less actively engaged participants whose scope for such agency cannot be a matter of their standing outside it as possessors of free will and hence as exceptions to the general rule. Rather it is a matter of their standing squarely within that order and responding to the greatest possible range of physical inputs, causal connections, sensory stimuli, perceptual cues, affective powers, formative encounters, or heightened modes of experience with respect to both bodily and mental attributes.

No doubt most philosophers will think that this list gives the Spinozist game away, ranging as it does across the whole gamut of human receptive, cognitive and intellectual capacities yet leaving those distinctions firmly in place (like that between the two 'attributes' of body and mind) despite its claim to have rendered them otiose or shown them up as mere products of a false, metaphysically deluded way of thought. However this invites the obvious response: that such an argument can carry conviction only for those already committed to the dualist way of thinking that has created such (on its own terms) insoluble problems for many who have espoused it whether in a hard-line, *echt*-Cartesian form or in a compromise version of the type widely canvassed in recent years. Among the proposals thus thrown up – one that has much preoccupied thinkers in the analytic mainstream – is the distinction between *reasons* and *causes* in accounting for human behaviour, a distinction that is taken to conserve the vital margin of freedom and autonomy by which human agents are set apart from the otherwise ubiquitous order of causal necessity.[39] This idea, with its primary source in Wittgenstein, has most likely exerted such widespread appeal in virtue of its appearing to resolve that problem without buying into anything like a Kantian metaphysics of morals based on the radical disjunction between the phenomenal and noumenal realms, or the domains of cognitive understanding and practical reason. Yet the proffered solution is really no such thing since it leaves completely open the obvious question – one pressed hard by determinists past and present – as to why talk of reasons (or of rational grounds, warrant, justification, and so forth) should necessarily be thought more conducive than physicalist or causal talk to the interests of maintaining sufficient room for the exercise of human autonomy or free will.[40] After all, no defender of moral autonomy would wish to decouple that idea from the precept that our actions, choices and decisions are arrived at under certain specifiable constraints, namely those that result from (what else?) our responsibly exercised powers of rational and ethical decision-making.

Still less should they wish to invoke some notion of ultimate randomness, chance, or probability – such as that which supposedly obtains at the microphysical level – by way of carving out a space for human freedom despite what once appeared, before the advent of quantum mechanics, as the implacable march of modern science toward the reign of all-encompassing causal explanation and hence of exceptionless determinism. Some philosophers – G. E. M. Anscombe surprisingly among them – have thought this to constitute a strong case and a means of enlisting scientific (i.e., quantum-theoretical) support for the idea that even physics has now passed beyond that old Newtonian determinist view which took its bearings from the strictly atypical instance of celestial mechanics and thus gave rise to all sorts of false dilemmas, not least in the ethical sphere.[41] However this is open to the double objection that it rests on just one, albeit orthodox yet deeply problematical interpretation of quantum physics, and that in any case it yields a result that no defender of moral autonomy could possibly wish to endorse. For it follows from the quantum analogy that our actions, choices and decisions are the upshot of purely random (or irreducibly probabilistic) events in the sub-neuro-physiological domain which must therefore be taken to preclude any form of conscious, deliberate, willed, or determined human agency and decision-making power.

No doubt my use of the word 'determined' is one that comes across, in this context, as a kind of arch philosophical pun since although deployed here with the primary sense 'involving a high degree of willed and self-authorized commitment' it cannot but evoke thoughts of that other sense ('compelled, constrained, or involuntarily performed') which of course carries just the opposite import. Yet the very fact that it has developed semantically in these two, as it might seem sharply divergent or flatly contradictory directions is an index to the types of problem that ensue from the deep-laid Cartesian presupposition that those senses could not possibly be conjoined so as to produce an intelligible monist conception of mind/body along basically Spinozist lines. That is to say, if it is apt to strike us as a pun – or, diagnostically speaking, as a curious symptom of our language being torn two ways by clashing intuitions – then this might be thought to constitute not so much a case for seeking more clearly to disambiguate the word but rather a case for regarding that supposedly confused and mischievous conflation of senses as a pointer toward the most significant truth to be had on this particular topic. Therefore, as a matter of priority, we had better look for ways of somehow rendering those senses mutually compatible and indeed strictly unthinkable in isolation – ways of undoing the entire post-Cartesian concordat in its variously hedged or qualified forms – in stead of continuing to prop up the dualist illusion by professing to find a semantic muddle (or perhaps, if theologically inclined, an instance of profoundly revealing paradox) in those two senses of the word 'determine'. For we should then be engaged upon a process of conceptual unravelling that might have the altogether beneficial upshot – unlike so many philosophic efforts in a negative-diagnostic vein – of loosening the grip exerted on our thought by notions of the mind/body 'relationship' or of the free will versus determinism 'problem' whose very phrasing betrays their source in that same rootedly Cartesian-Kantian paradigm.

V

We are now coming close (too close for comfort, some might say) to the ground staked out by Gilbert Ryle in his remarkably acute and prescient, though nowadays strangely neglected book, *The Concept of Mind.*[42] However, there is a crucial difference between the basically Spinozist conception that I am advocating here and Ryle's seemingly kindred attack on the dualist myth of the 'ghost in the machine', along with his programmatic suggestion that we translate intentional, belief-related, first-person inner-state-reporting and suchlike 'mentalist' ascriptions into a dispositional language that would find no place for that interior domain of purely private psychological or cogitative goings-on which he finds aptly figured in Rodin's sculpture 'The Thinker'. Indeed, it is just the difference required to counter the charge so often brought against Ryle's brand of 'philosophical behaviourism', namely that it ignores that entire range of (what we take to be) willed, deliberate, intentional, active, calculated, thoughtful, desirous, or again – in the ethical context – self-seeking or self-interested as opposed to genuinely other-regarding acts and choices. From this viewpoint there is simply no allowance, on Ryle's dispositionalist account, for those attributes that play an indispensable role in the understanding of ourselves and others as reflective moral agents or as thinking individuals capable of more-or-less mindful, attentive, intelligent, considerate, creative, resourceful, intellectually engaged (or whatever) ways of doing the various things that we do. To be sure, Ryle was well aware of this likely criticism and took pains to head it off – more explicitly in his work after *The Concept of Mind* – by advancing a theory that aimed to dispel the folk-psychological yen for mentalist notions like 'mindfulness', 'intelligence', 'consideration', 'creativity', and so forth, through the simple expedient of turning them into adverbial modifiers which then served to describe or to qualify the action concerned without invoking some mysterious inner realm of mind-states or first-person private occurrences.[43] Still his critics doubted that this could do the trick, amounting as it did either to a grammatical sleight-of-hand that offered only the semblance of a willingness to meet their objections or else to a covert means of smuggling back all those 'mentalist' predicates under cover of a merely linguistic or grammatical-taxonomic concession.

Where Ryle's argument lay open to these lines of attack was in adopting a form of philosophical behaviourism which, despite its express differences from the standard Watson-Skinner type approach, still seemed to take a starkly reductionist view of whatever it was that dualists supposed they were describing when they deployed such mentalist or folk-psychological terms. Thus his alternative, anti-Cartesian (as well as anti-Kantian and anti-phenomenological) 'concept of mind' was one that pressed so far in that direction as effectively to rule out any appeal to human perceptual, intuitive, cognitive, intellectual, or creative powers beyond the publicly manifest or outwardly observable. That is to say, Ryle remained sufficiently a behaviourist in the non-'philosophical' sense as to leave it unexplained how that alternative approach via the corresponding range of adverbial modifiers could open up a more than notional space for the exercise of just those capacities that most people (and indeed most philosophers) would regard as distinctively human. There is the same problem about Wittgenstein-influenced approaches which likewise – though usually without the same degree of argumentative vigour – put the case for an account of human actions, decisions, and thought-processes that dispenses with what they consider the otiose appeal to unverifiable since publicly unknowable 'private' states of mind.[44] Where this case runs into all the usual anti-behaviourist objections is by taking so passively conformist a view of what it means to act, decide, or speak in such a way that one's various instances of practical or verbal conduct will be sufficiently accordant with communal norms and hence properly intelligible, whether to fellow-members of the relevant community or indeed to oneself. For in so far as it fails to exhibit that degree of conformity – that compliance with the locally existent currency of sense-making values or beliefs – one's behaviour will be no less opaque to oneself than to those whose understanding (if not their assent) one presumably wishes to gain.

Thus, it is thought to be a direct consequence of Wittgenstein's remarks about 'private language' and 'following a rule' that we will always eventually be at a loss for criteria by which to decide issues of meaning, truth or correctness in such matters save by recourse to those same communal norms. Quite simply there is no making sense of any marked deviation from them, that is, any utterance, act, decision, or mode of justificatory reasoning that fails to count as such by the evaluative lights of some given cultural 'form of life' along with its cognate range of communally recognized language-games.[45] This is why, on Wittgenstein's account, the Cartesian idea of privileged first-person epistemic access is just as false when applied to our own 'inner states' – which we could not begin to understand or express without the benefit of a shared language – as when applied to our fellow human beings whose sayings and doings are more obviously subject to the need for some means of public manifestation if we are to gain any sense of their gist. Hence his joke that the idea of ascertaining the sense of one's words by checking them against some inner tribunal of meaning or intent is like the folly of buying a second copy of the daily newspaper so as make absolutely sure that what the first copy said was true.[46]

However, the apparent knock-down force of arguments like these – arguments that have been taken on board by a great many analytic philosophers, often with a view to holding the line against various strains of 'continental' thought from Husserlian phenomenology down – will seem nothing like so impressive if one considers how heavy are the costs to be paid in ethical and socio-political as well as in 'purely' philosophical terms. They betoken not merely an unfortunate failure but, on the part of some zealous advocates, a full-scale principled refusal to envisage how thinkers and agents might find themselves so deeply at odds with the mores, beliefs, or predominant values of their own time and place as indeed to create real problems of conceptual or ethical grasp for those more attuned to that cultural environment, but not ipso facto to render their motives inscrutably private or opaque. Nor is the case any more convincing when framed in the guise of a seeming dilemma, i.e., that the individuals concerned – whether scientists, artists, philosophers, or social reformers – must be thought *either* to have been engaged in saying and doing things quite beyond the ken of their fellow human beings *or else* (since that idea scarcely makes sense) not after all to have been so profoundly at odds with prevalent values and beliefs. That such arguments are often deployed in support of a conservative appeal to the priority of custom, tradition or received mores over the exercise of critical reason or independent thought is one indication of what is at stake in this tendency to raise the communitarian view to a high point of Wittgensteinian dogma.[47] Indeed it entails a veto on any such exercise in so far as potentially dissenting parties would either lack the means to communicate at all in the absence of a shared linguistic-cultural-ideological medium of exchange or else adapt their thinking to just such a medium and thereby negate or very strongly qualify their heterodox stance.

My main point here is that the kinds of false dilemma encountered in Wittgenstein, Quine, Ryle, Davidson, Rorty, and others are all versions – displaced or substitute versions – of the great central problem that has long been installed at the heart of philosophical enquiry, namely the free-will/determinism issue along with its various post-Cartesian (mind/body or mind/brain) derivatives. What they have in common despite some otherwise large differences of view is the commitment to a broadly naturalized conception of mind, knowledge, belief, language, ethics and philosophical enquiry as a whole which none the less remains partly in hock to that older way of thinking with its source in Kant and its diverse progeny in the post Frege-Russell analytic line of descent. The result, as I have said, is a strong tendency – visible in all those thinkers – to espouse some kind of compromise position that is supposed to yield the best of both worlds (a naturalized yet non-reductionist and adequately normative account) but which in fact brings along with it the same old problems in, if anything, a yet more intractable form since they become all the harder to locate and disentangle. For we are then left with a double liability engendered by the confluence of two ill-matched movements of thought whose attempted reconciliation merely serves to exacerbate the inbuilt problems that each runs up against when pressed beyond its own carefully defined or preferential comfort zone. On the one hand there is the Kantian epistemo-critical approach that has increasingly been shorn of its critical as well as its epistemological resources by the sundry strains of latter-day descriptivist, pragmatist, constructivist, framework-internalist, paradigm-relativist, or kindred approaches with their principal source in one or another version of the multiform 'linguistic turn'. What thus drops out is precisely that dimension – that space for the exercise of just such critical powers – which was very much a part of Kant's thinking even if his project ultimately failed through its fixing of a downright unbridgeable gulf between the natural (in Kantian terms, 'pathological') and the rationally or ethically accountable aspects of human thought and conduct. On the other hand, there is the naturalizing drive as expressed by a thinker like Quine which concludes by adopting a thoroughly reductive physicalist-behaviourist approach to epistemology coupled with an out-and-out holistic or contextualist theory of belief-formation, thus depriving itself of any normative resources beyond those that happen to prevail in this or that communal setting. Hence the otherwise strange convergence in recent thought – especially the 'strong' sociology of knowledge – of motifs deriving from Quine's brand of hard-line reductionist physicalism and Wittgenstein's (on the face of it) flatly opposed linguistic-constructivist or cultural-relativist approach.[48]

Nor is the resulting normative deficit made good by those various 'post-analytic' lines of thought which, as I have argued, provide no more than a means of getting around these problems by adopting a notional solution based on the idea – often with presumptive warrant from Wittgenstein – that they were never genuine problems to begin with. There is an obvious parallel here with compatibilist 'solutions' to the free-will/determinism issue which likewise propose that we avoid any clash between those rival views simply by accepting that 'causes' and 'reasons' are terms that figure in two distinct languages, discourses, conceptual schemes, or ways of talking about the springs of human thought and action.[49] Thus – to put it in the idiom that Rorty picked up from Davidson despite Davidson's intermittent doubts as to whether Rorty had got him right – the problem simply vanishes as soon as one sees that everything (or everything we claim to know about) is necessarily 'under a description', that is to say, subject to this or that mode of descriptive or explanatory treatment which effectively decides what shall count as an admissible or relevant statement concerning it.[50] In which case we can surely put an end to all those misconceived worries – such as the free-will/determinism issue or (in its latter-day, de-transcendentalized and often Wittgenstein-influenced variant) the dualism of reasons and causes – by simply accepting that these belong to two quite different descriptive registers and should therefore not be thought of as generating any kind of conflict. Yet like so many putative solutions advanced on Wittgensteinian or, more generally, on linguistic-relativist grounds it offers nothing more than a kind of emollient mental balm or a soothing assurance that we had best not bother our heads about all those problems heretofore conjured up by bad ('metaphysical') habits of thought. For this is just another version of that half-way naturalizing tendency in recent analytic (or post-analytic) philosophy that deploys the linguistic turn as a hedge against any stronger and – so it is assumed – unacceptably reductionist or hard-line physicalist approach.

If one were setting out to write a history of developments in analytic philosophy during the past three decades and looking for a focal point around which to construct the narrative then one might be best advised to focus on this series of attempts to steer a path between those two minefields where only the foolhardy nowadays wish to tread. The first is very plainly marked: 'Keep out: free-will = Kant = metaphysics = all that stuff about the noumenal domain and the transcendental subject as locus of a freedom beyond the furthest bounds of attainable knowledge or experience.' The second bears the equally dire inscription: 'Venture not here: strong naturalism = physicalism = determinism = failure to uphold the vital distinction between *causes* (whether conceived primarily in material or socio-cultural terms) and *reasons* (again of diverse sorts but united in appealing to an order of motives, meanings, intentions, purposes and justificatory principles that cannot be reduced to any kind of causal explanation)'. Of course there is a fairly obvious sense in which this way of framing the issue gets it badly out of kilter since it is not – or not mainly – a matter of the choice between Kant on the one hand and causal determinism on the other. Rather it is a matter of the stark dichotomy that runs like a fault-line throughout the entirety of Kant's critical project and which has since given rise to all those rifts and convoluted features of the philosophic landscape brought about by the unresolved dilemmas of that same project. Moreover – very largely in consequence of this and suchlike misdiagnoses of the problem – analytic philosophy has long been subject to a kind of compulsive-repetitive pattern whereby it has continually claimed to have come out on the far side of all such pseudo-dilemmas while in fact leaving them firmly in place under some new terminology or technical description.

This applies just as much to thinkers who avowedly reject every aspect of the Kantian legacy on account of its excess metaphysical baggage as to those, like P. F. Strawson and John McDowell, who argue (albeit from different standpoints) that there is something of value to be salvaged from Kant just so long as one is careful to lighten that baggage in currently acceptable ways.[51] What unites them – along with a good many others who would emphatically disclaim any direct allegiance or significant (even strong-revisionist) relationship to Kant – is their continued embroilment in the same sorts of problem that Kant diagnosed with such marvellous acuity in discussing the Antinomies and Paralogisms of pure reason, but which none the less retained their disturbing power within his own system of thought and reached out beyond that to exert a similar effect on various aspects and departments of the analytic enterprise. Indeed, I would go so far as to say that analytic philosophy of mind and language has devoted a large portion of its intellectual energies to the invention of various ways and means for avoiding any too plain acknowledgment of its still being caught up in those same paralogistic quandaries. To this extent there is something to be said for the post-analytic, even 'post-philosophical' recommendation of a thinker like Rorty, namely that the project has run aground on various artificial problems of its own bother-headed devising, and that by far the best thing for philosophers to do is close their professional/academic shop and get used to talking on a level with other participants in the wider cultural conversation. However where Rorty gets it wrong is in arguing from the fact that *so much* analytic business-as-usual is an unwitting re-run of Kantian themes with no record or prospect of advance to the more sweeping (and unwarranted) conclusion that *everything* subsumable under the description 'analytic philosophy' – or just plain 'philosophy' – had best be consigned to the scrapheap of pseudo-disciplines that have long outlived their moment of cultural usefulness. That there might be alternative resources available from a different philosophical quarter is an idea that Rorty is unwilling to contemplate not only on account of his attachment to the sweeping picture but also because those alternative resources are such as he is wont to denounce as yet more lamentable throwbacks to a thoroughly superseded 'metaphysical' world view.

What I have in mind here – as will surely have become apparent by now – is that range of realist and causal-explanatory modes of thought which have lately been enjoying a strong revival in epistemology, philosophy of mind, cognitive science, philosophical semantics, and other fields. The most significant aspect of these developments is their promise of affording a powerfully argued – as opposed to merely abstract or notional – means of closing the Kantian epistemological rift and the sundry related dualisms that have followed in its wake right down to the travails of current analytic philosophy. I would hope that the phrase 'powerfully argued' may carry some likewise powerful associative links with the case for causal realism that I have been arguing somewhat on-and-off in the course of this largely diagnostic or problem-oriented opening chapter. That is to say, it is the case for a fully fledged ontology of causal powers that aims to overcome those inherited dualisms – by now so deep-laid as to seem intrinsic to the very nature of thought – between mind and world, subject and object, concept and intuition, reasons and causes, or the active and passive ('spontaneous' and 'receptive') capacities of human intellectual-cognitive grasp. This latter pair of terms – 'spontaneous' and 'receptive' – has lately been taken up from Kant by John McDowell in place of the more familiar Kantian concept/intuition dichotomy and with a view to stressing what Kant puts forward as their strictly inseparable, mutually dependent, or reciprocally inter-involved character.[52] However, as I have argued at length elsewhere, MacDowell is no more able than Kant to prevent these terms from reintroducing all the same old problems and antinomies, and doing so to yet more disruptive effect on account of their gesturing toward something like a monist or quasi-Spinozist idea of mind and/or nature.

Of course it is no wonder that Kant should have drawn back from any closer engagement with claims of that sort given the risks – at very least of intensified censorship – that would surely have been incurred by anyone who laid their Spinozist cards more openly on the table.[53] Indeed, one might go so far as to suggest that the persistent hold of dualist conceptions from Descartes down has been strongly reinforced by the desire of many thinkers to maintain a self-protective distance between their approach to epistemological questions and a form of ontological monism with more or less overt materialist leanings. In which case, what I have described here mainly in terms of a post-Kantian predicament affecting large swathes of present-day analytic philosophy could just as well be viewed in a much longer perspective as the latest episode in a history of thought that has always been marked by this same constitutive dualism, whether expressed through the matter/spirit, soul/body, mind/brain, subject/object, or other such periodic variations on a common theme. To be sure, the attempt to press through and beyond such tenacious habits of thought – to achieve a genuine and working monism along something like Spinozist lines – is no longer subject to punitive sanctions by church or state on account of its materialist and hence atheistical or radically heterodox character.[54] Still it is fair to say that the constant re-emergence of dualist motifs in contexts where they are unexpected or downright unwelcome – such as McDowell's strenuous effort to sink the conceptually unsinkable difference between 'spontaneity' and 'receptivity' – is evidence of just how far that old veto has managed to infiltrate the discourse of epistemology and philosophy of mind. In fact, MacDowell is an emblematic thinker in this regard, one who has pretty much run the gamut of modern (especially post-Wittgensteinian) ways of thinking about such topics and then proposed the return to Kant – or a decidedly selective, partially naturalized reading of Kant – as if to a source very largely untapped or at any rate grievously misunderstood by earlier generations of Kantians.

What I think we should take from McDowell is not so much this straightforward back-to-Kant directive but rather the lesson that any hope of a better, less inherently problem-strewn approach to the mind/brain or free-will/determinism issues will need to look both forward and back from that whole long chapter of philosophic history that stretches from Kant to the latter-day linguistic turn. That is, it may well find its starting-point and model in that Spinozist strain of radical monism whereby 'mind' and 'nature' are properly conceived as joint attributes of one all-encompassing substance and also in the turn toward causal realism – together with a strong, ontologically grounded naturalism – that has been such a marked feature of recent work across a range of disciplines or subject-domains.[55] More to the point, this approach has tended to characterize just those areas of maximally fruitful or productive exchange between disciplines that had previously been thought of as requiring altogether different methods or working priorities. Thus, it has enabled a new-found sense of common purpose between, on the one hand, epistemologists or philosophers of science who endorse this view of nature as composed of certain intrinsic, i.e., kind-specific structures, powers and capacities which give a hold for the actively enquiring character of human investigative thought and, on the other, philosophers of mind and cognitive psychologists who can think of themselves as engaged upon the same project from their own distinct but closely related angles of approach. This has the further great advantage – for anyone not overly wedded to existing academic structures and assignments of special expertise – that it offers more room for collaborative effort not only among people in the broadly 'humanities' area but between those in the human, social, and natural/physical sciences.

VI

What has often got in the way of such overtures is the notion, again with its proximate source in Kant, that the natural sciences have their home in the realm of empirical-cognitive grasp where sensuous intuitions are 'brought under' adequate concepts, while there also exists – above and beyond that – the realm of ethical precepts and values wherein practical reason discovers its ultimate orientation. That these domains should not be allowed to overlap – that empirical science not presume to dictate in matters of moral judgment nor the guiding 'ideas' of pure practical reason be treated as though they might have some bearing on matters of cognitive warrant – is of course a main burden of Kant's argument in the First *Critique*. It is a doctrine that finds numerous (if metaphysically muted) echoes in later attempts to drive the wedge home or keep it firmly in place through repeated insistence on the irreducibility of certain cognate distinctions such as that between causes and reasons. Now there are signs that this dualist heritage has encountered such a mass of decisive objections from so many philosophic and natural-scientific as well as ethical quarters, that we are reaching the point where no amount of conceptual tweaking or piling-up of concessionary clauses can any longer disguise its failure to meet the most basic conditions of an adequate account. Chief among them is the requirement that certain no doubt distinctive attributes of the human animal – its singularly well-developed aptitude for a range of intelligent, complex, purposive, highly adaptable and context-sensitive modes of conduct – not be hived off into some special, philosophically ring-fenced protective zone or some causal-explanation-proof realm apart. Rather than resort to such philosophically as well as scientifically disreputable special-case options, philosophers should treat those human attributes as in principle open to the same sorts of explanatory treatment as have yielded such strikingly successful results in the natural-scientific domain.

Hence no doubt the notable upsurge of interest in Spinoza during the past decade, due in large part – as a survey of the literature will make very clear – to his having so remarkably prefigured the convergence of present-day philosophical, cognitive-psychological, and natural-scientific thinking on the claim that any adequate theory of mind must acknowledge the ultimate impossibility of conceiving mental events to occur in the absence of physical correlates.[56] Of course this still leaves room for philosophical manoeuvre, as for instance by viewing their correlation as a matter of token-token rather than type-type equivalence and thereby somewhat allaying the threat of wholesale physicalist determinism implicit in the claim to specify and locate the type of brain activity that causes – rather than merely underlies or accompanies – some given type of mental state or phenomenal experience.[57] Much the same purpose is served by currently favoured fallback concepts such as 'supervenience' or 'emergence' whose attractiveness, it seems, is that they provide a place of refuge for crypto-dualists or a buffer-zone for those who cannot but acknowledge the force of naturalist arguments yet cannot quite bring themselves to take such arguments at full causal-explanatory strength. Here again, Spinoza was the first to insist on the latter option as opposed to any dilute variant while taking stock of the challenge it presents – the seeming affront to various aspects of our cherished human self-image – with an intellectual honesty and clarity of grasp that have come into focus only now with modern developments in neurophysiology and cognitive science. Still, it should not be forgotten that Spinoza is equally ahead of his time (and also of ours) in seeing no necessity of outright conflict on logical, conceptual, metaphysical, or moral grounds between acceptance of the full-strength version and acceptance that we are indeed 'free' in a heterodox but well-defined sense of that term, a sense which moreover is the only one that matters for practical-effective as distinct from purely philosophical-speculative purposes. This is the view that human beings – like all other animate or sentient beings of whatever kind – are capable of exerting their powers in such a way as to achieve (or fall short of achieving) the utmost of their own distinctive or kind-specific potential.

Hence the Spinozist concept of *conatus*, that particular impulse, drive or desire whereby every creature endeavours to persist in its own mode of being, either at a relatively primitive (largely unchanging) level of organization or, as with humans, at a higher and more complex level where the mode in question is one that can preserve itself – not fall back into inert passivity – only through a process of constant change or continuing self-transformation. Hence also the cardinal distinction between active powers and passive or merely receptive dispositions which structures the entirety of Spinoza's thought and which he sees as providing the sole adequate or non-self-contradictory basis for a conception of mind as both integrally a part of the natural order where causal explanations hold sway and at the same time as capable of more or less effectively exerting its powers in and through that same order. This, in turn, requires that he distinguish between the concepts of *potestas* and *potentia*, both meaning 'power' but crucially divergent as regards the sort of power involved and its specific mode of action or manifestation. 'Potestas' denotes a strictly heteronomous form of power which bears upon the subject from outside or above and can be conceived only in coercive or privative terms as a form of physical, psychological, or socio-political constraint. 'Potentia' is the kind of self-actualizing power that results not from some Kantian exercise of absolute moral autonomy but rather from the freely-entered-into exposure to a maximal range of those causal stimuli that make for an enhanced or expanded capacity of active participation. So there is no room here for those strictly (on their own terms) insoluble antinomies that plagued the Kantian project of thought and have continued to vex philosophical discourse right up to and beyond the linguistic turn with its deceptive promise of deliverance from all such groundless 'metaphysical' pseudo-dilemmas.

It is among the great ironies of modern intellectual history that Kant should have been the first to propose a full-scale critique of metaphysics – of the various ways in which pure speculative reason was constantly tempted to overreach the limits of its proper domain and thus run aground on a range of self-induced quandaries – but also the thinker who managed to install such problems at the heart of both his own and many later philosophical endeavours. Indeed one could put a strong case that the travails of philosophy when faced with the mounting evidence from various scientific quarters have resulted in large part from this Kant-induced failure to take up Spinoza's project for a thoroughgoing monist account of mind as integral to – rather than 'emergent from' or 'supervenient upon' or loosely 'correlated with' – the order of natural causes, processes and events. As I have said, the trouble with these ideas is that they shy away from confronting the central issue, i.e., the issue as to whether or not there is room for non-naturalizable attributes, properties or predicates in a scientifically-informed as opposed to a folk-psychological conception of mind. In stead they seek refuge in the sorts of evasive or compromise formulation that give with one hand what they promptly take away with the other. Thus they avoid getting into direct conflict with the scientific-naturalist approach as they would by explicitly maintaining the thesis that a range of distinctively human or sentient attributes – such as free will, autonomy, intentionality, or qualia ('what it's like' to undergo certain kinds of sensory-perceptual-cognitive experience) – cannot in principle fall within the scope of any naturalistic or causal-explanatory account. At the same time, they seek to avoid imputations of reductionism or determinism by adopting a third-way idiom of 'emergence' or 'supervenience' that offers merely the illusion of a problem solved since it works by introducing a new term of art whose function is not so much to clarify the issue at hand as to shunt it off into a region of fuzzy semantics which can serve as a buffer to avert any direct physicalist threat.

Analytic epistemology and philosophy of mind have been stuck for at least the past three decades in what is starting to look very much like a dead-end dilemma. In brief, they have been striving to reconcile an acceptance that Descartes, Kant and company got it wrong about the mind/body or mind/brain relation, not least by conceiving it in terms of a 'relation' rather than a strict identity, with a continuing desire – one that no doubt goes deep into a certain received idea of philosophy's proper role – to preserve some conceptual space for a variously hedged or qualified dualist view. If it is ever to get on working terms with the sorts of challenge thrown up by advances in neuroscience as likewise by the more naturalistically-inclined branches of cognitive psychology then philosophy will need to jettison a good deal of the baggage bequeathed by its two most recent and deeply formative paradigm-shifts, that is to say, the Kantian-epistemological and the conceptual-linguistic-analytical 'turns'. For their single most decisive effect has been to place the Cartesian problematic squarely back at the heart of philosophical concerns, even – or especially – when thinkers since Kant have announced their intention to refine, reformulate, drastically revise, or simply abandon that old framework in the name of some supposedly less dilemma-prone since non-'metaphysical' approach.

Anyone who has followed these debates in recent years will know that such claims very often and oddly go along with a continuing active (not to say obsessive) interest in just the same issues that exercised those earlier philosophers, albeit now from a pragmatist, descriptivist, or quasi-naturalistic standpoint that purports to have emerged on the far side of all those superannuated problems but seems altogether incapable of leaving them alone. In which case, I have argued, its best hope of converting this desire into an operative programme is to adopt the kind of radically naturalised perspective that Spinoza was the first to envisage and which has nowadays arrived at the threshold of maturity with the advent of a neuro-philosophical approach to issues in epistemology and philosophy of mind. Only thus will it finally succeed in breaking with the deeply entrenched and remarkably protean dualist conception that has managed to retain such a singular hold on thinkers of otherwise diverse philosophical views. However, that paradigm has shaped so much of philosophy's agenda, from its dominant themes to the very vocabulary, syntax and field-specific idioms in which its debates are conducted, that the change here envisaged will require a transformation of quite extraordinary depth and scope. Meanwhile, what is more often on offer is a qualified naturalism that pays its dues – whether overtly or not – to various scarcely ignorable advances in neuroscience and cognitive psychology but which none the less seeks to mitigate their more worrisome effects by recourse to a likewise qualified (even, at times, somewhat shuffling and evasive) range of Cartesian assumptions. Among the most resourceful and conceptually adroit of these attempts to have one's philosophic cake and eat it is Harry Frankfurt's essay 'Freedom of the Will and the Concept of a Person', to which I shall now turn in Chapter 2.

Chapter 2

Frankfurt on Second-Order Desires and the Concept of a Person

I

Harry Frankfurt's 1971 essay 'Freedom of the Will and the Concept of a Person' is one of the best known recent contributions to philosophy of mind, moral philosophy, and – in the broad and I dare say proper if not currently received sense – cognitive psychology.[1] Indeed its main line of argument concerning the intimate link between our concept of human personhood and the distinctively human possession of 'second-order' desires or volitions has probably exerted as wide an influence and been subject to as many detailed re-workings as any other in the past half-century.[2] Nor is this at all surprising given Frankfurt's lucid and graceful yet pointedly analytic style, his regular choice of telling examples, and – more than anything – his having here tackled a complex of themes that not only go very much to the heart of recent philosophical debates but are apt to strike most reflective human beings from time to time.

Thus his essay seeks to articulate just what it is that makes such beings 'human' according to a certain criterion – or self-understanding – of genuine personhood that sets them apart both from non-human animals and also from those among their conspecifics who fall short of humanity in that regard. Moreover, it does so in an intellectual climate marked by the increasingly rapid accumulation and diffusion of new advances in just the areas (e.g., neurophysiology, genetics, and the more naturalistic varieties of cognitive science) that many people – philosophers and others – regard as a real or potential threat to any such conception. That is, those disciplines are viewed as laying siege to the basic attributes that are taken to mark human beings out as truly human since capable of actions, choices, decisions, or commitments which involve both an unconstrained exercise of will and the condition that such will be exercised in accord with the agent's best idea of what makes for good conduct or a life well lived by her own and other persons' ethical lights. Where Frankfurt's approach has proved so attractive is through the promise of offering a viable alternative to naturalism in its hard-line physicalist and (presumptively) determinist mode while not going along with any downright anti-naturalist or dualist creed that would fly in the face of so much presently accredited scientific knowledge.

Thus it offers a way of thinking about these issues which avoids any too direct invocation of Kantian ethical precepts or imperatives yet still leaves room – via the doctrine of 'second-order' desires – for a kind of self-transcendence through the bringing to bear of higher, more reflective or worthy volitions on those that belong to our unregenerate or animal natures. I should add that Frankfurt is not overly attached to this latter sort of talk, whether in its Kantian or its more traditional Christian-moralizing mode. All the same his essay falls in with such ideas at least to the extent of making it a requirement for the status of personhood (as distinct from mere membership of the human species) that the beings in question should possess that power of self-critical and self-evaluative thought which enables them to pass their first-order wants in review and, where appropriate, form the desire to cultivate a different set of wants or rank their existing set differently.

It seems to me that Frankfurt's essay cannot in the end make good on its promise to vindicate the claims of human autonomy or free will through this notion of our pre-reflective desires being subject to a higher-level process of quasi-juridical assessment as to their fitness or otherwise when measured against our best conceptions of the human good. For one thing, and most problematically, it fails to address the standard objection to all such ideas of progressive ascent from one to another level of increasing descriptive, explanatory, or justificatory power in whatever specific context of debate. Thus, the point is often made – most famously with regard to old-style logical positivist attempts to draw a firm line between first-order 'material-language' statements and second-order (metalinguistic) statements about those statements – that this procedure not only opens the way to a potential vicious regress but fails to establish any adequate formal (as opposed to pragmatic) grounds for supposing that such a line can be drawn with the required degree of conceptual or logico-semantic precision.[3] There is no need here to rehearse the manifold ways in which this difficulty first came to light and thereafter tended to re-surface with all the more disturbing or disruptive force in the wake of various claims to have resolved it on viable, that is to say, regress-blocking and clearly specified terms.[4] That account would be something like a potted history of the mainstream-analytical enterprise from its early stage of programmatic assurance, following Frege and Russell, that such distinctions could reliably be drawn to its wholesale demolition – as the orthodox version goes – in Quine's 'Two Dogmas of Empiricism', his famous attack on every last vestige of the analytic/synthetic dualism or the distinction between empirically verifiable 'matters of fact' and logically self-evident 'truths of reason'.[5]

Of course, this particular chapter of developments has to do primarily with issues in epistemology and philosophy of logic and language, rather than with issues in ethics and philosophy of mind such as occupy the main focus of interest in Frankfurt's essay. Nevertheless, as I have said, they do have a bearing on his claim that the possession of second-order desires or volitions is what makes for – indeed, what properly constitutes – the character of genuine human personhood as distinct from those various creaturely traits that mark human beings as just another, albeit highly evolved and psychologically complex kind of animal. Frankfurt himself concedes that there might be difficulties in this direction if it is allowed that 'a person may have . . . desires and volitions of a higher order than the second' (p. 91). For of course this idea brings along with it the possibility that the series thus envisaged might have no rationally assignable stopping-point and threaten not merely a notional regress – a problem for logically-minded philosophers – but a dissolution of authentic personhood in just the sense of that term that Frankfurt is seeking to establish. After all, if 'there is no theoretical limit to the length of the series of desires of higher and higher orders', then it would seem that 'nothing but common sense and, perhaps, a saving fatigue prevents an individual from obsessively refusing to identify himself with any of his desires until he forms a desire of the next higher order' (p. 91). However, he thinks, there is no need to entertain such extravagant worries since the stopping-point may always occur – and the threatened regress thus come to a merciful halt – at the stage of some decisive personal commitment which either exerts such a powerful grip on any higher-order volitions that they cannot but fall into line or else renders their existence altogether superfluous. In such cases it is a matter of indifference 'whether we explain this by saying that this commitment implicitly generates an endless series of confirming desires of higher orders, or by saying that the commitment is tantamount to a dissolution of the pointedness of all questions concerning higher orders of desire' (p. 92).

Nevertheless, these solutions have a somewhat makeshift or stopgap air, as if to ward off the regress objection by declaring it simply irrelevant for practical purposes or for anyone who knows 'from inside' what it is to have arrived at such a moment of decision. Indeed, they are reminiscent of patch-up attempts in other areas of thought like Russell's Theory of Types, i.e., his rule against self-predicative expressions or those that crossed the line between different orders of statement and thereby threatened serious trouble for the enterprise of set-theoretical thought and the attempt to place mathematics on a wholly consistent logical basis. Only thus, he believed, was it possible to ward off the kinds of damage that would otherwise be wrought by the paradoxes of self-reference or the potential for mischief contained in such expressions as 'the set of all sets that are not members of themselves'.[6] If all this seems pretty remote from the context of Frankfurt's argument – since concerned with issues of a purely formal or logico-conceptual as opposed to a subjective, experiential and ethico-evaluative nature – then one might pause to consider how often Frankfurt appeals to the language of first- and second-order desires or volitions, a language that cannot but raise certain questions of that same Russellian sort. After all, it is his leading contention that the criterion of genuine personhood is precisely the possession – ex hypothesi denied to non-human animals – of a capacity for freely-willed discriminative choice among the various first-order desires that are taken to constitute a separate realm of altogether 'lower' since unreflective or instinctually driven needs, cravings, instincts, appetites, and so forth. Then again, they might be construed as belonging to a somewhat more thoughtful stage of reflective ascent but only in so far as this is held subject to a further process of evaluation brought to bear from the next-higher level of deliberative judgment which thereby ensures – crucially for Frankfurt's case – that there remains the necessary leeway for free will and responsibly exercised choice.

So Russell-type problems do have a bearing here despite his confident claims cited in the previous paragraph that the regress objection can successfully be met just by seeing that it loses any force or point once we grasp that some higher-order volitions have a purposive strength and depth of commitment that renders them proof against any such regress-based or paradox-mongering objection. According to Frankfurt, it is possible to place a definite, non-arbitrary limit on the series of thoughts-about-thoughts-about-thoughts or desires-to-desire-to-desire (etc.) simply by grasping the salient fact that some of our decisions – those with this identity-shaping power – are such as to render that objection otiose, along with the related puzzle about self-predication that is apt to arise if one asks how any clear and subjectively relevant (i.e., other than purely logical) distinction can be drawn between those various levels. Thus '[w]hen a person identifies himself *decisively* with one of his first-order desires, this commitment "resounds" throughout the potentially endless array of higher orders' (p. 91). However, this leaves open the obvious question as to what room is left for the joint desiderata of freedom and responsibility – the constituents of full human personhood on Frankfurt's account – if that regress can be blocked only by an appeal to such a locus of self-determined and self-determining choice. For it then seems that this concept of a halt to the 'endless array' must also entail certain clearly-marked restrictions on the scope of that freedom if indeed it is to allow for any well-defined idea of personal responsibility. That is, there has to come a stage where the notion of choice as somehow both determined and determining – the former in so far as it pertains to lower-order desires, the latter in so far as those desires are subject to reflective evaluation – ceases to operate as the means of conserving a space for autonomous thought and becomes something more like the means of fixing a limit to the otherwise vertiginous range of possibilities thus opened up.

My point is that Frankfurt's use of the distinction between first- and second-order desires or volitions is one that inevitably leads to problems since it confronts thinking with a choice between (1) the kinds of potentially endless regress that ironists in the German Romantic tradition found utterly engrossing but which possess fewer charms for modern analytic philosophers, and (2) the kind of regress-blocking move which decrees that such complexities must have an end and that this involves the determination to make certain desires or volitions absolutely and completely one's own.[7] Yet it is then very hard to avoid the question implicitly posed by that seeming quirk of 'ordinary language' which enables us to say, as if without conceptual strain, that we are *determined to do* whatever we freely choose, or that our choice is a matter of *determination* to pursue some given life-plan, project, or mode of conduct itself undertaken – so the presumption goes – of our own spontaneous or unconstrained (hence responsibly exercised) will. It seems to me that Frankfurt's first-order/second-order way of framing these issues leads him into various, more or less nuanced or qualified statements of the thesis which all give rise to this unresolved – maybe unresolvable – conflict between the two senses of 'determine'. Thus on the one hand it signifies 'cause to think, feel, or act in a certain way through certain crucially *determinative* factors of a physical, environmental, social, or cultural sort', while on the other it carries the contrary or at any rate strongly opposed sense: 'decide, resolve, or *determine* to proceed in accord with one's best judgment or freedom to choose the most rational, desirable, advisable, or ethically responsible option'. Frankfurt takes it that metaphysical debates about free will versus determinism had much better be recast as debates about the relationship between first- and second-order desires, that is, the extent which the former may be influenced, guided, enhanced, reformed, or even subject to outright veto by the power of adjudicative challenge or review exercised by the latter. But he nowhere explains how that power of higher-level endorsement or rejection can itself be legitimized – thought of as rightfully binding – if its edicts have the kind of self-validating force that comes of their regress-blocking role and their function in defining what shall ultimately count as the motivating interest (or complex of interests) with which a person is so deeply identified that it becomes the very ground or criterion of personhood in their particular case.

II

This raises the question as to how there could exist that margin of freedom that Frankfurt considers indispensable if we are to make any sense of likewise indispensable notions such as those of will, decision, commitment, dedication, resolve, reflective or self-critical judgment, and responsibly exercised choice. For it would seem to place a drastic limit on any such freedom – rather, to exclude its very possibility – both in so far as lower-order desires are thought of as properly subject to being overruled without right of appeal and also in so far as higher-order volitions are themselves determined in both senses of the word. That is, they seem to figure on Frankfurt's account as products of human will – of the determination to follow through on some chosen course of action or mode of conduct – but a will that is constrained (along with all those lower-order desires that fall within its jurisdiction) by the fact of its so completely pervading the person's motivational mindset as to leave no room for deviation from the path marked out or determined in advance. In this respect Frankfurt's way of framing the issue can be seen as raising all the same problems as Kant's tortuous attempts to explain how the moral will can be conceived as an autonomous source of laws for its own proper guidance, or how the edicts, maxims, and imperatives of practical reason can be thought of both as issuing from and as bearing upon the subject in its self-legislative role.[8] Thus the word 'subject' in Kantian moral discourse is one that splits, like the word 'determine', into two quite distinct and indeed contradictory senses, with the first strongly linked to notions of the subject as locus of autonomy, agency, and the will to realize intentions formed through an exercise of independent thought while the second connotes the state of being subject – passively obedient – to laws handed down by some higher authority.

That this problem goes deep into the structure of Kant's entire critical project can be seen from the fact that it crops up again in relation to his epistemological arguments in the First *Critique.* More specifically, it concerns the gap that opens up between phenomenal (or sensuous) intuitions and concepts of understanding, a gap that is not so much closed or bridged as pasted over by Kant's notoriously vague talk of 'judgment' and 'imagination' as mediating faculties.[9] In the present context what it helps to bring out is the extent to which discourse on these topics since Kant has been marked – not to say hamstrung or hobbled – by a strain of deep-laid dualist thinking that constantly re-surfaces to vex any project aimed toward the final overcoming of all such 'metaphysical' residues.[10] With Frankfurt, it appears in the frequent signs of conceptual strain around statements to the general effect that, for all practical (including legitimate philosophical) purposes, the issue of free will versus determinism comes down to the question whether subjects or agents are capable – unlike non-human animals – to have the sort of will or will themselves to cultivate the sorts of first-order desire that they wish to have *qua* reflective individuals or persons in the plenary sense of that term. Thus '[j]ust as the freedom of an agent's action has to do with whether it is the action he wants to perform, so the question about the freedom of his will has to do with whether it is the will he wants to have' (p. 90). However, it is not at all clear how this shift to a second-order level of analysis does anything to ease the conceptual strain involved in understanding how freedom, autonomy or genuine volition can have any place in this notion of a will that is itself the outcome of a 'wanted' (hence willed) commitment, one that may either – at risk of vicious regress – be referred to some yet higher stage of wanting and willing or else taken as a terminal point and therefore as simply not open to reflective or critical-evaluative thought. According to Frankfurt, this problem simply doesn't arise since '[i]t is in securing conformity of his will to his second-order volitions . . . that a person exercises freedom of the will' (p. 90). But again, it is hard to see that there is room for freedom of will where such freedom is conceived as bringing the will into conformity with second-order volitions, or that those volitions can themselves be conceived as somehow introducing the necessary space for freedom to regain its foothold.

Of course it will be said that there are two sorts of 'will' in question here, the first- and second-order sorts, and that Frankfurt's whole purpose in writing the essay was to point up the crucial distinction between them precisely as a means of explaining how genuine personhood, autonomy and freedom differ from the mere unfettered pursuit of first-order wants and desires. However, this claim is more easily stated as a matter of abstract principle than convincingly borne out as a matter of finding room for those higher-level modes of volition, along with that vital margin of freedom in the absence of which there can be no making sense of the higher/lower distinction except as a merely technical device with no further moral or significant human implications. The main trouble lies in grasping what it could mean for somebody – let us say a rational, reflective, second-order-volition-forming individual who meets all Frankfurt's specified criteria for genuine personhood – to *freely wish* that her first-order desires should arrange themselves in this or that order of priority, or *freely prefer* that one such desire should take second place to another (conflicting) desire on well-considered moral, prudential, long-term beneficial, or other such grounds. After all, to the extent that those wishes are indeed well-considered and subject to reflective assessment of the kind that, in his view, constitutes the sine qua non of authentic second-order will-formation they are *for just that reason* not 'freely' arrived at – singled out for adoption from among some range of competing interests or conflicting priorities – as if through a rationally underdetermined or under-motivated act of choice. Rather, they must be taken to result from a more-or-less lengthy and complex process of rational deliberation whereby the subject achieves, or seeks to achieve, that measure of reflective equilibrium that would allow him or her to draw the appropriate conclusion and, where called for, translate it into action of a likewise appropriate sort. To suppose otherwise is to endorse the doctrine of 'doxastic voluntarism' according to which we can, often do and indeed often should choose or decide what's best to believe on the basis of preference or inclination as opposed to the basis of empirical evidence or rational-demonstrative warrant.[11] Among further objections to this idea – one espoused by latter-day pragmatists or 'strong'-descriptivists like Richard Rorty – is the standing temptation it offers to indulge a strain of Mary-Poppins-like wishful thinking as well as those other, more dangerous or sinister kinds of self-deception that Bertrand Russell anatomized in his response to William James concerning the latter's essay 'The Will to Believe'.[12] Also there is the case – of major import for the history of growing resistance to various forms of religious and political persecution – that since beliefs are properly arrived at through a process of persuasion by the best evidence, arguments or reasons to hand and therefore cannot (or should not) come down to a matter of wish-driven preference or choice therefore it is grossly unjust to punish people on account of their heterodox beliefs.

So there are problems with Frankfurt's central idea that one can keep open that vital space for the exercise of human freedom by making it a criterion of personhood that persons – as distinct from non-human animals or sub-personal human beings – should be capable of having second-order volitions and moreover of identifying with them in so deep or self-constitutive a way as to block the threatened regress from stage to stage through endless orders of reflection. What is chiefly problematic is just that equation between regress-blocking potential and fixity of purpose – or 'determination' in both senses of the word – as that which distinguishes genuine (mainly second-order) volition from mere first-order desire. For this inescapably poses the question with regard to doxastic voluntarism, that is, the question as to whether we can make any sense of the idea that beliefs are volitional, i.e., that they result from some exercise of will or decision to pursue one or other of the options among our preferentially ranked array of first-order wants or desires. That Frankfurt subscribes to some such idea – that in his view it is what makes room for freedom and sets persons apart from sub-personal (whether human or non-human animal) creatures – is strongly implied by numerous statements in the course of his essay. Thus, for instance, 'it is having second-order volitions, and not having second-order desires generally, that I regard as essential to being a person' (p. 86). Here he seems to be drawing a three-sided distinction between (1) non-human animals with first-order desires plain and simple, (2) human persons who have both first and second-order desires along with second-order volitions, and (3) that intermediate and in some way defective class of human beings – Frankfurt rather quaintly calls them 'wantons' – who are so much at the mercy of their first-order promptings, instincts, impulses, lusts, or whatever that they fall short of fully fledged personhood. Of the latter he writes that they 'are not persons because, whether or not they have desires of the second order, they have no second-order volitions' (p. 86).

Actually there is some looseness of phrasing or terminological slippage at points in the course of Frankfurt's essay, since earlier on he can be found positing a bipartite distinction between those 'many animals' that 'appear to have the capacity for . . . "desires of the first order", which are simply desires to do or not to do one thing or another', and the human animal who uniquely 'appears to have the capacity for reflective self-evaluation that is manifested in the formation of second-order desires' (p. 83). As the argument develops, Frankfurt comes to lay greater emphasis on the further – in his view cardinal – distinction between, on the one hand, persons who possess the reflective *and* the volitional capacity to turn their second-order desires into a fully motivating power of will and, on the other, 'wantons' in whom those second-order desires are too weak, confused, or indecisive to have any such effect. What this shows, I think, is his increasing sense of the above-mentioned problems with his basic first-order/second-order dualism. To begin with there is the problem about infinite regress to which one, albeit debatable solution is the citing of volitions (rather than desires) as the ultimate or buck-stopping locus for ascriptions of genuine second-order will. Then there is the problem as to why second-order desires should be thought of as any freer – or any less subject to various heteronomous drives and compulsions – than first-order desires. That the term 'heteronomous' imposes itself here as the best, most philosophically pointed or relevant term in this context is again a sure sign of just how deeply Frankfurt's thinking is caught up in the formal structure and, resulting from that, the various internal conflicts and antinomies of Kantian moral philosophy. Just as Kant conceives the 'autonomy' of practical reason in terms of its being held to account by a moral law of which it is somehow both author and compliant or non-compliant subject, so likewise Frankfurt appears to conceive volition as that which can somehow be called upon to reconcile the realms of desire (whether first- or second-order) and will (here cast in a jointly legislative, executive, and juridical role). This is indeed what he regards as the difference between 'will' in its everyday or normal philosophic usage as applied to human agents and 'will' in the distinctive sense that it acquires in the course of his own argument. As Frankfurt puts it, the notion of the will 'is not the notion of something that merely inclines an agent in some degree to act in a certain way', but is rather 'the notion of an *effective* desire – one that moves (or will or would move) a person all the way to action' (p. 84).

If we are here pointedly encouraged to think of 'persons' rather than 'agents' the reason is doubtless that Frankfurt sees – or suspects that readers will be apt to see – too close a relation between talk of agents or acts and the idea that first-order desires may carried right through to the point of direct implementation without any need for complicating detours through the realm of second-order reflection upon them. One is reminded of Hamlet's self-reproachful brooding on the contrast between his own state of chronic vacillation – 'sicklied o'er with the pale cast of thought' – and the purely impulsive, unthinking willingness of a soldier like Fortinbras to take any risk for the sake of some perhaps worthless or doomed venture.[13] Of course Frankfurt's point is just the opposite of Hamlet's: that in the absence of second-order reflection on the promptings of first-order impulse we should lack what most conspicuously sets human persons apart from both non-human animals and 'wantons', namely, the capacity for just that sort of self-aware, self-critical, and self-evaluative thought that may sometimes – as perhaps in Hamlet's case – exert an overly inhibiting effect on the power of decisive or resolute action. Thus, for Frankfurt, it consists in the reflective step back from a desire-driven will that all too readily translates into action, or in the pause for thought required by any such refusal to simply go along with the force of first-order instincts, impulses, or appetites. More than that, it consists in the difference between the kind of motivational force exerted by those first-order promptings even when subject to the rule of some purposive coordinating will and volitions of a higher, reflective-evaluative kind such that the will is held in check – or prevented from putting its purposes too quickly into practical effect – by a critical tribunal (call it the voice of conscience or of better judgment) which may or may not turn out to exercise its power of veto.

Here of course we are very much in Kantian country and obliged to fall back, as if necessarily, on legal or juridical idioms and metaphors. Such talk becomes well-nigh unavoidable if one thinks, like Frankfurt, constantly in terms of 'higher' and 'lower' faculties of mind and of these as possessing more or less warrant, authority, or power of command according to their place on a scale that is conceived in strongly hierarchical terms. Moreover, whether taken metaphorically or at face value, those terms connote something like an ordering of moral worth that runs from 'mere' (non-human animal or human subpersonal) desire, via the various intermediate levels of purposive will-formation, to the fully achieved capacity for weighing different possibilities and deciding between them on the basis of reflective and autonomous judgment rather than direct impulse or drive. My point is not so much that this smuggles in certain ideological values under cover of a 'purely' philosophical approach but rather that it shows how pervasive is the influence on Frankfurt's thought of those Kantian ideas about the nature, scope, and proper function of the faculties, along with their due degrees of role-specific subordination one to another. No doubt Kant's doctrine of the faculties is one that he would regard, like most philosophers nowadays, as metaphysically over-mortgaged and hence as meriting serious interest only in a suitably naturalized, de-transcendentalized, or scaled-down (e.g., Strawsonian) descriptivist form.[14] However it is not hard to make out the lineaments of that same Kantian doctrine in Frankfurt's leading premise that the truly definitive mark of genuine human personhood is the capacity to attain that measure of critical-evaluative detachment from our first-order wants which permits the formation of a second-order will superior to and exempt from the impulsions of rationally unconstrained or unreflective desire. For on this account, there is no escaping the idea of an ascent through successive, increasingly complex and more fully human orders of conscious and reflective thought, or again – perhaps more to the point – a descent from that level of achieved autonomous personhood through stages on the downward path to a level of non- or pre-human animal drive. What is at any rate clear from numerous passages in Frankfurt's essay is the fact of his subscribing to a uni-directional or top-down conception of that which fixes the essential difference between human and other sentient beings.

Frankfurt's essay was published in 1971 at a time when that distinction had not yet been challenged with anything like the range of arguments more recently brought against it by proponents of a radically de-anthropomorphized approach to these issues, among them Peter Singer.[15] All the same, there is a certain brisk assertiveness about his remark that 'we do in fact assume . . . that no member of another species is a person', or that 'one essential difference between persons and other creatures is to be found in the structure of a person's will', or again – more strikingly – that '[i]t does violence to our language to endorse the application of the term "person" to those numerous creatures which do have both psychological and material properties but which are manifestly not persons in any normal sense of the word' (pp. 81–2). These sentences all occur within the first two pages of Frankfurt's essay and take rise from his opening critique of P. F. Strawson's conception of the person (in his book *Individuals*) as just that type of entity to which can be ascribed both predicates that specify physical or corporeal characteristics and predicates that specify states of mind, intentions, beliefs, psychological conditions, and so forth.[16] What Frankfurt objects to about this conception is the fact that it seemingly admits to the category 'person' a variety of other (non-human) animal beings in relation to which, by his own lights, we should adopt a quite different attitude and not 'do violence' to language by sinking – or even by occasionally blurring – the relevant line of demarcation. Thus it is safe to suppose that Frankfurt would have raised a strong and principled voice of dissent had he come across Singer's proposal that certain non-human animals – among them but not exclusively the higher primates – should properly be accorded the basic rights that are taken to go with the possession of personhood on any unprejudiced understanding of the term, that is, quite apart from the largely irrelevant (as Singer sees it) issue of biological species-membership.[17] For Frankfurt, conversely, 'no animal other than man . . . appears to have the capacity for reflective self-evaluation that is manifested in the formation of second-order desires' (p. 83).

III

I think it is liable to strike the reader of these and other passages in Frankfurt's essay that his insistence on the point has an air of protesting too much, or of fending off doubts on this score – doubts concerning the security of just that prescriptive line of demarcation – that have been steadily gathering in strength since the time (actually well before Darwin) that science started to point towards an outlook of thoroughgoing naturalism with regard to human beings and their status vis-à-vis other animals.[18] Here I might instance some marvellously deft observations in Agamben's *The Open* concerning this pressure on received (whether religious or secular-humanist) ideas of human exceptionalism and the various ways in which thinkers of differing doctrinal adherence managed to evade or accommodate the challenge of an emergent evolutionary-naturalist world view.[19] It seems to me that Frankfurt's essay is a late offshoot of the same dilemma confronted by those who have refused to accept that world view at its full and, as they see it, humanly degrading or ethically debilitating force while none the less declining to seek refuge in the alternative appeal to a realm of anti-naturalistic (e.g., religious or Kantian) values that are taken to transcend any such grossly reductive physicalist approach.

That we are here addressing matters of a plainly metaphysical if not directly theological import is nowhere more evident than in a passage where Frankfurt stakes out the ground – or the region of conscious and self-conscious or reflective being – which belongs exclusively to human persons, or to those capable of attaining genuine personhood.

> The concept of a person is not only, then, the concept of a type of entity that has both first-order desires and volitions of the second order. It can also be construed as the concept of a type of entity for whom the freedom of the will may be a problem. This concept excludes all wantons, both infrahuman and human, since they fail to satisfy an essential condition for the enjoyment of

> freedom of the will. And it excludes those suprahuman beings, if any, whose wills are necessarily free. (p. 89)

This passage can be heard to echo any number of theologically inspired disquisitions concerning the proper human place – more precisely: the range of niches available to humans of various types – on the 'great chain of being' that was taken to run all the way from God, via the hierarchy of angels, to human beings, non-human animals, and thence on down through the sundry grades of vegetable or wholly inanimate mineral life.[20] Moreover, it seems to share their concern with defining what constitutes the human norm in contradistinction on the one hand to that which falls outside and below the norm and, on the other hand, to that which so far transcends it as simply to preclude all the conflicts and complexities of human existence.

Not that Frankfurt is in any way committed to a theocentric or religiously motivated outlook. After all, he is talking here about a certain well-defined range of issues in just those closely related areas – philosophy of mind, philosophical psychology, ethics, or metaphysics of the will from a broadly naturalized viewpoint – that need not (and nowadays for the most part do not) involve any overt or covert appeal to a validating ground beyond the compass of unaided human reason. Still there is there is a definite line of descent that runs from that old theocentric conception, through Kant's doctrine of the faculties and the complex, strictly ordered system of relationships between them, to Frankfurt's likewise strongly hierarchical conception of first- and second-order desires or volitions. Moreover this affinity is underlined by his idea that we most closely resemble non-human animals when we allow first-order desires to govern our conduct without the kinds of constraint and guidance afforded by reflective second-order volition, while conversely we attain to authentic personhood by engaging in just such higher-level processes of thought, self-criticism, and effective (i.e., practical even if difficult and sometimes temporally drawn out) will-formation. At any rate his use of 'wanton' – a censorious mode of description, even if deployed with a certain tongue-in-cheek archaizing tone – cannot but suggest that we take our intellectual-moral bearings from a scalar conception of the various orders of being with the human (unlike the animal or the divine) occupying a certain zone of the scale that is always potentially the site of a struggle waged between conflicting forms or forces of desire. Thus the two classes of 'wantons' and 'persons' are mutually exclusive if taken in the full, unqualified sense of each term since it is the having of second-order volitions, not just second-order desires, that qualifies a human being for personhood while the wanton is defined precisely as lacking any such capacity for moral character-shaping on the basis of rational self-evaluation combined with critical-reflective judgement and – crucially – a will to act upon any verdict thereby arrived at. What typifies a wanton is the fact that 'he does not care about his will', that '[h]is desires move him to do certain things, without its being true of him either that he wants to be moved by those desires or that he prefers to be moved by other desires' (pp. 86–7). What distinguishes persons, on the contrary, is that they make some desires truly and properly their own through a process of deliberative second-order thought that enables them to select and to will just those among the range presently competing for notice which qualify as worthy of adoption on all relevant, i.e., rationally and ethically acceptable grounds.

Here again Frankfurt reveals the Kantian lineage of his central concepts by insisting – as against any too rationalist or over-intellectualized account of these matters – that it is the exercise of will or capacity for such exercise that singles out some and not other human beings as eligible candidates for personhood rather than the exercise of rational thought if conceived in isolation from the active will. In his view 'a wanton may possess and employ rational faculties of a high order', since '[n]othing in the concept of a wanton implies that he cannot reason or that he cannot deliberate concerning how to do what he wants to do' (p. 87). What is missing from such deliberations when undertaken by the 'rational wanton', so Frankfurt maintains, is precisely that person-constitutive element of will. That is to say, it is the uniquely (though not universally) human capacity to take up a critical-reflective distance from our first-order desires and yet – where deemed appropriate – to adopt, endorse, or decisively confirm them as measuring up to our own best standards of moral or veridical warrant. So rationality, even 'of a high order', is not enough to justify the ascription of personhood in so far as '[w]hat distinguishes the rational wanton from other rational agents is that he is not concerned with the desirability of his desires themselves', and thus 'ignores the question of what his will is to be' (p. 87). Though Frankfurt doesn't say so, I suppose one could conclude that the upshot of such conduct – if pushed to a pathological extreme – is the Marquis de Sade's distinctly crazed but weirdly 'rational' imagining of the multiform perversities to which certain appetites or cravings can be carried through a rigorous working-out of their various possible or logically conceivable permutations.[21] Thus the rational wanton is inexorably driven by first-order desires and by the drive to maximize their means of fulfilment through application of rational decision-procedures yet is totally incapable of conducting any enquiry as to whether they are justifiable, that is, whether they can stand up to reflective or critical-evaluative scrutiny from a higher (second-order) viewpoint. 'Not only does he [the rational wanton – here for once the gender-non-specific masculine pronoun seems altogether appropriate] pursue whatever course of action he is most strongly inclined to pursue, but he does not care which of his inclinations is the strongest' (p. 87). Yet it also needs stressing, on Frankfurt's account, that rationality is a necessary if not sufficient condition for personhood rather than mere human-being or membership of the species homo sapiens. After all, 'it is only in virtue of his rational capacities that a person is capable of becoming critically aware of his own will and of forming volitions of the second order', in which case '[t]he structure of a person's will presupposes . . . that he is a rational being' (p. 87).

It seems to me that there are deep-laid conceptual tensions in Frankfurt's argument here and that these have to do with the underlying conflict between his basically Kantian conception of will as a matter of rational constraint upon the unregenerate promptings of desire and his semi-naturalised conception of human beings – even fully fledged 'persons' – as inescapably subject to just such promptings, however qualified or held in check by the countervailing agency of second-order desires and/or volitions. For although this idea of a dualism between lower and higher levels of human being is one with distinctly Kantian antecedents it departs very markedly from Kant in adopting the term 'desire' to describe not only the former but also the latter, i.e., that level of second-order reflective, self-critical, and sometimes desire-thwarting evaluative thought wherein human will is most authentically manifest. For Kant, such a usage would certainly have betrayed an attachment to some grossly inadequate conception of ethical values, a naturalistic conception that reduced them to merely 'pathological' products of human will in its lowest, desire-driven or appetitive mode.[22] To this extent he can clearly be seen to inherit a Christian-influenced tradition of thought which stresses the unregenerate nature of our post-lapsarian state and imposes a pitiless divorce between moral law as enjoined upon the subject by the autonomous, self-legislative power of practical reason and that other barely human (indeed near-animal) realm of 'inclination' where instinct and desire hold sway. At its most extreme, in the writings of Saint Augustine, this tradition has given rise to some strange and at times fairly comic theological-ethical contortions, as in Augustine's notion that before the Fall male erections were fully subject to control by the conscious and deliberative rational will rather than (as now) seeming to possess a perverse will of their own. (For the record: Augustine compares that erstwhile happy state to the capacity of certain people with unusual musculature to wiggle their ears with extraordinary skill or break wind to impressively sustained and musical effect.[23]) That a kindred way of thinking has managed to exert so powerful a grip across such a range of otherwise diverse disciplines, schools and periods of thought is doubtless a result of its fitting so well with Plato's famous conception of the human soul as a charioteer pulled aloft by one horse toward the heaven of contemplative reason but dragged down by the other to the realm of ignoble sensuous appetite or instinctual desire.[24] What Kant gives us in his practical philosophy – conceived as the cornerstone to his overall system or 'architectonic' – is essentially a doctrine of the faculties that seeks to make humanly intelligible sense of this starkly Manichaean vision.[25]

It is therefore not surprising, given all the well-known problems with Kant's remorselessly dualist view, that Frankfurt – like many philosophers nowadays – rejects any rigid application of the doctrine that opposes the dictates of rational (autonomous) will to the promptings of un-self-controlled (heteronomous) desire. One need not altogether go along with Jacques Lacan's scandal-provoking claim in 'Kant avec Sade' in order to see how very odd is the idea that a subject should be thought of as enjoying his or her greatest scope for autonomy of moral conscience or ethical choice at just the point where their inclinations are most strictly and remorselessly held in check, albeit at the bidding of a rational will that is somehow an expression of their own more principled or higher self.[26] Nor need one subscribe to an overly sanguine view of human nature in order to think that there is something wrong with Kant's Augustinian idea of human instinct, desire, and inclination as intrinsically corrupt or as always exerting a pernicious effect when allowed to interfere with the deliberative workings of 'autonomous' practical reason. Frankfurt is decidedly at one with the majority of present-day ethical naturalists in wishing to break with that hugely problematic legacy and abandon the idea of a sharp dichotomy between instinct, inclination, and desire on the one hand and reason, principle, volition, and will on the other. Hence, to repeat, his use of the same word 'desire' to describe both first- and second-order motivations. This suggests that the two orders are not so much locked in a struggle of absolute opposites or mutually exclusive drives and principles but engaged in a contest where they have at least enough in common for the issue and its outcome to possess real import at a humanly meaningful rather than a quasi-theological level. Moreover, Frankfurt's qualified naturalist sympathies show up in his account of how second-order desires relate to second-order volitions, that is, through an exercise of rationally motivated choice that determines just which of those maybe conflicting or as yet unclearly prioritized desires shall become identified with the subject's active will.

Still there are definite limits to Frankfurt's naturalism and also to his scope for throwing off that Platonic-Augustinian-Kantian legacy, given the marked persistence in his thought of other dualist themes which effectively replicate the same structure of assumptions. Chief among them, since bearing the greatest argumentative and philosophic load, is the distinction between first- and second-order desires, along with that between second-order desires and volitions. Thus volitions are taken – in distinctly Kantian mode – to fall within the category of 'desires' in so far as they belong to that range of motivating interests with which the subject identifies as a matter of active and personal commitment, but to stand outside (and potentially against) lower-order desires in so far as these latter embody the promptings of mere sensuous, instinctual, or creaturely gratification. This distinction is pressed home with maximum force in Frankfurt's comparison between two kinds of drug-addict, the one perpetually at war with himself over his constant struggle (and repeated failure) to break the habit which he knows to be destroying his life, while the other is subject to no such agonies of self-division and self-reproach since he has entirely given in to the habit. Of course this omits the case of the addict who has successfully managed to kick the habit, an omission that reflects Frankfurt's main interest in the workings of ambivalent, divided, or conflictual rather than straightforwardly efficacious moral will. The second kind of addict is a type-case of the 'wanton', one who 'does not prefer that one first-order desire rather than the other should constitute his will' (p. 88). Indeed, as Frankfurt puts it in an oddly ambiguous sentence, 'the wanton addict may be an animal, and thus incapable of being concerned about his will . . . . In any event he is, in respect of his wanton lack of concern, no different from an animal' (ibid.).

This passage leaves one in doubt – whether through deliberate or accidental looseness of phrasing – as to just what is meant by the non-human animal reference. That is, one is hard put to decide between the reading 'this is what happens in the case of mere non-human animals', and the reading 'this kind of blank indifference with regard to the relative value or worthiness of various motives and desires is what reduces human beings to the level of mere subhuman animality'. At any rate it is clear that Frankfurt adopts something very like the Kantian view of desire as inherently a product or expression of whatever in our unregenerate natures inclines us to act against the dictates of practical reason, that is, the maxims of virtuous conduct laid down by our own autonomous or self-legislative better selves. Hence the contrast he draws with the first type of drug-addict, one who sincerely hates his habit and despises himself for his failure to kick it yet is unable to find the requisite determination or strength of will. More precisely, his self-loathing and his genuine wish to pursue the alternative course are just not strong enough to subdue or vanquish the overwhelming first-order compulsion that impels him to continue in the erroneous path. In short, '[t]hese desires are too powerful for him to withstand, and invariably, in the end, they conquer him . . . [h]e is an unwilling addict, helplessly violated by his own desires' (p. 87).

As I have said, this is what locates Frankfurt's essay squarely within that mainstream tradition of Western ethical thought which identifies morality with some kind of check upon our natural, instinctual, or (above all) our merely 'animal' inclinations. Most often – and most notably in Kant – it has taken the form of a self-denying, self-thwarting, self-abnegating drive with its likeliest source in the kinds of ascetic imperative enjoined with varying degrees of rigour by religious (especially Christian) doctrines of salvation through mortification of the flesh.[27] What is so striking about Frankfurt's albeit more moderate, less Manichaean rendition of this theme is the extent to which it replicates the Kantian structure of argument along with all the conflicts, aporias, antinomies, and other such symptoms of conceptual strain to which that argument gave rise. Moreover, they are here expressed in terms of a veritable psychomachia, a drama acted out between contending desires, inclinations, impulses, motives, volitions, or exertions of will that often reads like an allegory with personified virtues and vices in the mode of *Pilgrim's Progress* or a medieval morality play. Such is the passage cited earlier where Frankfurt describes the unwilling addict as 'helplessly violated by his own desires', and also the following which I shall quote at length since it captures very well the way that this updated, semi-naturalized version of Kant tends to create its own dramatic plot-line with a suitably varied (if somewhat typecast) list of characters to move the action along.

> The unwilling addict identifies himself . . . through the formation of a second-order volition, with one rather than with the other of his conflicting first-order desires. He makes one of them more truly his own and, in so doing, he

> withdraws himself from the other. It is in virtue of this identification and withdrawal, accomplished through the formation of a second-order volition, that the unwilling addict may meaningfully make the analytically puzzling statements that the force moving him to take the drug is a force other than his own, and that it is not of his own free will but rather against his will that this force moves him to take it. (p. 88)

What is so curious about this passage is that it treats those statements not only as 'meaningful', i.e., as making good sense to the unwilling addict and to anyone who wishes to grasp his tragic predicament but also as possessing a strong claim to philosophical validity since, after all, they can be seen to articulate Frankfurt's own considered views on the matter. Thus, we are here presented with a story of mental desires, compulsions, conflicts, dilemmas, identifications, withdrawals, alliances, forces and counter-forces that seem to be envisaged almost after the manner of those ancient theories of consciousness or mind that posited the existence of homunculi, that is, Russian-doll like miniature beings who or which served to 'explain' the otherwise mysterious workings of our various sensory, cognitive, and intellectual faculties.

Of course, I am not suggesting that Frankfurt subscribes to anything like that doctrine in its primitive or mythical form. Rather, I am pointing to the fact that he tells this story – the story of the unwilling addict – very much in the style of a novelist who identifies so closely with the thoughts, feelings and self-interpretations of a given fictional character as to take theirs as the privileged narrative viewpoint in relation to which all events and other characters will ultimately fall into place. This is why, rhetorically and thematically speaking, it belongs very much to that older tradition where the novel first emerges from the morality tale or play and where there is not, as yet, any room to exploit those added possibilities of complex understanding that come of a more detached, ironic, or sceptically-inclined relation between author and character.[28] No doubt there is a sense in which the unwilling addict is here being put in his place for having manifestly failed to achieve that degree of practical wisdom that would allow him to act upon his better judgement by making it his own as a matter of effective will, that is, by identifying closely enough with his habit-kicking as against his habit-reinforcing desires. All the same it is clear from the various descriptions of the ongoing conflict or turmoil of motives that make up the unwilling addict's mental life that his is the chief point of reference both for Frankfurt's understanding of personhood in general and for what we are to think of his counterpart, the wanton, in whom no such struggle takes place and who therefore – it is argued – falls below the threshold for admission to the class of persons as distinct from mere human beings. Thus, in contrast to the passage cited earlier concerning the unwilling addict and his ventures into self-analysis, there is Frankfurt's description of the wanton who may well suffer the same kind of first-order conflict between desire to continue with the habit and desire to give up but who 'does not prefer that one of his conflicting desires should be paramount over the other' (p. 88). In such cases, '[i]t would be misleading to say that he is neutral as to the conflict between his desires, since this would suggest that he regards them as equally acceptable' (ibid.). On the contrary, what is lacking in the wanton is precisely that capacity to stand back and to assess, judge or critically evaluate his own first-order desires that would be requisite even to a subject who found himself unable to plump for one or the other of some sharply conflicting pair since the reasons and motives for each seemed to him so evenly balanced. As concerns the wanton, however, '[s]ince he has no identity apart from his first-order desires, it is true neither that he prefers one to the other nor that he prefers not to take sides' (p. 88).

IV

Hence perhaps Frankfurt's ambivalent remarks, as previously noted, with regard to what he thinks of as the hard-to-draw line between human and non-human (animal) wantons, since both strike him as manifesting the same incapacity for critical, reflective, second-order, revisionary, or self-evaluative thought. Just how hard-to-draw, on his account, comes out in a passage where the human wanton is described in terms that would seem to fit him more for placement in a zoo – or in one of those medieval bestiaries depicting human vices in animal form – than for the company of humankind. Thus,

> [t]he wanton addict cannot or does not care which of his conflicting first-order desires wins out. His lack of concern is not due to his inability to find a convincing basis for preference. It is due either to his lack of the capacity for reflection or to his mindless indifference to the enterprise of evaluating his own desires and motives. There is only one issue in the struggle to which his first-order conflict may lead: whether the one or the other of his conflicting desires is the stronger . . . . [I]t makes no difference *to him* whether his craving or his aversion gets the upper hand. He has no stake in the conflict between them and so, unlike the unwilling addict, he can neither win nor lose the struggle in which he is engaged. (p. 89)

This passage is remarkable for several things, among them its sheer determination to enforce the person/wanton dualism and its consequent refusal – somewhat against the logical grain of the argument here – to concede that such distinctions must at best be a matter of degree or of locating various points on a scale not only as between different individuals but also with respect to any given (not to say 'self-same') individual from one to another time or context. Where it raises logical problems is by presenting this as in some sense a struggle 'in which he [the wanton addict] is engaged' while at the same time working hard to persuade us that in truth there can be no such engagement in the wanton's case since he is altogether lacking in just those capacities which, if he possessed them, would earn his admission to the class of fully fledged persons. In short, '[w]hen a *person* acts, the desire by which he is moved is either the will he wants or a will he wants to be without', whereas '[w]hen a *wanton* acts, it is neither' (p. 89). Again there is something odd – logically as well as metaphysically perplexing – about the idea of *wanting* or *not wanting* to have (i.e., to have one's desires, thoughts and actions directed by) some particular exercise of will, or again, about this notion of the will to engage in or desist from this or that course of action as itself somehow subject to the prompting of our 'wants', preferences, or inclinations. For it is surely just the point of Frankfurt's cardinal distinction between first- and second-order desires that the latter should be thought of as strictly – indeed, by very definition – superordinate since this is what leads to their adoption in the form of definite commitments – or volitions – and hence their taking-up into the subject's will through an act of decisive identification with those that have proved most deserving of his or her allegiance.

So it is hard to make sense of his idea that the will must in turn be considered subject to 'wants', whether positive or negative, and therefore – it would seem – be thought to exert its authority only on condition of approval by the range of desires that predominate in some given subject at some given time in response to some given situation. Most likely my use of the word 'subject' in two different senses in the course of that last sentence will come across either as clumsy phrasing or as a kind of archly philosophical pun that strains the limits of semantic and conceptual decorum. However, that sense of strain is endemic to all thinking (at least, all mainstream philosophical thinking) about the nature, scope and limits of human epistemic and ethical responsibility, that is, the extent to which human beings can properly or intelligibly be held accountable for arriving at certain beliefs or convictions and putting them into effect through precept or practice.[29] This is partly, as I have said, a question of the logical regress that opens up whenever one seeks – in the manner first established by Plato and carried to its high point of elaboration by Kant – to specify a pecking order of the faculties. It is reinforced or brought home in epistemic and ethical terms by the way that such thinking makes it hard to avoid that problematical idea of the subject as at once the very locus of free will, personhood, or moral autonomy and the locus of subjection to precepts or dictates of which it is somehow (paradoxically) both author and passive recipient.

There is no escape here – in Frankfurt's naturalized or de-transcendentalized version of the Kantian doctrine – from the problem as to how one can save any plausible or non-contradictory account of what is supposed to constitute the essential difference between humans (or persons) as autonomous beings and non-human animals (along with wantons) as falling entirely under the rule of heteronomous compulsions or desires. Beyond that, it is a matter of the problems faced by any attempt to achieve a workable modus vivendi between the defence of free will, autonomy or Kantian practical reason conceived as distinctively human attributes and a moderately naturalized approach that would take on board the more salient aspects of present-day, scientifically-informed thinking and yet preserve space for a sufficiently robust conception of those same attributes. On most such accounts this would be a more expansive or accommodating space than is allowed for by genial physicalists like Dennett in his book *Elbow-Room: The Varieties of Free Will Worth Having*.[30] Still it would be nowhere near as large as that claimed by upholders of the strong-autonomist line who maintain – on various philosophic grounds – that the physical sciences (especially neurophysiology) are in the grip of a massive category-mistake when they suppose that the problem of consciousness or the issue of mind/brain identity could ever be clarified, let alone resolved, by any possible advance in the scope or reach of scientific explanation.[31] What emerges most strikingly from the passages that I have cited here, as likewise from a good many recent efforts in a moderating vein, is the fact that when thinkers strive to steer a path between contrary temptations or opposed sources of error they will often end up by ignoring or staving off rather than resolving the resultant dilemma.

I have suggested that these problems all bear witness to a deep-laid, culturally widespread and philosophically recurrent desire to fix a distance between the human and the non-human animal, or again (as Frankfurt would have it) between full personhood and whatever in our natures must be thought to fall short of that status through passive or unthinking enslavement to the rule of first-order motives and desires. His essay thus belongs to that multiform and venerable genre which Agamben has memorably traced in his book *The Open: Human and Animal*, namely the long series of attempts – in philosophy, theology, anthropology, ethnology, psychology, linguistics, and other fields – to carve out a unique (and uniquely distinguished) niche for homo sapiens despite, or because of, the growing sense that this distinction was under threat from developments spawned by some of those same disciplines.[32] What emerges from Agamben's survey is the way that thinkers were forced back upon ever more resourceful and inventive but also ever more elaborate, tenuous, wire-drawn or conceptually extravagant modes of argumentation so as to keep that niche open or hold that distinction in place. Not that Frankfurt's essay could fairly be described in any such terms, argued as it is in a clear-headed analytic style and with meticulous regard for the scope and limits of philosophic reasoning vis-à-vis other (e.g., psychological or natural-scientific) claims in this area. Still it is clear from the outset that this will be another contribution to the same legacy of thought, one that seeks to re-draw the human/non-human (or person/sub-personal human) line in a more naturalistic way, but which continues to insist that its drawing is prerequisite to any grasp of what it means to possess and exert that capacity for freely-willed autonomous choice that sets the former categorically apart from the latter in each case.

That is to say, like other recent ventures in this semi-naturalized and quasi-Kantian vein – among them John McDowell's *Mind and World* – the approach is very apt to hold out in one hand what it promptly takes back with the other.[33] Thus, we seem to be vouchsafed renewed access to a range of possibilities for staving off the threat of hard-line naturalism and determinism, possibilities that had once (not so long ago) been thought of as beyond the pale since they had been irredeemably tainted by association with Kantian transcendental idealism and its various (presumptively otiose) metaphysical commitments. In McDowell this has to do with refurbishing Kant's 'spontaneity'/'receptivity' dualism, taken to denote not a sharp and thus inherently trouble-making distinction like that between 'concept' and 'intuition' but rather a pair of mutually dependent or inseparably inter-involved capacities whose jointly active/passive synthesis is such as to resolve all the bad antinomies that have plagued philosophy from Kant on down. Chief among them are those of subject and object, free will and determinism, and of course mind and world, all of which – so McDowell firmly believes – can be laid to rest through the benign ministration of a Kantianism suitably filtered and revised (with the customary help from Wittgenstein) so as to coax it down from the transcendental-metaphysical heights and restore it to a properly naturalized sphere of communal practices or life-forms while at the same time deploying its normative resources to protect against any too strongly reductionist strain of philosophic naturalism. In Frankfurt, it takes the form of a first-order/second-order distinction as applied to human wants, desires, or volitions which preserves something – a structural analogue at least – of Kant's categorical distinction between the strictly 'pathological' realm of subjectively driven (especially sensuous) inclinations and the sovereign dictates of autonomous moral law. Here again, as with McDowell's Kant, that dichotomy is treated to a moderately naturalizing gloss whereby to eliminate its hapless since delusive reliance on a realm of transcendentally grounded moral precepts yet also to retain something of its role as a means of suggestively marking – rather than rigidly enforcing – the two distinct orders of desire.

Thus if Frankfurt, again like McDowell, takes a less extreme view than Kant concerning the difference between persons and non-human animals then this also applies, as might be expected, to their respective views of the difference between those first-order 'lower' desires that pertain most directly to the nature of our instincts, drives, or sensuous appetites and those second-order 'higher' volitions that require the application of critical-evaluative thought and the exercise of moral will. That is to say, Frankfurt is sufficiently of his own cultural time and place to make some allowance for the various ways in which a science-dominated naturalistic world view has raised obstacles to any such confident beating of the moral bounds, or at any rate made it far more difficult – philosophically or ethically speaking – to defend a full-scale anti-naturalist approach along strict Kantian lines. What has mostly taken the place of such efforts is just the kind of moderately naturalized account that finds expression in Frankfurt's essay, even if it appeared some decade before the wider movement in that direction to which McDowell's book bears witness. It is for this latter reason, I think, that Frankfurt writes with a vigour and a sense of breaking new philosophical ground that is very rarely apparent in more recent work on this and related themes. All the same it is worth looking more closely at some of the conceptual tensions or unresolved conflicts of aim and priority present in his thought. On the one hand, these have their ultimate source in the Kantian antinomies of pure and practical reason, the largely unacknowledged impact of which on analytic philosophy in its various departments or sub-divisions is a story that I have told in some detail elsewhere.[34] On the other, they also result from the encounter between that residual Kantian way of thinking and a naturalistic imperative which, even in its moderate forms, has put such ideas under growing philosophical strain. Indeed that strain has if anything been ratcheted up – rather than (as the received wisdom would have it) effectively dissolved or conjured away – through the sense that there must be some middle-ground position capable of easing these conceptual cramps by showing them up as nothing more than products of an old and nowadays discredited metaphysical world view.[35]

The effect of such therapeutic endeavours is very often to conceal from their own practitioners the extent to which they are still – and now less wittingly – in the grip of those same compulsive dualisms that are thought to have been left safely behind with the transition to a new, metaphysically unencumbered and sufficiently (though not overly) naturalized or 'de-transcendentalized' approach. It is here, I would suggest, that Frankfurt's essay has its special relevance and interest with regard to the way that these debates have gone since it appeared almost four decades ago. For there is much to be learned about philosophy's perceived situation vis-à-vis the natural sciences and its need to stake out some distinctive territorial claim in the fact that the essay has been so frequently cited and that his notion of second-order desires or volitions has become such a staple of attempts to vindicate the idea of free will as a marker of human (more precisely: of personal) status and identity. Along with that goes the precept of an absolute or principled distinction between human persons and those other non- or sub-human animals whose lack of this second-order capacity for self-critical, reflective or evaluative thought about their own first-order desires is what marks them as belonging to a lower rank on the scale of sentient being. It seems to me that there are large problems with Frankfurt's attempt to make good this case, and moreover, that those problems have a lot to do with his and his readers' wish to maintain that decisive margin of the truly, irreducibly, or properly human that would allow us to keep determinism at bay and hang on to our sense of moral privilege despite and against sundry present-day (mainly natural-scientific) threats and encroachments.

Indeed those problems come through with unignorable force in the last few paragraphs of his essay where Frankfurt takes stock of what he hopes to have achieved and enters a number of surprisingly large caveats in that regard. These caveats appear to be prompted by his desire to place the maximum possible distance between his own views on the free-will/determinism issue and Roderick Chisholm's concept of agent-causation, namely the idea that 'human freedom entails an absence of causal determination', that 'whenever a person performs a free action . . . it's a miracle', and moreover that 'a free agent has "a prerogative which some would attribute only to God: each of us, when we act, is a prime mover unmoved"' (p. 93).[36] Frankfurt is confident in rejecting such notions as a mere recrudescence of religiously-inspired and philosophically as well as scientifically discredited ideas about human beings as uniquely placed in the divine order of things by virtue of their absolute exemption from the otherwise universal laws of deterministic cause and effect. Besides – as he very reasonably comments – 'why, in any case, should anyone *care* whether he can interrupt the natural order of causes in the way that Chisholm describes?' (ibid.). Yet it is clear that Frankfurt does have concerns of this kind and, moreover, that his whole line of argument here is very largely driven by the desire, motive, volition, or will to carve out just such a space of freedom for the exercise of human autonomy as against all manner of perceived encroachments from the natural-scientific quarter. My point in listing those various candidates for the role of carver-out is that they each have a definite place in Frankfurt's account of how human beings – or persons – manage to buck the scientific (determinist) trend and also that they each mark a certain point on the scale that runs from a naturalistic to a broadly Kantian or autonomist conception of human agency. What is most striking about these last few paragraphs is the extent to which Frankfurt backs away from any strong autonomist claim while none the less continuing to fend off the combined (at least as he perceives them) threats of naturalism and hard-line physicalist determinism.

Thus the paradoxes at this stage come thick and fast, to the point where it is remarkably hard to discern just what is meant by the various terms that make up Frankfurt's philosophical lexicon. 'Whatever his will', we are told, 'the will of the person whose will is free could have been otherwise; he could have done otherwise than to constitute his will as he did' (p. 94). But in that case clearly there must be some superordinate constituting agency – some 'want', to use his own expression – that is so placed within the overall structure or economy of motivating drives as to rank higher than will in terms of its directive or governing power. This interpretation seems to be supported by Frankfurt's going on to say that 'the assumption that a person is morally responsible for what he has done does not entail that the person was in a position to have whatever will he wanted' (ibid.). So far as I can see, the only way to make logical sense of this claim is to take it as involving the huge concession – huge, that is, for anyone who wishes to defend the principles of human free will or autonomy – that we can only will what we 'want' to will, or again (in negative terms) that our will to do or to refrain from doing this or that is itself constrained and potentially subject to veto by the force of some other, more powerful desire that overcomes any reasons, precepts, or guiding principles that the will may muster on its own behalf. All the same, 'the assumption that a person is morally responsible for what he has done does not entail that the person was in a position to have whatever will he wanted' (ibid.). Here again it is hard to know what Frankfurt means since this sentence seems to push the spiral to a higher stage where not only is the person's will subject to wants that may ultimately thwart or frustrate it but those wants are themselves subject in turn to various possible 'positions' – or scope-restrictive predicaments – in which the person concerned is unable (and moreover, by the logic of the case, unwilling) to act upon them.

Thus one is puzzled to grasp what can possibly distinguish the willing from the unwilling drug addict as Frankfurt describes their respective and, he thinks, decisively different cases. After all, they are both stymied – prevented from acting in their own best interests, everything considered – by a crucial failure of will even if that failure, according to Frankfurt, pertains to different levels or stages of will-formation. On his account it occurs either (as concerns the wanton) through a sheer lack of second-order desires with sufficient strength or tenacity of purpose to become second-order volitions or else (as concerns the unwilling addict) through the fact that those second-order desires fall short of what is required to transform themselves into the kind of full-strength, self-defining, or properly personal volition that could constitute a subject's chief motivation for behaving in this or that way. Still, it might be asked – from a more naturalistic or less residually Kantian standpoint – just why we should accept this a priori dualism between, on the one hand, conflicts of motive or allegiance that belong to the presumptively lower sphere of first-order desires and, on the other, conflicts of a more elevated sort which, even where they issue in a failure to adopt the preferable path, none the less bear witness to a higher capacity for suffering such deep-laid dilemmas. For it will then look more like a rearguard attempt to shore up the Kantian-autonomist defences against the threat of a consistent naturalism that would find no philosophically legitimate room – that is to say, no other than face-saving or self-image-protective reason – for any such resort to a priori notions of fully achieved as distinct from merely 'wanton' and to that extent quasi-animal states of human being.

Things get even murkier when Frankfurt remarks, as if by way of clarification, that '[t]he willing addict's will is not free, for his desire to take the drug will be effective regardless of whether or not he wants this desire to constitute his will' (p. 94). But if the issue of his wanting or not wanting to remain in thrall to the addiction is indeed a matter of indifference in this sort of case – if the fact of his carrying on with the habit is enough to place the two possibilities metaphysically, ethically, or psychologically on a par – then it seems to contradict Frankfurt's thesis concerning the crucial distinction between persons, including motivationally torn or self-divided persons, and that other class of beings (wantons, non-human animals, humans falling short of personhood) whose defective status he has explained precisely in terms of that difference which he here lets go with remarkable ease. '[W]hen he takes the drug', Frankfurt remarks, 'he takes it freely and of his own free will.' (p. 94). This claim is very odd – involving what seems a sizable affront to received philosophical ideas as well as to the dominant narco-physio-psychological conception of addictive behaviour – and looks even odder when Frankfurt goes on to say that he is inclined 'to understand the situation as involving the overdetermination of his first-order desire to take the drug', and that '[t]his desire is an effective desire because he is physiologically addicted'. By now it would appear that the willing and the unwilling addicts are in truth – as per the sentence quoted at the start of this paragraph – distinguished only by a nominal shift from one to another distribution of emphases between the three terms 'desire', 'want', and 'will'.

Moreover, paradoxically enough, Frankfurt can then proceed on the basis of just this stipulative re-jigging of semantic bounds to take what amounts to a strong voluntarist line on addiction – or on 'wantonness' of any kind in so far as it issues in bad, irresponsible, recidivist, or self-destructive behaviour – since, after all, his argument has now worked itself around to a point where the Kantian structure of assumptions has pretty much collapsed under pressure from the naturalizing drive that is also a prominent aspect of Frankfurt's thinking. This is why he eventually declares in favour of what seems in ethical as well as in social, political, and legal terms quite a hard-line doctrine of accountability that would – if carried into policy and practice – leave precious little room for pleas of reduced responsibility, mitigating circumstance, or social/cultural deprivation as grounds for special-case treatment. Thus, if the addict's desire is effective on account of his physiological addiction, nevertheless 'it is his effective desire also because he wants it to be' (p. 95). More specifically, '[h]is will is outside his control, but, by his second-order desire that his desire for the drug should be effective, he has made this will his own. Given that it is therefore not only because of his addiction that his desire for the drug is effective, he may be morally responsible for taking the drug' (ibid.). In which case, quite simply, there is no getting the confirmed addict off the moral hook since in the end – as the upshot of all these semantic shifts – it is the addict *ipse*, whatever his degree of psycho-physiological enslavement, who must turn out to be 'willing' not only in the negative sense of passively enduring his addiction but also in the positive sense of actively wanting or desiring to continue with the habit. For we have now reached the stage in Frankfurt's dialectic of the faculties where there would seem as much reason to assert 'Human beings are responsible for everything they do' as to assert 'Human beings are responsible for none of their actions'. And we have reached that stage precisely on account of his essaying a normative conception of personhood that is distinctly Kantian in its basic approach to issues of autonomy, free will, and motivation but which backs away from any full-scale acceptance of the metaphysical doctrines whereby Kant sought to uphold that conception. What results is a semi-naturalized account of the relations between desire, want, volition, and will that ends up by shying away from the determinist conclusion and thus embracing a full-strength voluntarist or autonomist outlook according to which, even if someone's will is 'outside his control', nevertheless 'by his second-order desire . . . he has made this will his own'.

V

This is, to say the least, an extraordinary conclusion and one that should I think give us pause in reflecting on the kinds of hidden liability to which philosophy is subject when it tries to bring off that particular trick of retaining those elements of Kantian thought that suit its argumentative purposes while emphatically rejecting the rest of Kant's transcendental-metaphysical system.[37] The consequence of this is a curious hybrid doctrine which produces what might very well be thought the worst of both worlds: a theory that combines the unyielding rigour of a deontological ethic (i.e., a doctrine of autonomous or self-legislative practical reason whereby the person is held absolutely responsible for each and every act) with a partially naturalized view of human desires, wants, and volitions which effectively excludes any possible room for the exercise of such autonomy by placing will in a subordinate role to the promptings of want and desire. Such is the upshot of Frankfurt's argument – borne out by the various passages cited above – even though it goes drastically against what he takes to be the main gist of his essay, that is, the vindication of a strong conception of human choice and responsibility in firm opposition to determinist or hard-line naturalistic approaches. I would guess that this conflict of aim with outcome is one chief reason for the odd statement, in his closing paragraph, that his understanding of free will 'appears to be neutral with regard to the problem of determinism', despite the 'innocuous appearance of paradox' in the statement – one that Frankfurt seems happy to endorse – that 'it is determined, ineluctably and by forces beyond their control, that certain people have free wills and that others do not' (p. 95).

I think this is not so much a paradox, or, if so, not so much an 'innocuous' paradox but rather something more like a reductio ad absurdum of Frankfurt's whole line of argument, leading as it does to a flatly contradictory statement at least on his own expressly voluntarist principles. The same applies to his subsequent claim that '[t]here is no incoherence in the proposition that some agency other than a person's own is responsible (even *morally* responsible) for the fact that he enjoys or fails to enjoy freedom of will' (p. 95). Here again one has to say that there is a very basic and, given his own assertions elsewhere, a self-contradictory tension between this and Frankfurt's emphatic commitment to the idea of personhood as crucially consisting in the power of second-order volitions to monitor, check, subdue, deflect, modify or simply overrule the promptings of first-order desire. 'Perhaps', as he writes in a final twist to this sequence of paradoxical musings, 'it is also conceivable . . . for states of affairs to come about in a way other than by chance or as the outcome of a sequence of natural causes' (ibid). It is hard to conceive what this might mean – what the alternative possibility might be – if not some version of the Chisholm-style appeal to agent-causation that Frankfurt had roundly rejected just a couple of pages before. I can think of no text that more strikingly demonstrates the kinds of dilemma that philosophers are forced into when they attempt to bring off this kind of balancing-act. Thus a great many of them are still very much in denial – and casting around for some such halfway plausible compromise solution – when it comes to confronting the massive challenge to traditional (including received philosophic) modes of thought represented by advances in the natural-scientific domain.

Indeed it is no exaggeration to say that philosophy's self-mage and sense of its right to exist as a self-respecting discipline of thought is closely bound up with something very like Frankfurt's conception of autonomous personhood as involving the exercise of second-order, reflective thought and judgment. Nor is this at all surprising, given what philosophers typically take to be their proper sphere of interest, if not – any longer – their uniquely privileged realm of special expertise. What his essay brings out with particular force is the extent to which, in this respect at least, mainstream philosophy has tended to articulate the sorts of attitude that many people take toward developments in neurophysiology and other branches of the natural sciences which threaten – or which might well be taken to threaten – the belief in human autonomy. That is to say, it gives carefully worked-out arguments for a view of the relation between desire, wants, volitions, and will (along with all their internal divisions) that in its own way strikingly replicates the sorts of ambivalence that many scientifically-informed people are liable to feel when trying to square their everyday conceptions of what is distinctive about human personhood with the kinds of thinking that increasingly appear to dominate the view from those scientific quarters. It seems to me a fair inference from the rate and direction of various advances in the relevant fields of research – neurophysiology, cognitive psychology, and related areas – that naturalism in its strong rather than in any qualified or hedged-about guise is the only outlook that makes sense in a way that is consistent with those advances and not prone to self-destruct on the kinds of paradox or flat contradiction that emerge in the course of Frankfurt's essay.

No doubt it is the case for various reasons – moral, legal, political, social, and cultural – that the ideas of autonomy and free will are deeply built into our elective self-image as human beings, even if (as naturalists are apt to argue) this is chiefly on account of their conducing to a more stable, less conflict-ridden mode of social co-existence and hence a better prospect for species flourishing by way of natural selection. Philosophy has long been a bastion of such thinking with its sundry suggestions as to where the crucial difference is supposed to reside, from Plato's 'other place' (*topos ouranos*) of ideal forms to Descartes' realm of 'clear and distinct ideas', Kant's domain of synthetic a priori intuitions and concepts as distinct from the deliverances of pure reason, and Husserl's notion of transcendental phenomenology as aimed toward a region of pure eidetic essences.[38] Thus, at least until the middle years of the last century, the majority of philosophers were firmly opposed to any naturalistic approach that would threaten the values of autonomy and selfhood conceived as intrinsically beyond the furthest reach of any physical or causal-explanatory account.

Of course there were notable exceptions, beginning with the ancient Greek atomists and carrying on through a widely-spaced yet continuous and indeed – given the strong (sometimes lethal) kinds of disincentive ranged against them – impressively persistent line of materialist or radically anti-dualist thinkers.[39] Nowadays this situation has changed to the point where those who reject the mind/brain identity thesis are forced very much onto the back foot or may feel themselves compelled to endorse some version of the dualist argument on intuitive or purported a priori grounds. This typically involves either the appeal to qualia ('what it's like' to perceive colours, listen to an oboe, suffer toothache, and so forth) or else – in Searle's somewhat harder-headed but still residually dualist version of the case – the presumptive truth that minds *just are* self-evidently marked by a special quality or attribute (that of intentionality) that could not possibly be realized by means of any inorganic, e.g., silicon-based hardware or support system.[40] The first line of argument, like so much present-day philosophy, trades on a purely linguistic point abut differing modes of talk or conceptualization and the problem – maybe the impossibility, though this is again a linguistic-conceptual matter – of translating the mental-experiential-phenomenological into the physical or neurobiological without significant remainder. The second has the severe disadvantage, albeit often masked by the over-confidence that comes of a priori conviction, of simply taking for granted – as if it were self-evident – the absurdity involved in supposing the idea of 'artificial' (i.e., non-humanly-embodied) intelligence to be anything other than a category-mistake of the most blatant kind.[41]

Cartesian dualism finds a kind of muted, shame-faced or last-ditch expression in Colin McGinn's 'mysterian' thesis that even though minds may be brain-dependent or brain-identical in some ultimate sense it is a sense that we'll never be able to fathom because we are just not bright enough to figure it out.[42] One striking feature of this case is that it totally reverses the Kantian idea of a noumenal domain – that of the 'thing-in-itself' – that is capable of being conceived in the abstract through a speculative stretch of pure reason and must be so conceived if we are to render our knowledge and experience metaphysically intelligible but which cannot, on pain of creating insoluble dilemmas, be ascribed to the remit of knowledge whereby sensuous intuitions are 'brought under' concepts of understanding.[43] For McGinn, conversely, what lies beyond the range of human conceptual grasp is that ultimate material (i.e., neurophysiological) nexus where goings-on in the brain are somehow – through a process which remains, to us, utterly and forever incomprehensible – transmuted into goings-on in the mind. It strikes me after some fairly extensive reading-around in the literature that a good proportion of it – that which strives to discover some via media between the claims of hard-line reductive physicalism and an appeal to the phenomenological self-evidence of 'what it's like' to undergo this or that kind of sensory or subjective experience – amounts to a just series of further variations on the theme first sounded by Frankfurt in his 1971 essay. For it does little more than reiterate the notion of phenomenological ascent from level to level of a consciousness presumed a priori capable of achieving such ascent (since defined precisely through and by that capacity) whilst none the less conceived as being in some sense identical with its physical hardware in order to head off any imputation of lingering dualism or subjectivism. Hence the frequent talk nowadays of 'emergent' or 'supervenient' properties, conceived as strictly *dependent upon* and even (in some rather mysterious sense) *identical with* their physical support-system or means of instantiation yet by no means simply *reducible to* it in the manner proposed by hard-line physicalists or central-state materialists.[44]

I think it is fair to say that 'consciousness', along with its other more specialized cognate terms such as 'intentionality' and 'qualia', has become an *explanandum* that all too often serves as its own *explanans.* In other words it is a name for whatever is both supposed to require some account beyond reach of our present-best (maybe best-attainable) physicalist understanding and taken to constitute precisely the reason, self-evident to consciousness itself, why such understanding must fall so manifestly short. Moreover this involves just the kind of purely notional ascent – along with the attendant risk of vicious regress or mere circularity – that can likewise be seen in the earlier attempts of logical positivists and empiricists to construct a model of 'material' (or empirical) as distinct from higher-order 'formal' (or logical) languages.[45] That programme had two main objectives: to hold a firm line between evidence and theory, in conjunction with that between context of discovery and context of justification, and also – as I have said – to head off any problems of self-reference like those which Russell famously encountered and purported to resolve with his Theory of Types.[46] That the same sorts of problem tend to arise with efforts to conserve some unique, privileged, or distinctive role for human consciousness vis-à-vis what non-human animals are thought to enjoy in the way of mental life or 'internal states' is, I think, one sure indication that such efforts are misconceived.

This applies all the more when they are placed in the service of other, more ethically loaded claims for the qualitative difference between human and non-human animal modes of sentient experience and hence – so the argument often runs – the error of supposing that there exist any rules, codes, or obligations that should regulate 'our' treatment of 'them' in a properly responsible way. Here again such thinking finds a precedent in Kant's idea that the error in question is the sort that typically arises when people mistake the promptings of 'mere' sentiment or creaturely sympathy for the demands of strict moral duty or the maxims and imperatives of practical reason.[47] It is this Kantian conception that is invoked – whether explicitly or not – by those who would draw a categorical line between human and non-human modes of awareness or sentient existence. Thus when philosophers mount a case along these lines it typically involves some further dichotomy – as for instance between conscious and self-conscious states or self-conscious states and those that can or could find expression in articulate or propositional form – whereby to shore up that supposedly vital distinction.[48] Yet this begs the question not only as regards the privilege attached to language or speech by thinkers from Aristotle down and open to challenge on the grounds, once again, of inherent circularity but also as regards its frequent deployment as a means of reinforcing traditional attitudes concerning the intrinsic superiority of human over non-human creatures. Such attitudes – whether grounded in religious or secular versions of the exceptionalist doctrine – are always liable to work out as a pretext for humans to treat other animals pretty much as they see fit.

To this extent the main issue is still what it was for Descartes and his contemporaries, namely the question as to whether those other animals are like or decisively unlike 'us' in the most salient physical, perceptual, psychological, social, and (at least arguably) ethical ways. Descartes' answer – that non-human animals were cunningly contrived machines and that merely to raise such a question was both absurd and impious – is one that would find few takers nowadays, at least in anything like so extreme or unqualified a form. On the other hand, there are still adherents to a basically Cartesian outlook who would seek to maintain the crucial distinction, albeit (most often) with obligatory gestures toward a scientifically respectable, quasi-naturalistic approach that would draw the line firmly at any avowal of Cartesian substance-dualism. Indeed, current versions of property-dualism – for a long time the favoured fallback position in debates of this sort – are more than likely to come hedged around with all manner of caveats to the general effect that any dualist (or mentalist) talk therein to be found is quite compatible with some version of the mind-brain identity thesis, perhaps via an appeal to Davidsonian 'anomalous monism'.[49] All the same it is not hard to make out the lineaments of that same old Cartesian position in many recent, philosophically au courant efforts to carve out some well-defined niche for what's thought distinctively or uniquely human about the human animal. Thus, for instance, the traditional view looms large when Peter Carruthers sets out to explain how the really important difference is that between the kind of self-awareness that non-human animals (or some of them) perhaps have and the kind of expressible, i.e., articulate, conceptually mediated and therefore speech-apt second-order consciousness enjoyed by human beings and – so he argues – by human beings alone.[50]

It is clear enough what response this would draw from thinkers, such as Peter Singer, whose chief concern is with the effect of such anthropocentric or 'speciesist' attitudes on the wider culture wherein they translate into various practices that are found unacceptable – or downright abhorrent – by those of a contrary persuasion.[51] For them, the relevant point is still best made by Jeremy Bentham's vigorous rejoinder that 'the question is not, Can they reason?' nor, Can they talk? but, Can they suffer?'[52] After all, it is scarcely deniable even by unabashed upholders of the exceptionalist thesis in its strong form that any appeal to rationality, language, or other such presumptive indices of human uniqueness must be made from (what else?) a human viewpoint which is sure to embody just that range of values and priorities. Moreover, it is here – in the area presently criss-crossed by numerous debates in epistemology, cognitive psychology, linguistics, anthropology, ethnology, and the social sciences – that it becomes most difficult (maybe impossible) to disentangle issues concerning the scope and limits of human vis-à-vis non-human animal modes of existence from issues concerning the ethical dimension of 'our' dealings with 'them'. Thus Bentham's point is subject to dispute not only by those, like Carruthers, who take the attributes of language and reason to constitute a definite and privileged mark of the human but also – less explicitly – by those who want to stake out a middle-ground, quasi-naturalist position that would still leave room for some scaled-down version of the same anthropocentric view. It is this approach that Frankfurt's essay exemplifies to most striking effect, along with all the symptomatic problems and stress-points – among them the constant proneness to various forms of circularity and vicious regress – that tend to characterize such projects.

Indeed one might go farther and assert that the problems in question are sure to arise when philosophers allow their naturalist commitments to be more or less qualified or held in check by the conjoint effect of two near-related but distinct motivations. One is the doubtless very deep-laid desire among most human beings to find something specific or unique to themselves that sets them decisively apart from other animals. The second is that other, more specialized or intra-disciplinary incentive which leads them (philosophers) to situate precisely this question – that of consciousness *qua* supposed distinguishing feature of humanity – at the centre of their various epistemological, ethical, linguistic, and (in so far as they are nowadays willing to endorse the description) metaphysical concerns. As I have said, this preoccupation extends well beyond the company of those who would defend a strong, even neo-Cartesian doctrine concerning the absolutely privileged status of human conscious, self-conscious, reflective, or ethically autonomous being. It is just as central to the arguments of many who would have no truck with that fully-fledged dualist way of thinking and who would see it as the product of an old, scientifically under-informed conception of mind. For them – the majority of present-day philosophers – such ideas are simply unsustainable when confronted with the range of scientific (i.e., neurophysiological and cognitive-psychological) evidence currently on offer. Such evidence is often assumed to create large problems for the neo-Cartesian appeal to a realm of mental, intentional, or phenomenological experience conceived as intrinsically exceeding the limits of any brain-based (no matter how fine-grained) causal-explanatory account. Even so the very persistence of these debates and the fact that philosophers feel constantly obliged to rehearse them – to beat the bounds of physicalism from both sides – is a sure sign that they have not let go of that nagging preoccupation.

VI

What unites these thinkers across some otherwise large divergences of view is the underlying notion that they are able to discuss this topic in a meaningful way only on condition that, as conscious (or self-conscious) creatures, they must possess the kind of understanding that cannot be fully accounted for in downright physicalist or central-state-materialist terms. It would not, I think, be difficult to show that this basically anti-naturalist premise – or residual Cartesian assumption – is surreptitiously at work in a good few of those seemingly hard-headed approaches to the mind-brain issue which lay claim to a physicalism-compatible view while none the less endorsing the received, philosophically sanctioned idea that there is more to consciousness than could ever be cashed out in such reductive terms.

It seems to me that this gets the emphasis precisely back-to-front and that the best way forward is that marked out by those previous major advances in the natural sciences, from Galileo down, that have come about mainly through the willingness to break with anthropocentric or intuitively self-evident habits of thought and to go with the best scientific theories or hypotheses to hand. No doubt there remains a puzzle – one much touted by anti-naturalists and upholders of the various (albeit heavily qualified) dualisms that are nowadays doing the rounds – as to how that outlook can possibly accommodate the sheer variety of human subjective, perceptual, ideational, phenomenal, or other such inherently first-person-indexed, i.e., irreducibly experiential modes of knowledge and awareness.[53] However that puzzle is better thought of by analogy with our still perceiving, again as a matter of intuitive self-evidence, that the sun rotates around the earth – rises in the morning and sinks below the horizon every evening – or with our continuing to 'see' certain well-known instances of visual illusion under their illusory rather than what we know to be their true, dimensionally accurate or (in some cases) geometrically demonstrable properties, ratios, or shapes. Once thought of in this way the 'problem of consciousness' begins to look more like the pseudo-problem of explaining how phenomenal qualities like heat reduce to physical quantities like mean kinetic energy of molecules, or again, how our intuitive feel for 'analogue' (continuously varying) properties like smoothness, softness, roundness, gradualness and so forth, can possibly be registered or represented by discrete or 'digital' patterns of neuronal firing.

Of course, if one concludes that these questions are wrongly framed or involve some kind of basic category-mistake – and are hence incapable of finding any other than a vacuous or misconceived answer – then one risks being lumped together with McGinn and the 'new mysterians' as believing that the problem is inherently too deep or complex for us to get our heads around it. In fact, this is just the opposite kind of position since it holds that the problem is mainly a result of our clinging to commonsense-intuitive ideas of what it means to be somehow at the focal point of those various sensory, perceptual, epistemic, subjective, affective, or phenomenological states. Much better, I suggest, to take a lesson from various chapters in the history of science which tell how a good many such states – initially those at the sensory-perceptual end of the scale but increasingly those of a high-level cognitive or epistemic character – have been subject to ever more detailed kinds of scientific, i.e., physically specified descriptive and explanatory treatment. Thus the most productive, least mystery-mongering approach is one that views (say) the qualia issue by analogy with cases like the explanation of perceived temperature in terms of mean kinetic energy of molecules plus an incomplete though rapidly expanding knowledge of the sensory and neurophysical processes involved. At least it holds out more promise of advance – of progressive reduction in the range of phenomena considered beyond reach of any such naturalistic account – than the appeal to some quasi-dualist notion of phenomenal experience as either qualitatively sui generis or else, in the current mysterian mode, as quite conceivably having some ultimate physical explanation but one so complex as to far surpass our limited powers of scientific or indeed philosophical grasp.

In fact this case very often amounts to a point-for-point reversal of the science-first argument which in stead gives pride of place to the standard range of philosophic puzzles about consciousness, qualia, subjectivity, or 'what it's like' to undergo various sorts of phenomenal experience, and which takes their philosophically recalcitrant nature as evidence enough that they are humanly – hence scientifically – forever insoluble. It strikes me that, despite the intensive focus on these issues during the period of almost four decades since Frankfurt's essay was published, the philosophical discussion has moved on rather little while the science has achieved some very striking advances in the scope and depth of its conceptual as well as its descriptive and explanatory grasp. No doubt there will continue to be many philosophers who reject a priori the very idea that advances of that kind could ever resolve such intrinsically hard and distinctively philosophic issues. After all, there are those who reject any notion that the progress of scientific knowledge with regard to microphysical structures, properties, or dispositions should properly be taken to require some adjustment to our sense of whether Locke was justified in denying the possibility of ever advancing from 'nominal' to 'real' essences or definitions, or again, whether Hume was justified in his scepticism concerning the validity of causal explanations. If one can reasonably take the view that things don't move on philosophically in quite the same way that they move on scientifically – and therefore that it not completely off-the-point or just a sign of disciplinary obsolescence when philosophers raise problems from Locke or Hume – still this affords no justification for the idea that an impasse in philosophic thought should be taken to block or to close off the prospect of future scientific advance.

This is just another case of speculative reason's proneness to take what Alexander Pope called 'the high priori road' and discover putative grounds for asserting or denying what can only be determined by other, less perfectly self-assured but altogether more reliable investigative means. It is the kind of error that philosophy is especially (perhaps constitutively) prone to by way of reasserting its status, autonomy, or continuing claim to serious attention despite the extent to which other branches of enquiry – pre-eminently the physical sciences – have seemed to make ever greater inroads on its once all-encompassing area of competence or special expertise. All the same it is a notion that philosophy must learn to live without if is not to become either the last redoubt of beliefs that are all the more entrenched and dogmatic for their superannuated character or else, as debunkers like Rorty would have it, at best just another entertaining but otherwise pretty much irrelevant voice in the ongoing cultural conversation. In Chapter 3 I shall move beyond the diagnostic phase of asking what's wrong with halfway attempts to escape this sorry predicament by placing its traditional range of concerns on a moderately naturalistic footing. In particular, I shall raise the possibility – all the more attractive in view of current failures to make much progress or get much clearer concerning the 'hard problem' of consciousness – that philosophers might well shift their focus of interest away from the *cogito* and its various surrogates without thereby giving up on the central, most challenging and (I would suggest) most constructively debatable issues in philosophy of mind.

Chapter 3

Deflating the Cogito: Thought, Knowledge, and the Limits of Consciousness

I

It will be my contention here that a great deal of philosophical confusion might be avoided by distinguishing more clearly between *thought* and *consciousness*. That confusion has regularly cropped up when philosophers have sought, like Descartes and Kant, to find some grounding for knowledge and truth in the a priori certitude that thought has access to a *cogito* plainly since consciously possessed of just such a self-validating ground, or when they have reasoned from the certain (apodictic) consciousness of this or that thought-content to the indubitable truth or validity of the content in question. It is likewise to be found – or so I would suggest – in the sorts of argument typically adduced by present-day thinkers of an anti-reductionist persuasion who take the existence of intentional, phenomenal, or (what this claim very often comes down to) of consciously accessed mental states as evidence for various controversial theses in philosophy of mind, among them that which rejects any claims for machine intelligence or strong AI.[1]

Of course there is a risk that my counter-proposal regarding the need for a firmer sense of that distinction might seem guilty of the opposite error, i.e., the fallacy of misplaced concreteness whereby I accord categorical status to a notional dichotomy that in truth cannot play anything like such a load-bearing argumentative role. After all, so it might very well be maintained from a Wittgensteinian or Austinian viewpoint, philosophers are sure to come off the tracks if they insist on making clear-cut distinctions or enforcing those distinctions through terms of art (such as 'consciousness' and 'thought') which ordinary language more sensibly allows to carry on performing their communicative role with a tolerable margin of overlap or boundary-blurring between them.[2] And indeed, there is a sense in which any deployment or discussion of these terms will necessarily involve a stipulative element, that is, a certain preferential usage whereby they are defined in keeping with the line of argument currently being advanced. Still, it is worth remarking that some such usages are better in the fairly straightforward way of sharpening the focus on points at issue and thereby enabling debate to proceed with a stronger sense of aim and intent. It seems to me that the thought/consciousness muddle has had just the opposite effect and worked its mischief not only where there is evidence of some gross confusion but also in contexts where there is no such appearance of conceptual or terminological laxity. Moreover, it has been all the more damaging for the fact that this is a real as opposed to merely nominal distinction, or one which picks out two salient attributes of human cognitive process that have to be conceived as qualitatively distinct and hence as requiring a different mode of treatment or analysis.

Indeed, I would go so far as to suggest that thought and consciousness are sometimes mutually exclusive in so far as the former term applies most aptly to mental goings-on that *must necessarily* transpire without the accompaniment of conscious awareness since they involve a going-beyond – or a movement of cognitive advance – that *by very definition* cannot possibly occupy the spotlight zone of conscious awareness. Thus, acts of thought in the proper, as distinct from the loose (whether everyday or incautious philosophical) sense of that phrase are acts that intrinsically elude or transcend the supervisory grasp of conscious awareness or the power of mind to pass its own procedures in reflective review. At a first approximation, thinking is the sort of mental activity that typically goes on when mathematicians prove theorems, or when physicists devise a hypothesis with new and far-reaching explanatory powers, or when artists figure out some hitherto undreamed-of way to break with existing painterly conventions. Or again, it is the sort of activity that must be involved – 'must' as a matter of phenomenological as well as conceptual necessity – when competent speakers of this or that language do what they customarily do and talk straight ahead in (more or less) grammatically well-formed sentences about various topics which they manage to communicate in (more or less) the words and the manner they intend. That last case-in-point of the thought/consciousness distinction is perhaps most useful for those who are neither mathematicians or scientists – who don't have any direct experience of what those relatively specialist types of thinking might entail – but who grasp as a matter of hands-on knowledge that most often we can make perfectly intelligible sense without having launched upon the speech-act concerned with any kind of conscious or deliberative pre-comprehension.

This is not to say, in *echt*-Heideggerian mode, that 'language speaks us' rather than us speaking language on account of the depth-hermeneutic dimension that always exceeds any merely subjective, expressive, or intentional (first-person indexed) power of utterance.[3] Nor is it to take the line of those – a rather motley bunch including post-structuralists, Wittgensteinians, Foucauldian discourse-theorists, Rortian 'textualists' or strong-descriptivists, et al. – who would urge that language (in whatever most-favoured sense of that term) goes all the way down to our deepest conceptual or ontological commitments, and therefore that objects of consciousness, like states of consciousness, are so many products of the language through which they achieve discursive currency or public circulation.[4] Rather, it is a question of thought doing what it normally (in the context of 'ordinary language') or sometimes more creatively and unpredictably does, that is, maintaining a head start over any consciously accessible awareness of its own procedures. As William Empson once said, you don't talk in conscious or deliberate accordance with the rules of grammar – still less the largely unformalizable 'rules' of contextual relevance, pragmatic adjustment, conversational implicature, and so forth – any more than you play ball-games well through an expert knowledge of the laws of dynamics.[5] Even here, with regard to one of the most everyday human accomplishments, there is a marked tendency – as witness the debate around Chomskian transformational-generative grammar – to confuse thought and consciousness, or that which pertains to our normal competence in producing and interpreting grammatical speech-forms and that which belongs to our reflective capacity for acquiring a sense (a linguistic-theoretical or philosophic sense) of what must be supposed to go on in that process.[6]

There is a case to be made that Chomsky himself added to this confusion by aligning his linguistic project with that high-rationalist tradition in epistemology and philosophy of mind which had its main sources in Descartes and the Port-Royale philosopher-grammarians.[7] As I have argued elsewhere, the main consequence of that avowal has been to create the misleading impression that his project stands or falls on a claim concerning our 'knowledge of language' that would have that knowledge be somehow a matter of conscious or reflective grasp.[8] This in turn has offered its critics a chance to observe how utterly implausible it is that so fluent yet complex, spontaneous yet highly structured, remarkable yet perfectly normal an activity should be thought of as involving any such input from processes of conscious – let alone self-conscious – awareness.[9] After all, it is among Chomsky's leading claims that linguistic competence in his sense of the term (that is, the capacity for language-production and interpretation in accord with certain depth-structural and depth-to-surface transformational rules) is a highly 'modularized' or self-contained human endowment which exhibits a striking degree of autonomy with respect to otherwise large differences of culture, education, literacy, or 'intelligence' as defined by whatever currently favoured metric. Although he has more recently opted for a much simplified replacement of the original TG model with his 'government and binding' approach Chomsky still seems committed to that same Cartesian or ultra-rationalist philosophy of mind, if anything strengthened by a further infusion of neo-Platonist lore concerning the dependence of linguistic uptake on our possessing a large stock of innate ideas.[10] It is for the same reason that he has always emphasized the need to draw a sharp methodological line between those syntactic and structural-semantic aspects of language that invite (indeed require) such treatment and those broadly 'pragmatic' aspects having to do with context, occasion, cultural environment, communicative strategy, etc.

This goes along with his clearly stated commitment to taking as the linguist's primary topic-domain all and only those constituent features of language that are capable of specification in internalist, i.e., speaker-relative but strictly trans-individual terms. That is, we are here concerned – as might be expected, given Chomsky's rationalist allegiance – with just such presumptive universals of human cognitive-linguistic aptitude as lend themselves to structural analysis by way of an innatist account. Yet there remain some deep-laid, perhaps irresolvable tensions in his constant appeal to that tradition in philosophy of mind that gives pride of place to the conscious (and self-conscious) mind, sometimes – as with Descartes – in response to the threat of hyperbolic sceptical doubt or as a cornerstone of various, mostly foundationalist epistemologies. What creates those tensions, through a conflict of priorities visible in thinkers all the way from Descartes to Husserl, is a double and contradictory commitment. On the one hand is the idea of *consciousness* as that which secures its role by appeal to some mode of reflective, apodictic, or phenomenological self-evidence. On the other – also there in Descartes but less prominent or well-known since less dramatic in its impact – is the idea of *thought* as that which can dispense with any such idea of privileged first-person epistemic access since it finds its transcendental (strictly so defined) guarantee in the mere fact that 'thinking goes on', rather than by appealing to the *cogito* as punctual and self-validating source of knowledge. 'Transcendental' here bears the sense first attributed to it by Kant in his First *Critique* and has to do with the 'condition of possibility', i.e., the presuppositional *sine qua non* for whatever it is that is answering before the Kantian tribunal of critical reason.[11] Thus – to tie my discussion in with Kant's most basic (if notoriously problematic) example of transcendental inference – just as our a priori intuitions of space and time are conditions of possibility for any particular act of spatio-temporal grasp so likewise the *cogito* is itself reliant for whatever demonstrative power it may possess on the logically prior existence of forms or modalities of thought that have no indexical reference to any specific thinker or thought-act.

It is not so much in the *Meditations* but in other, more methodologically oriented texts such as his *Rules for the Direction of the Mind* and *Discourse on the Method*, that Descartes makes due allowance for the way that thought may surpass or very often elude the spotlight beam (if such it is) of conscious awareness.[12] In this respect the conflict of priorities involved is closely analogous to the problem with Chomskian linguistics. That is, his programme involves the joint but incompatible premises (1) that language at least in its grammatical aspect is highly modular or 'encapsulated' as indeed it must be if we are to talk straight ahead and make sense in real time, and (2) that the existence of this language-capacity – typified, as Chomsky often remarks, by its singular 'creative' power to produce any number of new grammatical sentences from a small stock of component parts – must be conceived by direct analogy with just that strand within the rationalist tradition that likewise places maximum stress on the autonomous (which here implies consciously directed and self-reflective) character of human thought. Another clear sign of this is the way that Chomsky argues for a link – at very least a strong elective affinity – between his rationalist view of human language as a product of our highly evolved cognitive resources and his ethico-political views as an anarchist believer in the freedom, right and obligation of individual citizens to exercise their faculties of reason and moral conscience over and against the various abuses of government or state power.[13] If he has sometimes seen fit to enter certain caveats or disclaimers in that regard, mainly to dissuade political opponents from confusedly attacking his linguistic theories on ideological grounds, it is none the less a significant aspect of his thinking and one that has often played a powerful motivating role for those of a sympathetic or left-dissident persuasion. However my point, philosophically speaking, is that it offers yet another case of the kinds of problem that arise when consciousness and thought – as they should be distinguished according to my recommendation here – are either invoked pretty much interchangeably or else in a way that entails or implies the dependence of the latter on the former.

It would not be difficult to multiply examples from the high tradition of modern post-Kantian epistemology and philosophy of mind, especially – though this will not be the main focus of my interest here – the 'continental' (i.e., mainland-European) line of descent. The confusion is nowhere more evident than in Kant's vain attempts, in the *Critique of Pure Reason*, to put asunder what he himself had all too promiscuously joined and to sort out the various bewilderments created by his failure to draw that cardinal distinction with adequate clarity and care. Hence the great morass of the 'Transcendental Aesthetic' with its somewhat desperate recourse to the role of examples as the 'go-kart of judgment', and to 'productive imagination' as a 'mysterious art' that is 'buried in the depths of the soul', yet without which we could know nothing whatsoever even though 'nature' prevents us from knowing anything at all concerning it.[14] Hence also Kant's problems in trying to explain how the transcendental subject, that merest of abstract or notional entities, can somehow maintain its necessary role as guarantor of personal identity and sole source of assurance that our world of space-time appearances will continue to hang together despite being emptied of every last attribute that might lay his argument open to the charge – again, the fallacy of misplaced concreteness – that he brings against Descartes' delusory appeal to the *cogito* as ground of indubitable truth. In each case, as throughout all the mazy convolutions of Kant's doctrine of the faculties, what seems to have produced this sorry predicament is the failure to distinguish with adequate clarity between consciousness and thought, or those dimensions of human cognitive process that depend for their very existence on our having epistemic, phenomenological, or reflective access to them and those dimensions that inherently elude such access since it is in their very nature to exceed or outrun the capacities of conscious awareness.

Of course this will appear grossly unfair to Kant, since he spends so much time and intellectual effort in beating the bounds between those various faculties – sensuous intuition, conceptual understanding, imagination, reason, practical reason, pure practical reason, judgment in its various modes – which must not be confused (he is very anxious to stress) on pain of creating all manner of serious and far-reaching confusion. Still, it is arguably the case that Kant's proliferation of faculties and sub-faculties has less to do with the intrinsic demands of conceptual rigour or with curbing the high gyrations of reason in its purely metaphysical or speculative mode and more with his failure to take due account of the thought/consciousness duality. This offers a helpful gloss on many of those intractable problems that Kant bequeathed to modern philosophy. Among them is the whole vexed issue of synthetic a priori knowledge in relation to mathematics and the formal sciences where his arguments to that effect have the unfortunate upshot of requiring that the truth or validity of hypotheses, theorems and axioms alike must be a matter of correspondence between the thoughts wherein they achieve articulate (propositional) form and the apodictic witness of human consciousness. It was this supposition that famously received its quietus with the discovery, shortly after Kant, that alternatives to Euclidean geometry were thinkable not only as abstract or logical constructions divorced from the realm of conceptual (synthetic a priori) possibility but also as genuine working hypotheses with – as it later turned out – a strong claim to getting it right as applied to the at first highly counter-intuitive domain of General Relativity.[15] Then again, there is the realism/anti-realism debate which finds its most pointed set-piece example in the clash of views between mathematical realists (or Platonists) who maintain that truth is objective, recognition-transcendent, or epistemically unconstrained and anti-realists who argue that this is strictly nonsensical since we cannot, by very definition, have knowledgeable warrant to assert the existence of truths that purportedly transcend the utmost limits of human knowability.[16]

II

I have put the case elsewhere that anti-realism is itself incompatible with any remotely adequate account of how mathematical knowledge makes progress or how conceptual advances come about through the discovery (not the invention or creation) of truths that must already have existed – although beyond reach of demonstration – throughout the time when an adequate formal proof was as yet unforthcoming.[17] However my chief point here is that this whole debate takes wing from the mistaken assumption – most evident among the anti-realists but often partially conceded in the reasoning of their realist opponents – that if truth-claims are to have any proper or intelligible warrant then they must entail some demonstrable reference to the epistemic conditions under which they can or might be subject to proof or ascertainment. Yet this is just to say, with whatever nuances or qualifications, that the truth-value of any given thought must at some stage depend on its being provable, recognizable, or manifestable within the scope and limits of the cognisance of conscious subjects whose epistemic powers become the measure of truth-aptitude.[18] These are just two of the long-standing quandaries that might be adduced in support of the claim that the Kantian inheritance has worked great mischief through its failure to observe that cardinal distinction between consciousness and thought. One could readily extend the diagnosis to developments in the mainland-European line of descent where, for instance, it seems to play a sizable role in the various deep-laid problems confronted by Husserl's enterprise of transcendental phenomenology.[19] In particular, these concern the kinds of problem that arise when he attempts to reconcile the two dimensions of genesis and structure, or mathematical knowledge as that which develops through the history of human investigation under certain contingent epistemic limits and mathematical truth as that which transcends any such account since it belongs to a realm of absolute ideal objectivity where those limits are strictly irrelevant.[20] Again, what creates the apparent dilemma – as with the realist/anti-realist debate – is precisely that sinking of the difference between thought and consciousness that has also created such confusion in the 'other', analytic camp.

Moreover, it is one that has large repercussions for some of the most seemingly advanced or cutting-edge debates in present-day philosophy of mind and cognitive science, especially those that focus on the so-called 'hard problem' of phenomenal experience. This is the problem of somehow bridging the presumed explanatory gap between what we know (or might eventually come to know) about the workings of the human brain through neuroscience together with its various allied disciplines and, on the other hand, just 'what it's like' to be a sentient human being and perceive the colour red, listen to an oboe, smell a perfume, taste a peach, or touch some particularly soft or coarse material.[21] Where philosophers divide most sharply is on the question as to whether that bridge might ever be constructed through the acquisition of increasingly detailed and fine-grained knowledge of neurophysiological structure plus whatever is needed in the way of more advanced theoretical and causal-explanatory grasp. To a rough approximation this ranges thinkers on opposite sides of the 'mind = brain' debate, or sorts them into those who uphold and those who reject some version of the claim that sensory, perceptual, subjective, or phenomenal states of mind can – or might ultimately – be brought under (so as to avoid the more tendentious phrase 'reduced to') physical states of the brain. On the one side are grouped outright physicalists, eliminativists, central-state materialists, and others who consider it merely self-evident that minds cannot possibly exist apart from brains and therefore that qualia (or phenomenal, 'what-it's-like') modes of experience must be capable of a full and conceptually adequate explanation in the terms of some future, perhaps far-off but attainable stage of scientific progress. On the other are those who insist, on the contrary, that there is all the difference in the world between *any* description or specification of brain-states no matter how detailed or refined and the character of *any* first-person-indexed experience that by its very nature eludes such treatment since it has to do with phenomenal, i.e., subjective or qualitative states of mind that simply cannot be described without remainder on a physicalist account.[22]

As I say, this is a broad-brush picture of the various, often subtly differentiated positions adopted by contributors to the current debate. Still one might reasonably argue – perhaps after a lengthy perusal of the relevant analytic sources – that the debate has itself become overly focused on just such specialized intra-disciplinary topics (mostly involving the analysis, refinement or rejection of other contributors' terms of address) and is now in real danger of obscuring the main issues. Thus much of it turns on highly technical, often semantic considerations regarding the usage (or 'logical grammar') of certain concepts that are taken to bear a large weight of implicit argument in respect of the mind/brain identity thesis. Of course such discussion is well worth carrying on – and indeed performs an essential service – where it is a matter of getting straight about concepts that are ill-defined or ambiguous and may hence give rise to various sorts of confusion or category-mistake. However it is hard to read widely in the recent literature without being struck by the drastic narrowing of focus that has occurred since Gilbert Ryle first introduced that approach in his sweepingly iconoclastic approach to philosophy of mind.[23] It may be said – by those who believe that analytic philosophy has achieved real advances during the past 50 years through just such a process of progressively refining its conceptual terms and distinctions – that Ryle's more dramatic anti-mentalist claims would have benefited greatly from acquaintance with those later developments. Still, the contrast doesn't work out so straightforwardly to the credit of present-day work since so much of that work is devoted to unpacking concepts that already play a prominent role in discussion of the 'hard problem' and which often serve as convenient units of currency in a highly regulated, i.e., professionalized, philosophic discourse. Indeed, this seems like another example – one of many in the analytic field – of what G. E. Moore was first to describe as the 'paradox of analysis', or the fact that any discourse aspiring to the condition of the analytic statement (or taking the business of conceptual unpacking as a chief aim) will either reduce to a series of vacuous tautologies or else turn out to have smuggled in other, more substantive theses by the back door.[24] At any rate, there is reason to conclude that much recent analytical philosophy of mind has boxed itself into a corner where even the most perceptive or inventive thinkers are unlikely to produce any significant progress, let alone any breakthrough advance.

However, although it is undoubtedly a part of the problem, this large-scale *déformation professionelle* is not my chief concern in this chapter. Rather, I want to focus on that particular aspect of it which shows up in the failure to distinguish with sufficient clarity between consciousness and thought. More precisely, what has often occurred is an attempt to collapse that dualism in one or the other direction along with a tendency to grant it surreptitious re-admission and have it work all manner of mischief. I have already offered several analogous cases – from Descartes and Kant to Husserl and Chomsky – of the different forms that this liability can take in thinkers across a wide range of periods and philosophical/disciplinary affiliations. With each of them it issues in a version of the false dilemma that would end up *either* by confining thought (in all its formal, heuristic, inventive, or creative aspects) to the scope and limits of consciousness *or else* by so defining the nature of consciousness as to make it a more-or-less surrogate term for the property (whatever it is) of human minds that enables such thought-processes to occur and which also offers us, when thinking about them, our sole possibility of grasping their content and character. Hence perhaps the odd situation noted by Michael Devitt in his book *Ignorance of Language*: that a great many expert and highly intelligent linguists and cognitive psychologists have expended vast amounts of intellectual energy in trying to understand the sorts of things that go on in the everyday human business of linguistic-communicative utterance or uptake and yet – he concludes – made rather little genuine progress despite having produced a likewise vast amount of detailed descriptive, theoretical, and (on its own terms) explanatory data.[25] This does rather tend to suggest, in confirmation of my general thesis, that the kind of linguistic know-how involved in the production and comprehension of well-formed grammatical utterances is not – or not to any large or reliable degree – the kind that lends itself to introspective knowledge or conscious (much less to self-conscious or reflective) awareness.

Devitt takes issue with the Chomskian idea that linguistically competent speakers and listeners must be taken to possess an operational *knowledge* of various formally specified rules, protocols, underlying structures, depth-to-surface transformational procedures, and so forth. It is this kind of knowledge, however defined or wherever located, that the theory proposes as the sole adequate means of describing and explaining those speakers' otherwise near-miraculous powers of creative (though none the less rule-governed) linguistic communication. Just how conscious that knowledge is supposed to be, or just how accessible to reflective awareness, is not at all clear from a reading of the linguistics and cognitive-science literature. This may well have to do with a certain need to equivocate – plainly put: to fudge the issue – if the thesis is not to run up against objections concerning its downright implausibility as a matter of human mental-psychological capacity. However Devitt's arguments do a good job in bringing out the way that language and the thought-processes involved in linguistic utterance and uptake belong to a region of cognitive psychology that normally operates well beyond the spotlight of consciousness. The same applies, or so I would suggest, to many other human activities or processes of thought where the very condition of achieving some genuine advance is that thinking go beyond what lies within the scope of conscious, i.e., reflectively available knowledge. Thus, when philosophers ignore or downplay this distinction – when they revert to some however qualified version of Descartes' inaugural mistake – they make it impossible for themselves to explain just how it can be that thought is propelled into new regions of, e.g., mathematical or scientific understanding before the advance or discovery in question has reached the stage of explicit or communicable grasp. Moreover, they oblige themselves to 'explain' consciousness in terms that make sense or that pass analytic muster according to standards – those most at home in the formal sciences – which entail the exclusion, or at any rate marginalization, of anything that cannot credibly be treated by means of conceptual analysis. Either way, as I have said, there is a narrowing of focus that works to obscure one salient fact about the thought/consciousness relationship. This has to do with its character of essential non-coincidence or the impossibility of just that punctual, mutually transparent and reciprocal relation between them which many would consider to have had its last outing with Descartes but which none the less lingers on in various forms across a wide swathe of present-day philosophical debate.

Here it is worth entering a brief disclaimer since some of what I've had said might appear to place me in the wrong (or at any rate by me unintended) company. Thus, my argument so far – that consciousness and thought are two very different and sometimes (especially at moments of creative or innovative thinking) mutually exclusive capacities of mind – should not be construed as in any way endorsing the so-called 'new mysterian' thesis with respect to our chances of ever solving or even getting reasonably straight about the 'hard problem'. This thesis was first advanced by Owen Flanagan and taken up in a more systematic way by Colin McGinn who asserts as a matter of logical truth that, since we don't have access to other people's minds but solely to our own, irreducibly subjective or phenomenal modes of experience we are not and could not possibly be in any position to achieve a scientific (i.e., an objective or non-perceiver-relative) understanding of such experience.[26] Moreover – what chiefly distinguishes the 'new' mysterians from earlier thinkers who likewise took a sceptical (sometimes religiously motivated) line on the question – this problem is one that will always and inevitably lie beyond hope of human solution because we are just not bright enough to get our heads (or our powers of deliberative thought) around it. For McGinn, this is a claim about the scope and limits (more crucially the limits) of human intelligence that no scientific, technological or even philosophic-theoretical advance could ever come close to annulling since it pertains to the peculiar nature of the problem which in turn results from the likewise peculiar – indeed strictly sui generis – nature of conscious or phenomenal experience. So there is, on the face of it, a marked resemblance between McGinn's thesis and my own in so far as we both hold that consciousness eludes any specification in terms derived from or best adapted to the conceptual analysis of thought. However, McGinn would by no means support the other main prong of my argument, namely that thought – in its more radically progressive or creative-exploratory mode – may sometimes surpass the utmost limits of conscious or self-aware understanding. Thus he is happy to leave the 'hard problem' very firmly and immovably in place (since on his account thought-based theories of consciousness are sure to fall lamentably short) but not at all keen to entertain the idea that consciousness might itself prove wanting when it comes to those kinds of intellectual, mathematical, scientific, or other advances in thought that inherently elude conscious grasp at their time of first occurrence since by very definition they go beyond existing or pre-established modes of awareness.

This dissymmetry in McGinn's approach is yet another example of the failure to appreciate that thought and consciousness, although crucially distinct, are not to be treated either as contenders for first-place honours in the cognitive stakes nor as ranked higher or lower on a scale of creativity, innovative power, or capacity for breakthrough achievements in this or that field. With McGinn, as so often with writers who take the 'hard problem' as an ultimate challenge, there is a strong implication that what cannot be explained by any means at our disposal is *ipso facto* of a 'higher', i.e., to us mere mortals inaccessible and hence by very definition transcendent (= surpassing our human, all-too-human cognitive limits) order of truth. This is where the 'new mysterians' might justifiably be seen as latter-day exponents of an old, superficially updated mysterian doctrine with regard to the creaturely limits of human understanding and the folly – moreover the arrogance and hubris – of presuming to comprehend what lies beyond our powers of comprehension. Against such notions the point needs to be made very firmly that if consciousness poses a tough, perhaps unanswerable problem for thought, then it is equally the case that thought – even (or especially) where it exceeds the utmost powers of conscious grasp – is a problem for any philosophical outlook grounded in some more-or-less qualified version of the Cartesian consciousness-first doctrine. What most often drops out in the analytic treatment of this issue is the very possibility that thought might be engaged in the business of *creating concepts*, that is, not just (as the standard practice would require) teasing out the logical grammar or the subjacent implications of concepts that already have their place within some existent discourse but rather coming up with new concepts that surpass currently prevailing ideas of conceptual possibility.

It is a safe bet that this claim – one that finds eloquent expression in the writings of Gilles Deleuze – will strike many readers, at any rate those of an analytic persuasion, as having a strongly paradoxical and even an oxymoronic character.[27] However that impression is one that comes not from any intrinsic absurdity about the idea, but rather from the deep-laid assumption among many analytic philosophers that conceptual analysis can have to do only with concepts that exist and have their rightful role within a first-order discourse (most often though not always that of one or another natural language) to which the analysis should properly stand in a critical, elucidatory, or meta-linguistic relationship. Of course they don't all apply this rule in the hard-line fashion of the logical empiricists, and indeed, there are branches of the analytic enterprise in its broad-church characterization – such as Austinian 'ordinary-language' philosophy – where the line becomes extremely hard to draw.[28] This is partly a matter of principle (since these thinkers reject that *echt*-analytic approach as radically misconceived) and partly a matter of the natural-linguistic qualities – the deliberately, even studiously down-to-earth yet often quirky or mischievous tone – that typifies Austin's writing. Nevertheless, the appeal to 'ordinary language' is itself another way of enforcing or at any rate endorsing the same basic distinction. That is to say, it works here by insisting that such language is perfectly adequate, correct, or sufficient for all practical-communicative purposes and therefore that philosophers in the 'high'-analytic tradition descending from Frege and Russell – those who take it to require conceptual elucidation or correction by means of logical analysis – are getting things completely back-to-front.[29]

Thus, the two main approaches that have somehow managed to cohabit, albeit uneasily, within the broadly-defined 'analytic' fold have this much at least in common: that they each (though for opposite reasons) seek to place a curb on philosophy's pretensions to devise new concepts and thereby extend, refine, or inventively/creatively redeploy the resources of natural language. To simplify matters very drastically: where the purebred analytics impose that curb through their insistence on meta-linguistic rigour and a clean separation between first- and second-order (natural and formal) languages the Austinians do so through a contrary insistence that everyday utterance cannot be bettered in point of communicative subtlety and power and therefore that philosophy has no business proposing such corrective measures. What is firmly ruled out in both cases is the idea – one that Deleuze takes as the hallmark of any philosophy worth its salt – that conceptual analysis is perfectly compatible with, and moreover strictly indissociable from, the kind of creativity that challenges the sense-making norms or the codified standards and protocols of some given (i.e., received or familiar) discourse.

It seems to me that that this sedulous beating-of-the-bounds in analytic philosophy is yet another version of the old prejudice according to which thought is defined as in its very nature constrained by the requirements of conceptual-analytic rigour and hence as having no licence to assume a creative-exploratory role that might take it far out on innovative paths that stretched those existing concepts to the limit and beyond. It seems to me that the current impasse in debates about the mind/brain problem – and especially the various claims and counter-claims concerning qualia, phenomenal states, or 'what it's like' to experience this or that sensation – has a great deal to do with this inherited set of problems. With some parties it takes the form of a refusal to concede that thought (here equated with the scope of analysis conceived in basically reductionist terms) might somehow transcend its inherent limitations as that which *by very definition* cannot surpass or even hope to encompass the domain of phenomenal experience let alone that of conscious, self-conscious, or reflective awareness. Then there are those like McGinn who take a rather similar defeatist line but who argue in terms of endemic human incapacity – of our lacking the requisite intellectual powers to get our heads around the mind/brain problem – and hence view it as an epistemological barrier rather than as pointing to any kind of deeper-laid ontological problem. As I have said, their position needs distinguishing from that of the 'old mysterians' (the majority of thinkers up until roughly a century ago) who took it that consciousness under its various extant descriptions – mind, soul, thought, intellect, often used more or less interchangeably – was by very definition beyond reach of any explanatory account since it had to do essentially with that which distinguished human beings made in God's image from the rest of the physical, including the non-human animal, creation. Philosophers from Descartes and Kant to the present have often been drawn to this basically exceptionalist idea of the human, i.e., the idea of consciousness or mind as a vital support for the further (sometimes but not always theologically motivated) conviction that there is – must be – something in the nature of what it is to be a sentient and sapient human being that places such beings in a realm intrinsically beyond the scope of any explanation couched in scientific, physical, or causal-explanatory terms.

As hardly needs saying, the new mysterians have no wish to trade on this particular nexus of philosophic and thinly-disguised theological arguments. All the same there is something distinctly crypto-theological about McGinn's programmatic insistence that philosophers who presume to come up with theories, hypotheses, or even interesting new ideas about the 'hard problem' are thereby exhibiting their unfortunate failure to grasp why those notions can never be anything more than exhibitions of rationalist hubris. Here again, it seems to me that what lies behind this oddly dogmatic emphasis on the limits of human understanding – an insistence all the odder for its coming at a time of such impressive advances in neuroscience, cognitive psychology, and allied disciplines – is a failure to recognize that thought has resources of its own that don't (or don't necessarily) go by way of conscious awareness and that may none the less offer insights into the nature and workings of consciousness. That is to say, the new mysterians have effectively turned a puzzle into a full-scale obstacle to thought by assuming or decreeing that 'consciousness" denotes the locus of certain intractable or (to us) downright insoluble problems that can always be relied upon to flag up the limits of human ratiocinative brainpower. However this presupposes what I take to be a dogma or a largely unexamined article of faith, namely – to repeat – the idea that consciousness must of its very nature (or by very definition) subtend and encompass whatever thought is able to achieve by way of elucidatory reasoning or description/explanation of all that transpires in our various states or activities of conscious, phenomenal, or subjective (first-person-indexed) awareness.

III

My point is much akin to that which Spinoza made when he remarked that, just as the mind has powers that we are as yet unable to conceive or envisage, so likewise the body has powers of which – in what he took to be our 'common sense'-Cartesian way of thinking – we remain in ignorance precisely on account of the prejudice that credits mind with the possession of autonomous agency and thought while the body is supposed to languish in a state of subjection to brute physical or causal forces.[30] Spinoza's great design was of course to overcome this dualist prejudice by developing a radically monist approach according to which mind and body were two 'attributes' of the selfsame substance that in turn encompassed the various 'modes' or particular instances/manifestations of this strict isomorphism between thought and objects-of-thought.[31] If recent philosophers of mind and cognitive psychologists have been keen to recruit Spinoza to their cause – most often in support of some more-or-less qualified physicalist or mind/brain identity thesis, as for instance with Donald Davidson's 'anomalous monism' – then it seems to me that his heterodox approach to these questions holds equally pertinent lessons for anyone seeking to clarify the relation between thought and consciousness.[32] For it is a main plank in Spinoza's case against Cartesian dualism that just as mind and body cannot be conceived as utterly distinct substances that somehow mysteriously interact – or cannot be thus conceived without creating all manner of logical, scientific, and metaphysical absurdity – so likewise we are apt to run into philosophic trouble if we envisage the wrong, mystery-inducing kind of relationship between consciousness and thought. Or again, in more affirmative Spinozist vein: just as the mind and the body each have powers that we are unable to know or even guess at given our highly restrictive conception of both, so likewise thought has a whole vast range of as-yet undiscovered capacities – including that of conceptual creativity in Deleuze's sense of the phrase – that could take it into realms far beyond those where it is presently most at home.

Descartes seems to exercise a strangely ambivalent or a double and contradictory influence over recent debate in this field. There is more than a vestige of Cartesian dualism – along with the nowadays well-nigh ubiquitous turn against notions of Cartesian privileged epistemic access – in the McGinn-type defeatist or mysterian argument that thought (here conceived in conceptual-analytic or causal-explanatory terms) is by its very nature predestined to fail and merely advertise its own constitutive limits when it seeks to get a hold on the nature of consciousness or to answer the 'hard problem'. Paradoxically enough, this argument combines a standard post-Wittgensteinian rejection of anything like the apodictic 'way of ideas' as a source of veridical insight with the presumption that consciousness, whatever it is, must have something about it (for phenomenal states the quality captured by Nagel's 'what it's like' to be a bat, see red, hear the sound of an oboe, etc.) that locates it in a realm altogether beyond the compass of discursive or articulate thought.[33] And indeed – as I have said – we do need to distinguish clearly between, on the one hand, those peculiarly first-person modes of experience or self-intimating knowledge that properly count as instances of consciousness and on the other those mental operations, whether conscious or not, that involve some ratiocinative process whereby thought is enabled to achieve certain kinds of cognitive advance or some more-or-less ground-breaking extension to the scope of its conceptual, intellectual, or creative-imaginative powers.

Such advances may occur across the widest range of disciplines and activities, from mathematics and the physical sciences, through philosophy in its various forms and genres, to literature, music, and the visual arts. In each case it is possible – as often remarked by those who have accomplished some such signal advance – for thought to come up with various theories, hypotheses, proofs, explanatory conjectures, formal procedures, and so forth, that strike the thinker as having been discovered or invented through a process of conceptual creativity (to cite Deleuze again) which somehow transpires beneath the threshold of conscious or reflective awareness. Since testimony of this sort has come from every quarter of the sciences and arts it seems to offer strong prima facie evidence that thought has resources which, when deployed at full cognitive stretch, are apt to run way of ahead of what is accessible to any version of the *cogito* as conceived by philosophers from Descartes to Husserl. Moreover, as I have said, it would seem self-evident as a matter of conceptual-ideational genesis that any truly creative or innovative thinking must always involve a leap beyond what is presently available to consciousness in the form of received scientific, philosophical, or common-sense-intuitive knowledge.

Yet there is no good reason to conclude – along with an otherwise motley company of dualists, idealists, anti-physicalists, mental-monists, pan-psychists, and of course new mysterians – that this inbuilt disparity or structural non-coincidence between thought and consciousness requires us to acknowledge the latter as somehow surpassing all the cognitive or epistemic limits of the former. Nor should we feel driven to accept – in much the same philosophic company – any notion of thought as confined to those aspects or dimensions of human understanding that are capable of adequate representation in non-intentional or non-subjectivity-involving terms, and are hence a fortiori disqualified from offering an answer to the 'hard problem' or even laying claim to a useful capacity for conceptual clarification in that regard. Where these debates have regularly run into a brick wall of mutual incomprehension is in the fixed belief *on both sides* that the issue is one of a peculiar depth, tenacity, or conceptual elusiveness which therefore involves a drastic choice between one or other of two supposedly option-exhausting alternatives. Thus, the hard problem is one that must either (as central-state materialists or eliminativists would have it) require dissolution by showing it to be just a pseudo-problem conjured up by our delinquent 'metaphysical' habits of talk or else be conceived (by dualists and anti-reductionists of various stripe) as an issue like no other except perhaps the freewill/determinism problem to which it is in any case close kin. On this account, shared by those new mysterians who deny the capacity of thought to get any purchase on the problem of consciousness, we have to acknowledge the existence of a strictly sui generis realm – that of qualia or other such 'what it's like' phenomena – which lies absolutely and by very definition beyond reach of thought in its conceptual, representational, or (most of all) its causal-explanatory register.

However we are forced to this philosophically defeatist conclusion only on the prior and to that extent unwarranted premise that consciousness just isn't like other subjective phenomena for which scientists have good working definitions ('"heat" = "mean kinetic energy of molecules"', etc.) which cut out any reference to 'what it's like' for sentient beings but which don't for that reason attract the charge of evading the central issue. The idea that philosophers (scientists even) might yet come up with some breakthrough conjecture with regard to the nature and workings of consciousness will seem evasive in just that way only to anyone who takes it as self-evident that the hard problem is indeed beyond any hope of conceptual figuring-out, but not (of course) to anyone who takes a more confident view or who prefers to keep an open mind on the question. Thus, they will make allowance for the standing possibility, no matter how remote at present, that thought – scientific thought in particular, but also more broadly the kind of thinking that takes its lead from previous advances in the scope of human conceptual-explanatory power – might well encompass some further advance that effectively removes sentient or conscious phenomena from the range of those deemed permanently off-bounds for science or intrinsically incapable of any such mistaken since grossly 'reductive' treatment. After all, this way of thinking productively or creatively about problems that as yet cannot be brought into the daylight realm of fully conscious awareness or lucid apprehension is one with a good many major achievements to its credit, including quite a few that would once have been considered way off the scale of practical (or even conceptual) possibility.

Of course when philosophers refer to the 'hard problem' of consciousness – in whichever way precisely they construe that problem and with whatever salient differences of view concerning its ultimate solubility – they must be taken to imply that this is a problem unlike all those others that science has managed to resolve. More specifically, it is treated as one that requires something more than the kind of radical, paradigm-transforming, yet still in the long term scientifically intelligible shift of world view whereby (in Quine's words) 'Kepler superseded Ptolemy, or Einstein Newton, or Darwin Aristotle'.[34] What that label typically serves to flag up is the idea that any adequate characterization of consciousness or phenomenal experience in scientific, theoretical, objective, analytic, or causal-explanatory terms – assuming (most often *concesso non dato*) that such an account might be possible – would involve so great and far-reaching a challenge to the whole set of working assumptions built into those terms as to constitute a far more drastic revolution in thought than any among the range that normally figure (for thinkers like Quine and Kuhn) as major episodes of the kind.[35] That is, it would entail such a wholesale redefinition of what should be taken to satisfy the relevant criteria – a redefinition so drastically skewed toward the register of subjective, phenomenal, or conscious (first-person indexed) warrant – that the very idea of 'scientific' validity would by then have undergone a sea-change into something utterly remote from its previous usage. At any rate this seems to be what various parties to the current debate have in mind, irrespective of their particular take on the best way for philosophy to go once it has grasped the sheer enormity of the challenge involved. Such is supposedly the unique dilemma to which the 'hard problem' gives rise. Thus one can *either* have a hard-headed theory of consciousness that takes its main bearings from neuroscience and cognitive psychology and which thus plays the problem down as misconceived *or else* a theory with its roots in phenomenology or some such consciousness-friendly approach which makes a virtue of respecting the peculiar nature of the problem but which by the same token – so opponents would assert – foregoes any claim to resolve it or even achieve greater clarity regarding that nature.

On this point there is general agreement across a range of otherwise sharply contending (e.g., dualist, epiphenomenalist, panpsychist, or new mysterian) positions. Where they converge is in assuming that something has to give – or be given up – in the clash of views between, on the one hand, a naturalistic or science-led outlook that is ultimately premised on the descriptive-explanatory closure of the world and all its furniture (minds included) under the laws of physics and, on the other, an anti-naturalistic or consciousness-led outlook that rejects any such claim as just a flagrant category-mistake or a further bad instance of philosophers' befuddling infatuation with science. It is on just these *tertium non datur* grounds – the purported non-existence of any third way or alternative solution to the dilemma as posed in these terms – that new mysterians can erect their case for treating the whole debate as misconceived and futile since plainly incapable of moving beyond its present deadlocked condition. However, as I have said, this is much better seen as merely a product of the old Cartesian prejudice that continues to exert its spell even – or especially – on thinkers who believe themselves to have thrown off any such lingering influence through a good strong dose of the Wittgensteinian anti-private-language therapy that is meant to serve as a lifelong inoculation against delusory ideas of privileged first-person epistemic access. Here as so often in Wittgenstein's reception-history it is a case of the physician who not only failed to heal himself but who managed – albeit despite his express intentions – to spread all manner of sceptical doubt (via the rule-following 'paradox' and other such obsessional topics of debate) the effect of which has been to reinforce that deep strain of more-than-methodological solipsism that forms such a striking component of much post-Cartesian epistemology.[36]

It seems to me that this effect is further strengthened by the widespread tendency either to blur the thought/consciousness distinction or else to raise it into a full-scale conceptual or definitional barrier that flatly interdicts any philosophic commerce between those hypostasized faculties of mind. The former most often results in a failure to make adequate allowance for the various ways in which thought may exceed the bounds of presently attainable conscious reflection and thus stake out an exploratory path that points beyond the kinds of self-locked dilemma that have typified the discourse of first-person, apodictic, or phenomenological enquiry from Descartes and Kant down to Husserl. The latter tendency – that which exaggerates the depth or breadth of this explanatory gap between the powers of thought and the nature or quiddity of phenomenal consciousness – has much the same effect (though for just the opposite reason) since it treats consciousness as something so purely and self-evidently sui generis as offer no explanatory hold for those capacities of mind that depend upon thought in its discursive, conceptual, or representational modes. What is lost from view in each of these narrowed perspectives is the power of thought at its most inventive and rigorous, creative and consequent, or speculative and highly disciplined – none of which adjectival pairings should by now strike the reader as involving a downright contradiction in terms – to come up with solutions to deep-laid problems that could not have been solved by any amount of reflective self-communing in the 'Cartesian theatre' of the solitary, self-sufficient *cogito*.

Of course there are very few philosophers nowadays who would grant even heavily qualified credence to that Cartesian idea of epistemology as a salvific operation where truth, knowledge and mind-independent reality are somehow held secure against the ravages of sceptical doubt by appealing to the self-evidence that I (the thinking subject) can be known to exist and moreover to provide such ultimate security in and through the very act of thought. Such supposed grounds of apodictic certainty were already highly suspect to Kant when he reproached Descartes for making such inordinate claims and purporting to derive such large, far-reaching, or metaphysically substantive conclusions from the meagre – since purely formal or place-holding – role of the transcendental subject as a condition of possibility for knowledge and experience in general.[37] However as with Kant, so likewise with Wittgenstein: the warnings have fallen not so much on deaf ears but rather on ears that were all too receptive in so far as many have taken them philosophically to heart in a way that tends more to compound than to remedy or alleviate the original problem. Thus, just as the Kantian aftermath witnessed a perpetual re-emergence of all the same problems under different, periodically shifting labels – right down to the whole range of present-day disputes between realists and anti-realists, externalists and internalists, objectivists and constructivists, or those who think of truth as recognition-transcendent and those conceive it as epistemically constrained – so likewise (to repeat) Wittgenstein's therapeutic efforts to wean us off dependence on foundationalist or first-person privileged notions of knowledge and truth may be seen to have miscarried, or succeeded only in the pyrrhic sense of having managed to generate a vast amount of wholly inconclusive and inordinately self-occupied commentary. And just as, in Kant's case, the problems arose in large part from the confusingly complex relation and the frequent unremarked boundary-crossing between consciousness and thought, so in Wittgenstein's the followers are still bogged down in the same sorts of confusion despite having taken the 'linguistic turn' with its much-advertised capacity simply to dissolve rather than deludedly seek to resolve such philosophic (pseudo)-problems.[38]

If the issue about consciousness and phenomenal awareness should count as one of these – and for Wittgenstein certainly it did – then the desired result is surely not to be had through his particular, problem-displacing or disguising rather than problem-confronting kind of linguistic therapy. What is rarely noted by his faithful exegetes is the extent to which all those inherited quandaries or dilemmas that received their most elaborate statement in Kant's First *Critique* may be seen as still exerting a profound effect on Wittgenstein's efforts to coax us down from the metaphysical heights or (the same exercise in a slightly different key) up from the sceptical depths. Nowhere is this more obvious than in the whole rather footling and off-the-point debate about 'following a rule', inspired jointly by Wittgenstein's explicit remarks on that topic and his cognate, equally bafflement-inducing remarks about the impossibility of 'private language'.[39] This confusion is chiefly a consequence of the failure to acknowledge how thought – prototypically, in this context, mathematical or logical thought – may occur and achieve various constructive, sometimes revelatory results in the absence or without the active participation of conscious, let alone reflective or self-conscious awareness.[40] Hence Wittgenstein's altogether misleading emphasis on the state or the mental-psychological condition of 'certainty' as that which provides a kind of bedrock guarantee that doubting must have an end, that radical scepticism is simply unsustainable, and that at this point – as he famously puts it – 'my spade is turned' since it is pointless to ask for anything 'deeper' in the way of knowledge, objectivity, or truth.[41]

No doubt it will be said that 'certainty' according to Wittgenstein is not a matter of any individual mind-state but rather a matter of those shared (everyday or expert) practices and life-forms that alone provide the relevant criteria for judging what is properly to count as a right, correct, consistent, or valid as opposed to a wrong, incorrect, inconsistent, or invalid way of proceeding in some given case. However the idea of communally-sanctioned 'agreement in judgment' as the furthest one can possibly get by way of veridical warrant or justification is so patently unable to make the least sense of concepts such as truth, knowledge and progress in the formal and the natural sciences that it has to be considered a reductio ad absurdum of this entire line of argument.[42] And if the point requires reinforcement then one need look no further than Saul Kripke's thoroughly perverse attempt to press Wittgenstein's sceptical reasoning to the limit and then embrace the communitarian 'solution' as the best on offer despite frequent protestations that he finds the whole exercise and its upshot incredible to the point of manifest absurdity.[43]

This is, I would suggest, yet another case of how philosophical arguments are apt to go off the rails if they don't take sufficiently clear-headed account of the various ways in which thought may exceed, surpass, or elude the register of conscious awareness. What seems to have driven Wittgenstein, Kripke and their sundry bother-headed exegetes into this particular dead-end is the utterly mistaken although, it would seem, philosophically tempting idea that knowledge in mathematics and the formal sciences has to be defined in terms of some individual or collective mind-state rather than with reference to the way things stand as a matter of objective, recognition-transcendent truth. Of course that idea has a motivating interest of kinds, namely the belief – held most firmly by card-carrying anti-realists and taken on board to a large extent by response-dependence theorists and others of a more moderate mind – that objectivist conceptions of truth are strictly nonsensical or self-defeating since they place such truth by very definition beyond the furthest bounds of humanly achievable knowledge or ascertainment.[44] Hence no doubt the strong appeal, for some, of theories that bring the whole issue epistemically back down to earth by replacing truth-talk with knowledge-talk and then replacing knowledge-talk with talk of 'warranted assertibility'.[45]

From here it is but a short distance – at any rate for those of a Wittgensteinian persuasion – to defining assertoric warrant as a function of 'certainty' on the part of individual subjects and 'agreement in judgment' as the best they can reasonably hope for in such matters since the sole and ultimate standard of truth, knowledge, rational warrant, or formal (e.g., mathematical or rule-following) correctness is what passes muster by the shared lights of some relatively specialist or broader community of shared interests, values, and beliefs. This is chiefly in response to the usual, well-practised anti-realist challenge which says that we can *either* have objectivist truth with no possibility of ever getting to know it *or* 'truth' as epistemically constrained, and hence redefined in just such a way as to bring it always and infallibly within range of human knowability.[46] Those who have taken a leaf from Wittgenstein's book will naturally go for the latter option and simply add that the constraints in force are those that characterize some given, communally sanctioned range of 'language-games', practices, or 'forms of life' which effectively decide what shall or should count as true, valid, or correct to the best of anyone's acculturated mode of understanding. What is so odd about all this – quite apart from its utter failure to explain how present-best belief may often fall short of knowledge and present-best knowledge often fall short of some more deeper, more complex, or encompassing truth – is that the communitarian appeal goes along with a notion of intersubjective 'agreement in judgment' which would itself seem to be in flagrant breach of the Wittgensteinian veto on any idea of mental states or other such 'private' tribunals as a source of validating warrant. That is to say, its various deliberations and verdicts are best thought of as issuing from a sort of 'common mind' (to borrow a useful phrase from Philip Pettit) where what goes on by way of collectively performed checking, assessment or evaluation must be envisaged as the communal equivalent of what goes on when individual subjects undertake the same kinds of operation.[47]

Thus we are really no further along than those presumptively deluded types – like Wittgenstein's character who buys a second copy of the daily newspaper just to make sure that what the first one says is true – who suppose that we can check the truth of our statements, the meaning of our sentences, or the validity of our truth-claims by introspecting (in however 'transcendentally' disciplined a fashion) or by testing for a due correspondence between their manifest purport and that which we properly or veridically meant to say.[48] However, like so many of Wittgenstein's striking analogies and parabolic fictions, this is really just a species of philosophical red herring which sets up a patently absurd line of argument (the idea of truth as whatever matches one's deepest intuition or most firmly-held conviction) and then wields it as a stick with which to beat any other line of argument which has it that truth may reside somewhere outside or beyond the sphere of communal warrant. Moreover, as I have said, his alternative proposal – what Kripke rather lugubriously calls the 'sceptical solution' to the 'sceptical problem' about rule-following – in fact falls plump into Wittgenstein's self-laid trap since it locates the ultimate court of appeal in that state of 'certainty' that comes of simply carrying on (thinking, acting, reasoning, judging, calculating, continuing a numerical sequence, and so forth) in accordance with the rules or conventions that constitute some given, communally sanctioned form of life.

In short, we are here left with the worst of both worlds: a conventionalist doctrine that really amounts to fully fledged cultural relativism whatever its adherents' protests to contrary effect and a bottom-line appeal to 'certainty' as the final criterion of knowledge and truth which likewise cannot be redeemed from its inherently subjective (or 'private') character by insisting that any intuitions thus acquired have the backing of communal assent. For of course that assent will count for nothing at all – will simply not register as such – unless and until the individual thinker, agent, or rule-follower comes to *experience for herself* the kind of certainty that results from her complying with the relevant norms or accepted modes of thought and judgment. So there is something deeply inconsistent, even contradictory, about Wittgenstein's insisting on the one hand that certainty plus communal agreement is the end-point of enquiry and an adequate (indeed the sole available) answer to scepticism, and on the other that any foundationalist idea of apodictic or first-person privileged epistemic warrant is a hopeless last resort that will surely resurrect all the same old problems in a yet more intransigent form. It seems to me that the decisive wrong turn is taken – by Wittgenstein and others inclined toward that way of thinking – with the notion of truth (or some surrogate term) as epistemically, hence linguistically, and hence (by the kind of reasoning that drives so much of present-day thought) culturally or communally constrained. Moreover, as I have argued, this error results from the wider tendency – that which unites all Descartes' children, philosophers and non-philosophers alike – to place undue emphasis on consciousness (its nature, its constitutive scope and limits, and its role in the mind's more advanced creative-investigative powers) and thus fail to reckon with the various ways in which thought may outstrip or overtake the best efforts of conscious reflection.

IV

Where my argument differs from that of the new mysterians is in taking itself *not* as a standing rebuke to the claims put forward by some of the more sanguine cognitive psychologists or philosophers of mind but rather as a recommendation that workers in the field refrain from premature, dogmatic, or under-evidenced verdicts either way. Also, it urges then to take more account of the fact – since here we do have a great deal of evidence, albeit from other disciplines or regions of enquiry – that, so far from being a laggard with respect to consciousness, thought has capacities for finding things out or for going beyond some existing state of knowledge that are simply unavailable to conscious awareness. Indeed this makes sense in a fairly straightforward, even obvious way if one considers how hard it is to break with such received or acculturated habits of mind, and how deep is the connection between that ingrained resistance to change – or attachment to sundry inherited modes of scientific, philosophic, metaphysical, or everyday common-sense belief – and our conscious inclination to think and behave in accordance with those same prevailing norms. If the act of discovery/invention/creation is one that by very definition involves the capacity to step outside the confines of previous knowledge or practice then it is likeliest to come at just those moments – or in just those states of mind – when consciousness retreats to the point of disinhibiting certain heterodox thoughts that would otherwise be subject to veto by the pressures of adherence to existing beliefs.

Most philosophers are willing to accept such ideas when applied to the creative arts, whether by way of the appeal to some kind of (Freudian or other) unconscious process or by way of the fairly traditional notion that artists have access – for better or worse – to sources of innovative thought beyond anything licensed by more restrictive (or reputable) forms of conceptualization. Many would likewise endorse the claim that progress very often comes about in the formal and the physical sciences through imaginative leaps or powers of creative-exploratory thought that intrinsically elude conscious grasp at their time of discovery or first formulation. Here the most obvious explanation would be that they find no place marked out in advance as offering a focus for the kind of mental activity that we are apt – in however folksy or philosophically misleading a way – to describe as having something (an object or idea) 'in our mind's eye'. After all, the history of scientific progress is in large part the history of just those rationally, logically or empirically motivated breaks with the naïve self-evidence of intuition or sensory perception that enabled such otherwise improbable shifts as (to repeat the Quinean litany) the upheavals of thought by which 'Kepler superseded Ptolemy, or Einstein Newton, or Darwin Aristotle'.[49] More to the point: it is crucial to maintain a due sense of the thought/consciousness distinction and of thought's capacity to outrun the powers of conscious or reflective awareness if we are to find some rationally accountable alternative to the Quinean-Kuhnian doctrine of wholesale ontological relativity along with its twin methodological props, i.e., the 'theory-laden' character of empirical data and (its symmetrical counterpart) the 'underdetermination' of theory by evidence.[50] This leads in turn – irresistibly so, if one accepts Quine's ultra-pragmatist view – to the notion of scientific paradigm-change as a matter of more-or-less radical adjustment at this or that point in the overall 'fabric' of belief, an adjustment brought about by some empirical anomaly or localized conflict between theory and evidence that might have been resolved in a different, equally 'rational' way had scientists come at the issue with a different set of foregone commitments and priorities.

Where such thinking goes wrong, as it does most conspicuously in Kuhn's case, is through confusing the various processes of *thought* that achieved these revolutions in knowledge – processes that often occurred without the innovator's fully comprehending the nature of his own discovery, and that sometimes (as with Kepler) had to wait some time before its import really struck home – with issues concerning the phenomenological, psychological, or (inter-)subjective aspect of scientific paradigm-change. That is to say, it is when consciousness comes into the picture and is allowed to play some role in deciding what may or may not have been the mind-set that enabled (or hindered) some major advance that there tends to develop a relativistic or at any rate strongly anti-objectivist outlook. At this point – often with further support from Wittgenstein – the conclusion is drawn that scientific 'revolutions' or breakthrough 'discoveries' are chiefly a matter of seeing things under a different, maybe radically transfigured aspect depending on the viewpoint (or, more specifically, the range of cultural-theoretical-conceptual-metaphysical-ontological assumptions) that effectively decide how things appear to this or that observer/investigator.[51] It seems to me that this is yet another instance of the problems that typically arise when philosophers (or philosopher-historians of science) substitute the no doubt interesting but, from a scientific standpoint, irrelevant question as to what sorts of conscious state may have characterized some given episode of theory-change for the far more important – indeed scientifically and philosophically central – question as to how thought achieves certain kinds of advances against considerable odds of received belief or entrenched prejudice. The latter leads on to issues concerning the scope and limits of consciousness along with the question whether consciousness (or the effort to achieve and maintain conscious surveillance of the mental processes involved) might have a retardant rather than a stimulant effect on the more innovative modes of formal, natural-scientific, philosophical, or artistic thought.

Here it is worth adding – lest 'artistic thought' be deemed out-of-keeping with those other disciplines, or even, by some, as a kind of oxymoron – that one further benefit of distinguishing more clearly between consciousness and thought would be to challenge certain deep-laid philosophic prejudices with respect to the arts and their claim (or lack of it) on our interest as thinking or rational beings. Thus it might throw much-needed doubt on the idea, going back to Plato, that if the sources of artistic or creative inspiration lie beyond the compass of conscious, reflective, or lucid self-awareness then they must have to do with irrational promptings that don't in the least merit such interest and could even be deemed inimical to the wider intellectual and moral good.[52] To this extent it helps not only to defuse a longstanding and damaging misapprehension but also to explain what happens (for instance) in a truly attentive reading of poetry or the experience of listening to music with one's perceptual, cognitive and longer-range structural powers of apprehension at full re-creative stretch. For such experiences can sometimes – *contra* Plato but more in keeping with Aristotle's treatment of the topic – be a stimulus to modes of thought that are none the less demanding and intensely focused for their taking place outside the zone of fully conscious or perspicuous awareness. It seems to me, after many years' listening and much reading-around among the music critics and analysts, that what mostly happens when a piece of music criticism or analysis 'clicks' is that it gives a consciously accessible or – very often the same thing – verbally articulate form to understandings that were already 'there' in one's previous auditions of the work and that may have involved a highly complex and demanding process of thought, albeit a process that had not yet arrived at the point of explicit summation or review.[53] For a preternaturally brilliant, perceptive and intelligent statement of the equivalent case with respect to poetry the reader could do no better than consult William Empson's great book *Seven Types of Ambiguity*.[54]

In short, I would want to claim that a clearer distinction between consciousness and thought might bring large benefits to our often confused or falsely polarized ways of conceiving other relationships such as those between science and art, the rational and the non-rational, discursive and non-discursive understanding, or concept and intuition. It is not that these dichotomies are just plain false – since clearly they mark a significant and closely-related series of attempts to get a hold on certain salient differential features of our cognitive lives – but rather that they are skewed in various ways by the over-emphasis on consciousness as a normal (if not requisite) concomitant to any mental process that genuinely qualifies for the title of thought. At its most misleading this goes along with the residual Cartesian concept of mind that associates consciousness with a kind of internal spotlight whose zone of illumination extends just so far as to capture those mental contents or processes with a claim to that privileged (since epistemically self-validating) title. This latter idea has come in for quite a drubbing from Dennett, Rorty and others but receives perhaps its best, most pointed, though not altogether dismissive critique from Kathleen Wilkes when she remarks that, after all, '[t]he technique of attending to one's experience as such is a sophisticated and rarely-used one', and moreover that the 'spotlight' model or metaphor is a bad fit in the case of sensations since 'we seem to regard it as one and the same thing to have a sensation and to be conscious of that sensation – illumination and illumined coalesce'. In fact, she continues, 'the inner spotlight idea, if it grips at all, looks more appropriate to conscious *self*-consciousness, where we can make sense of the idea that a conscious thought takes as its object a mental event of some kind; the metaphor here is not wholly misleading, and presumably lies behind the idea of *introspection*'.[55]

This strikes me as exactly right and as helping to locate one major source of the difficulties that occur when thought is identified too closely with consciousness, and consciousness itself very often elided with something like a Cartesian appeal to the *cogito* as that which gives the reflective subject its indubitable grounding in the pure self-evidence of rational thought. Indeed Wilkes's essay presents a strong case for regarding consciousness as so inherently multiform or heterogeneous a topic with so many likewise endemic doubts as to how it stands in relation to other (e.g. perceptual, phenomenal, cognitive, or conceptual) capacities of mind that it should best be ignored for scientific and most philosophical purposes. Thus on the one hand, not all phenomenal states are 'conscious' in any useful or meaningful sense of the term while on the other, just as problematically, there are manifold ways in which states of consciousness may seem to possess a certain quite specific phenomenal content and yet fail to do so through various kinds of sensory illusion or perceptual error. Nor is there much mileage to be had from the appeal to qualia or, after Nagel, to the usual set-piece examples of 'what-it's-like' phenomena as proof that there exists a first-person realm of conscious and self-conscious experience irreducible to any scientific or causal-explanatory, e.g., neurophysiological account. After all, so far is consciousness from being a hallmark or sine qua non of phenomenal qualia – assuming their existence for the sake of argument – that in fact a large proportion of our sensory-perceptual experience takes place beneath the level of conscious awareness. Such cases range all the way from extreme instances like that of blindsight, through sundry forms or techniques of subliminal processing, to the well-known puzzlement of motorists who often find that they have covered long stretches of road and no doubt responded to numerous ambient stimuli without any memory of having done so.

Besides, there is the problem – for anyone canvassing this as a means to secure the anti-physicalist or anti-reductionist claim – that the appeal to qualia as prime candidates for the role of immediate, directly introspectible, self-intimating modes of experience has reckon with that large body of present-day philosophic thought which rejects any version of the 'myth of the given', of knowledge-by-acquaintance, or the 'way of ideas' as affording access to the phenomenological *Ding-an-sich*. On the contrary, as thinkers from Kant to Wittgenstein, Sellars, Davidson, and McDowell have been keen to stress: all presentation is representation since everything – or everything in so far as we can possibly perceive, conceive, recognize, or know it – falls under a description and therefore cannot count as providing any such rock-bottom epistemic, apodictic, or existential guarantee.[56] It seems to me that Wilkes is right to suggest that we treat 'conscious' as a second-order concept, like 'intelligent', which properly applies not to this or that particular state, activity, process, strategy, or mode of conduct but rather to a whole, more-or-less diverse range of such first-order items that have to link up in certain ways – to form the kind of pattern that merits ascription of the predicate 'conscious' – if its usage is to be justified. Just as talk of 'intelligence' tends to create the philosophically misleading (as well as psychologically and socially disastrous) idea that it denotes a uni-scalar dimension whereby persons can be tested and shown to possess 'it' in various degrees, so likewise talk of 'consciousness' produces the false impression that it has to do with some distinctive though remarkably hard-to-define mental state that applies in varying degrees to sundry human (and maybe non-human animal) modes of perceptual, phenomenal or cognitive experience.

In the same way, just as (following Ryle) we may prefer to use the adverbial form 'intelligently' so as to acknowledge the range of thoughtful activities – across the whole spectrum of human endeavour – that merit that description, so likewise 'consciously' much better captures the vast range of ways in which human (and maybe non-human) beings can manifest an alert, attentive, careful, adaptive, sensitive, and intelligent manner of proceeding in their various, inseparably physical and mental activities.[57] Thus, in Wilkes's words,

> we presuppose a whole slew of psychological ascriptions – to do with perception, motivation, belief and desire, misperception, illusion, recognition *etc.* – when an ascription of consciousness makes sense; conversely, where some set of first-order mental statements are appropriate, then the 'fact' of consciousness follows automatically – there is no further question to ask.[58]

All of which – along with the difficulty (perhaps impossibility) of finding any set of necessary and sufficient conditions for what counts as a conscious state – leads Wilkes to conclude that consciousness is simply not a fit candidate for scientific treatment. Moreover, although she doesn't spell this out so clearly, her argument carries a strong implication that its fitness for a leading role in philosophical debate is very far from well-established despite its having drawn so much in the way of detailed attention from philosophers of mind and philosophically-minded cognitive psychologists. It is not so much that consciousness *qua* 'hard problem' is beyond our native human wit to figure out or make sense of in rationally accountable terms since, despite its elusiveness at present, one main lesson of the history of science is that major discoveries or insights may be just around the corner even though unsuspected – or as yet beyond epistemic reach – by the currently most advanced workers in some given field. Rather it is a question of the problem being largely irrelevant for anyone whose chief interest is either in offering a scientifically useful, i.e., well-defined and testable hypothesis concerning the nature of mind or the interaction of mind and world, or else in arguing a philosophically cogent (conceptually precise and metaphysically substantive) case with respect to that same issue.

V

This leaves us with the second-order task of explaining why the 'hard problem' has exerted such a hold on philosophers from Descartes right down to the time – roughly the period post-1960 – when it started to become increasingly clear that the emergent combination of neurophysiology and cognitive science was about to present a sizable challenge to just about every received way of thinking about these issues. The response, or one common factor across an otherwise large range of responses, was to assume that at least there existed a 'problem of consciousness' irrespective of whether one took that problem to be solvable by the best (scientific or philosophical) methods to hand or took it inherently to elude such treatment. Where philosophers typically go wrong, I would suggest, is in supposing that any commitment in this regard must come down to a three-sided choice between (1) deeming consciousness a real and intimately knowable though strictly sui generis phenomenon that holds out against 'reductionist' (i.e., scientific or causal-explanatory) treatment, (2) viewing it as open to investigation by a fairly standard range of analytic methods like any other reputable topic-domain, or (3) concluding – in new-mysterian style – that whatever its nature it is just not the kind of thing that we poor limited human knowers can possibly get our heads or our rational-cognitive-discursive faculties around. However this leaves open the further possibility (4) that consciousness may be the name for a real, that is, subjectively salient characteristic of the human way of being-in-the-world which no doubt figures centrally for us human beings as a matter of common-sense-intuitive outlook but which should not, on that account, figure centrally in the quest for more adequate knowledge of ourselves within a larger, more philosophically as well as scientifically encompassing framework. After all, as I have said, a good proportion of major scientific advances from Copernicus down have involved just such a radical break with the kind of world view that had previously appeared a matter of straightforward sensory or phenomenal self-evidence, especially when backed by the authority of doctrinal (i.e., most often theologically grounded) truth.

That the problem of consciousness is intrinsically or at any rate *by us* insoluble – rather than at present just happening to reside beyond our best efforts to solve it – strikes me (to repeat) as more like an item of barely transmuted metaphysico-theological belief than a viewpoint capable of adequate defence on philosophical, let alone scientific or science-compatible grounds. Once again, I would suggest that this prejudice comes largely from the tendency to take consciousness as the hallmark of advanced, refined, conceptually adequate or philosophically sophisticated thinking and hence to assume that any thought bereft of such conscious *rapport-à-soi* must to that extent always be lagging behind in the quest for lucid self-knowledge. Of course when the issue is posed in these terms – terms that very plainly betray their Cartesian inheritance – then indeed it must seem self-evident that only a reflective consciousness trained intently on its own workings can have a genuine claim to describe, comprehend, or elucidate those workings. Although few contributors to the current debate would endorse this as anything like an accurate statement of their view it is, I think, a fair enough summary account of the kinds of presupposition – not to say prejudice – that have given rise to that debate, chief among them the idea (whether upheld or rejected) that the hard problem is hard precisely because it concerns a mode of knowledge attainable only on the basis of first-person acquaintance or privileged epistemic access.

As I have said, that idea still exerts its hold on a great many otherwise sharply conflicting approaches to the question, not least because it is there at the very heart of the dispute between those who think the problem humanly insoluble, those who seek a scientific or at any rate science-compatible solution, and those who believe that any prospect of advance must lie in an altogether different (e.g., phenomenological) direction. Moreover the residual Cartesian influence is sometimes liable to produce some odd crossings-over or strained alliances between these positions. For instance, one may well take the point of Daniel Dennett's numerous well-aimed sallies against notions of the mind as a 'Cartesian theatre' where thoughts, beliefs, memories, desires, sensations, etc., are passed in lucid spectatorial review, along with his proposal that we practice a form of 'heterophenomenology' whereby we should strive so far as possible to treat our own (supposedly) 'private' mental goings-on just as we would those of other people whose modes of conscious or phenomenal experience we can often interpret with considerable skill but without any such privileged 'inside' knowledge.[59] Yet it is hard not to feel, despite the frequent acuity and brilliance of *Consciousness Explained*, that the book's very title – for all its deliberate chutzpah – gives the game away in advance by running together two terms of art (or, very often, of fighting talk) from disparate quarters of debate. Thus it succeeds not so much in its avowed aim, i.e., that of overcoming or finessing the presumed gulf between science and subjectivity but rather in drawing attention to the fact that, if consciousness poses anything like the kinds of problem associated with it by other, less sanguine types, then no amount of 'explanation' in Dennett's basically physicalist (hardware-plus-software) terms is going to meet the standard range of *qualia*-based and suchlike subjectivist ripostes.

There is the same basic problem with his idea of heterophenomenology as a bridge between consciousness in some phenomenologically familiar sense of that term and consciousness in so far as it might figure in a scientific world view premised on the ultimate closure of everything (minds or phenomenal states-of-awareness included) under some possible even if as-yet scarcely conceivable physical description. After all, as becomes very clear in Husserl's Fifth *Cartesian Meditation* and in the travails of many commentators since, there are large problems in the path of anyone who seeks to reconcile the first-person-centred apodictic mode of phenomenological enquiry with an approach that not only concedes the existence and the various claims upon us of other people but also purports to derive this knowledge from a likewise apodictic, transcendentally grounded process of reflection.[60] I shall not here labour this (as it might seem) marginally relevant point except to say that it helps to explain, from the perspective of a different philosophical tradition, why difficulties arise for a thinker like Dennett who adopts the idea of heterophenomenology as a putative means to keep consciousness squarely in the picture as a suitable, indeed prime topic for debate but who wants to avoid any hint of subscribing to notions of first-person privileged access or other such left-over props from the old Cartesian theatre. What gets in the way of that ambition is simply the fact that any talk of consciousness in a philosophic context where the word inevitably flags up a Big Problem is sure to carry along with it a lot of the same inherited baggage that has hobbled this discussion throughout its long history to date. More specifically, it is sure to come up against the issue between those who maintain and those who deny that consciousness can indeed be 'explained' in some set of terms that would take the debate onto new and more productive ground by eliminating the appeal to qualia or other such intrinsically private, hence (some would say) explanatorily vacuous entities or mind-states.

Dennett seems often to be tugged in both directions, and to keep his balance – or his argument from veering too plainly in one or the other – mainly by a good deal of nifty argumentative footwork and a readiness to shift ground (to emphasize the phenomenology or the claims of rapidly advancing neuroscience and cognitive psychology) as and when required. Still it is a difficult position to sustain for very long and one that can never achieve any kind of secure or lasting stability since it is subject to all the unresolved (most likely unresolvable) conflicts of aim and priority that have typified the mind/body debate from Descartes down. Thus one finds William James, in his *Principles of Psychology*, performing very much the same kind of regular two-step shuffle as he strives to do justice to the 'stream of consciousness' along with all its qualitative nuances and ultra-fine shadings of phenomenological content while also trying to accommodate the widest available range of current scientific findings.[61] This is the main reason why Wilkes concludes her own survey of the field by suggesting, as against Nagel and those of a similar mind, that '[t]here seems no particular reason to suppose that science has, or fails to, grapple with the question of "what it's like for the *X* to be an *X*" – especially if this is said to be inexpressible and ineffable even within the far richer conceptual resources of the vernacular'.[62] Thus, it is not that the problem is real and pressing but just too hard for us human knowers, nor that a solution is (or might be) within reach given some decisive advance in the relevant scientific fields, nor again that it is merely a pseudo-problem that can best be dissolved through some well-judged ministrations in the late-Wittgensteinian or other such therapeutic mode. Rather it is a matter of seeing that the problem as typically presented – how could there possibly be room for consciousness or 'what it's like' in a world view premised on the explanatory closure of physics, that is, the subsumption of everything, consciousness included, under some ultimate (whether or not humanly knowable) set of physical-scientific laws? – is simply not one that need engage cognitive scientists or those whose proper interests lie elsewhere.

I have argued that this point might not have been so easily missed had philosophers kept a more careful eye on the thought/consciousness distinction, or the capacity of thought to run well ahead of what consciousness is able to bring to mind with anything like the clarity or force of apodictic self-evidence. That thinking is carried on to some highly productive, creative and potentially problem-solving ends in the absence of any such lucid conscious grasp is a claim that would I think be accepted by most philosophers except the very few who would nowadays subscribe to a fully fledged Cartesian view of the matter. All the same it is one that often seems to get forgotten – or not to count for much – when the 'hard problem' comes up for debate. In that context it tends to be supposed that if consciousness cannot *bear witness to and for itself* in a way that involves a self-reflexive appeal to the nature and modalities of conscious awareness then the resultant story must, for precisely that reason, be radically deficient or patently missing the point. But of course this supposition is tantamount to ruling that thought – or the kinds of thought-process that don't exhibit that elusive character – *simply cannot* find out anything of genuine relevance or interest with regard to the qualitatively unique or sui generis phenomenon of consciousness. I would argue, conversely, that here as in many fields (at any rate those with some definite idea of what would count as a genuine advance) the best hope lies in overcoming the human tendency to anthropomorphic projection, or the proneness to stick with intuitive – most often sensory-based – modes of perceptual self-evidence as a substitute for the more recalcitrant business of framing adequate, scientifically valid or properly explanatory concepts.

Of course that analogy would be roundly rejected by those parties to the current debate who treat the appeal to qualia or kindred 'what it's like' phenomena as a touchstone of first-person conscious experience and therefore as a standing refutation of any claim to describe or explain consciousness 'from outside'. For them, the idea that such appeals are on a footing with the pre-Copernican common-sense-intuitive conviction that *of course* the Sun revolves around the Earth – since everyone can plainly see that it does – is just the kind of error typically made by hard-line physicalists or eliminativists, such as the Churchlands, whose blithely dismissive talk of 'folk psychology' is (they would say) nothing more than a smokescreen erected to conceal the poverty of an out-and-out reductive scientific approach when applied far beyond its due remit.[63] Still one may doubt that physicalism could ever make good on its more aggressive or imperialist claims while nevertheless harbouring equal doubts as to the intelligibility – let alone the explanatory usefulness – of any alternative based on the idea of a subjective-phenomenal *je ne sais quoi* that absolutely eludes any specification except some more-or-less evocative rehearsal in first-person-privileged (hence ultimately private or incommunicable) terms.

In this connection Wilkes has a nice passage which makes the point against radically first-person perspectives – and, more generally, against attaching too much importance to the 'hard problem' – through a comparison with maps as a means of finding one's way around some unfamiliar terrain or transport system.

> Just as a subway map often has an ancillary arrow stating 'YOU ARE HERE', so the notional, complete psychophysiological-physical picture of the world could be supplemented for its users: 'THIS IS YOU', from which the way things are experienced from that position may be inferred. The addition YOU ARE HERE to a map is, evidently, not the cartographer's business, and such maps could not be sold at bookstores; similarly, the addendum THIS IS YOU to a completed psycho-physical account is not a proper part of the scientist's concern. But a fixed map in a subway station on which the station master could stick his arrow, or an objective psycho-physical description to which someone could *give* the subjective 'fix' – and both would be heuristics, assisting the user to interpret the maps – could each supply the point of view: from here, Victoria is four stops east, from this perspective, red and green would be indistinguishable (*i.e.*, the perspective of a colourblind man).[64]

What is so useful about Wilkes's analogy is that it helps to refocus attention away from the stand-off that has mostly characterized this debate, i.e., that between physicalists, central-state materialists, or hard-line eliminativists on the one hand and subjectivists, phenomenologists, or qualia-theorists on the other. Instead it points us towards the more tractable issue as to whether consciousness and its various phenomenal surrogates might perfectly well retain their salience as items of our everyday common-sense world view while pretty much dropping out of the picture for scientific and at any rate some philosophical purposes. It would then be more straightforwardly a question of deciding – without prejudice either way as concerns that old chestnut, the 'hard problem' – which parts of philosophy may have a use or a positive need for continuing talk of consciousness and which parts can much better do without it or without the involvement of load-bearing claims and assumptions regarding it. While there would still no doubt be room for disagreement over just where the line should fall in the case of some disciplines or subject-areas – those (like philosophical psychology or philosophy of mind) where any decision is sure to reflect certain foregone commitments and priorities – there would also be a useful sharpening of the line with respect to others (especially epistemology and cognitive science) where the Cartesian conflation of consciousness and thought has exerted such a powerful distorting influence. In particular philosophers would find themselves much better placed to achieve clarity concerning the way in which thought can exceed the compass of conscious awareness or the extent to which consciousness is by its very nature slow to catch up with the kinds of development – quite possibly including developments with a strong bearing on the 'hard problem' itself – that involve the mind's adventuring (as Wordsworth said of Newton) into new and strange 'seas of thought'.

All in all it seems a good bargain to strike, given the extreme unlikelihood of any remotely convincing solution to that problem. Moreover, as the thought/consciousness decoupling helps to show, this is not by reason of our human thick-wittedness or native incapacity to stretch our minds around it but rather by reason of its simply not being the kind of problem to which we can or should expect to find such an answer. No doubt this is one main reason for the current, very marked growth of interest in Spinoza among thinkers – philosophers, psychologists, and cognitive scientists included – who want to frame these questions in an altogether different way, or (as some would take to follow directly from the kinds of argument mounted here) move them off the agenda once and for all.[65] In Chapter 4 I shall look at some of the ways in which philosophy has yet to catch up with Spinoza's radically monist conception of mind and body as joint 'attributes' of a selfsame substance which appears to us under that dualist aspect, or (in the favoured analytic idiom) under that preferential description, only on account of our limiting perspective as human cognizers inveterately prone to such Cartesian errors and illusions. On the other hand – crucial to Spinoza's case for re-thinking the *cogito*, as likewise to latter-day proposals in that regard – it is possible to gain a cognitive standpoint beyond those 'common sense' limits by adopting just the kind of non-parochial or non-anthropocentric outlook that has enabled the achievement of a great many major scientific advances from Copernicus and Galileo to Darwin. Thus Spinoza helps us, by pointed contrast, to grasp how Kant mistook the character of his own critical project and created all manner of subsequent confusion when he described that project as a veritable 'Copernican revolution' in the history of thought. Above all – and most important for our purposes here – it is Spinoza's great merit to have thought his way through and beyond that other tenacious dualism that already ranged typecast 'rationalists' against similarly typecast 'empiricists', and that is still being played out (mostly to damaging since dialogue-blocking effect) in numerous sectors of present-day philosophic debate. Although naturalism, rather than empiricism, is now more usually taken to characterize the chief opposition to rationalist claims in epistemology, ethics, and philosophy of mind there is a clear continuity with the situation as Spinoza found it and as it also struck Kant later on, albeit with consequences – in the latter case – that have often tended to deflect attention from Spinoza's utterly different way of conceiving these issues.

However there are numerous signs that philosophy in the analytic mainstream has at last moved beyond the conceptual-linguistic 'turn' that was Kant's chief bequest to his epistemologically-minded heirs, disciples and opponents alike. Among the questions that find themselves squarely back on the agenda for serious treatment are those of a downright metaphysical character having to do with the way things must necessarily stand vis-à-vis the world and all its furniture, ourselves included, rather than with the particular range of operative concepts and categories that we human knowers may bring to it. That is to say, it is now widely accepted that philosophy can and should pursue just the sorts of enquiry into matters concerning the nature and structure of real-world entities or properties – along with matters concerning mind and its likewise non-mind-dependent place in the order of things – which were ruled strictly off-bounds by Kant and his even more metaphysics-averse successors. The story of how this reversal of the veto occurred is one that has been well told recently by Timothy Williamson who hails it as a large-scale recovery of intellectual nerve, or a true 'revolution' in the older (etymological) sense of a full-circle return to neglected though essential aims and priorities.[66] On his account the shift came about as a consequence of rising dissatisfaction after decades when philosophers sought to avoid such issues through the false belief that a due attentiveness to language and conceptual precision must entail a confinement to language and concepts as philosophy's sole legitimate concern. 'Although philosophers have more reason than physicists to consider matters of language or thought, philosophy is in no deep sense a linguistic or conceptual enquiry, any more than physics is.'[67] If Kripke and his followers did most to contest that previous orthodoxy with regard to the physical object-domain of entities, species, historically existent persons, and (prototypically) natural kinds then by the same token it is the naturalizing thinkers who have done most to advance that realist cause with regard to the metaphysics of mind.[68] All the more reason – I shall now proceed to argue – for looking to Spinoza not only as a notably prescient advocate of the mind/body or mind/brain identity thesis but also as a singular case of what rigorous thinking can achieve despite and against the pressures of received (in this case dualist or consciousness-based) doctrine.

Chapter 4

Catching Up with Spinoza: Naturalism, Rationalism and Cognitive Science

I

If there has always been a 'New Spinoza' this is no doubt because his thinking so strongly resists assimilation on any of the terms laid down by every mainstream school of European philosophy from Descartes to the present. Thus his work has very often been taken up by radicals or dissidents – those who approach it with a view to transforming the discourse of ontology, metaphysics, epistemology, ethics, politics, or aesthetics – while always leaving something unaccounted for, or something that is consequently thought to require a likewise radical critique.[1] This pattern of response goes a long way back – historically as well as philosophically speaking – to the earliest stages of Spinoza's reception when his writings became a chief zone of engagement in the struggle for freedom of conscience and belief, or for emancipation from the dictates of religious (whether Christian or Jewish) orthodoxy.[2] Later on it assumed the same kind of salience for the quarrel between idealism and materialism, or – at its most extreme – between the Romantic (German and English) idea of Spinoza as a 'God-intoxicated' mystic and his underground reputation as an out-and-out determinist, materialist, and atheist.[3]

Perhaps the most remarkable thing about this reception-history was the fact that both parties to each dispute could cite chapter and verse from Spinoza's texts, and moreover buttress their respective readings with a good show of exegetical care and argumentative rigour. It is the same with those recent or present-day schools of Spinoza interpretation which are often sharply at odds with each other on basic points of method, doctrine and principle yet which likewise manage to put up a strong textual-documentary as well as philosophical case. Thus, for instance, thinkers such as Althusser and Balibar – 'structural Marxists', as the label went – could very plausibly appeal to Spinoza by way of support for their rationalist account of the relationship between lived experience, ideology, and the process of 'scientific' concept-formation[4] while others, like Gilles Deleuze, could just as plausibly invoke him as the chief source or elective precursor for a kind of radical process metaphysics grounded in the notions of desiring-production and 'molecular' or 'deterritorialized' energy-flows.[5] More than that, the impact of his work was clearly visible across a swathe of developments in hermeneutics, critical theory, and the human and social sciences where Spinoza's philosophico-historical critique of revelation and scriptural warrant was among the most crucial early contributions to the project of secular enlightenment thought.[6]

Some years ago I wrote a book about Spinoza that put the case for his pervasive yet under-acknowledged influence and tried to sort out some of these multiple, often closely intersecting yet sometimes wildly divergent lines of intellectual descent.[7] In particular, I traced the conflict of interpretations which started out with his double role as arch-heretic or vilified atheist on the one hand and, on the other, purveyor of a knowledge – a mystical-intuitive mode of comprehension – beyond all the limits and endemic shortcomings of plain-prose reason. This conflict has been repeated in various displaced or surrogate forms over the past three centuries of often intense and heated debate around Spinoza's thought. Nowadays it appears in the clash of priorities between those in the 'analytic' camp who regard him as having some useful (if often misleadingly formulated) things to say about issues in metaphysics, epistemology or philosophy of mind[8] and those of a 'continental' persuasion who tend more often to emphasize Spinoza's politics or what they see as the basically political nature of Spinozist ethics, ontology, and psychology.[9] Even so this fails to capture the full complexity of the situation since there is something in common – philosophically if not politically speaking – between the analytic drive for conceptual clarity and precision and Althusser's claim for Spinoza (in company with Marx) as having achieved a decisive epistemological 'break' with the currency of common-sense or ideological belief.[10] Indeed, the very fact of his having spawned so diverse and complex a reception-history is one measure of Spinoza's extreme singularity and his way of holding out against classification according to such ready-made categories. French thinkers in the wake of structuralism, Deleuze especially, have on the whole been more concerned to emphasize this aspect of Spinoza's thought as a part of their campaign against the grip of conceptual abstraction or 'totalizing' systems of whatever kind, not least Althusserian Marxism.[11] However it has also left a strong impression among his analytical commentators through their sense of his standing quite apart from – and posing a sizeable challenge to – some of the most rooted assumptions of mainstream philosophic thought.

So for all their marked, even drastic differences of interest, idiom, and dominant agenda the 'two traditions' are at any rate largely agreed in their perception of Spinoza as a thoroughly anomalous and (to say the least) provocative thinker. While to some, this has been cause for unqualified celebration – in particular those, like Deleuze, who enlist him on the side of radical difference or heterogeneity – in others, it has provoked a very mixed response and sometimes taxed their exegetical patience to the limit and beyond. Here I am thinking chiefly of Jonathan Bennett's approach in the mode of Russell-style 'rational reconstruction' where he offers a patient, detailed, and often admiring account of Spinoza's *Ethics* until he gets to the 'third kind of knowledge' – *scientia intuitiva* achieved through the 'intellectual love of God' – at which point all this patience suddenly runs out and his commentary gives voice to a sense of bafflement and downright exasperation.[12] Thus, taking to task a comment by the more indulgent Stuart Hampshire: 'I contend that instead of implying that Spinoza has brought us "beyond the limits of literal understanding" and that this is acceptable because it is inherent in his chosen topic, we should say openly that Spinoza is talking nonsense and that there is no reason to put up with it.'[13] As for Frederick Pollock and his claim that these passages are 'among the most brilliant endeavours of speculative philosophy', and moreover that they 'throw a kind of poetical glow over the formality of [Spinoza's] exposition', Bennett is quite unable to contain his indignation. Thus: 'when a commentator as shrewd as Pollock is reduced to such babbling by his desire to praise the final stretch of the *Ethics*, that is further evidence that this material is worthless. Worse, it is dangerous: it is rubbish that causes others to write rubbish'.[14] Still, as I say, even these sharply conflicting valuations bear witness to the sheer singularity of Spinoza's thought and its power to solicit uncommonly intense and deeply felt modes of response, whether as an unprecedented challenge or a scandal to received ideas. In this context we might recall Derrida's etymologically pointed use of 'solicit' (from the Latin *solicitare*) with the sense of challenging and summoning-forth but also of shaking to the very foundations.[15] What unites these otherwise disparate approaches is their willingness – albeit very often within certain clearly marked limits – to accept the possibility of a thinking at odds with those dominant conceptions that have shaped the self-image of reputable philosophic discourse.

If commentators once joined battle over the issue of Spinoza as atheist and radical materialist versus Spinoza as nature-mystic and proto-Wordsworthian pantheist they now more often take sides over matters of ontology, epistemology, or philosophy of mind and language. Or again, they divide with respect to the question whether these are indeed (as analytic philosophers would have it) the core issues in Spinoza's thought or whether – on the dominant 'continental' view – they must ultimately take second place to his ethico-political concerns. Thus, as things stand at present, it is hard to imagine (say) followers of Althusser, Balibar or Deleuze entering into some kind of constructive dialogue with philosophers whose main points of reference are the commentaries offered by analytic thinkers like Bennett, Donald Davidson, or Alan Donagan.[16] Yet, in truth, the Spinoza who emerges through Althusser's structuralist-Marxist reading bears a closer resemblance to Bennett's Spinoza – the rationalist thinker of 'adequate ideas' as opposed to the delusions of imaginary common-sense belief – than to anything that finds room in Deleuze's (for want of any better description) radical-empiricist account. And again, despite obvious differences of idiom, what Deleuze has to say about Spinoza's doctrine of the affects and his notion of *conatus* as the inbuilt drive toward self-preservation and fulfilment on the part of every living organism finds a close parallel in readings from a very different quarter that likewise place chief emphasis on his treatment of the positive and negative emotions as the basis for any rational account of knowledge as conducive to human well-being. Among the latter can be counted Antonio Damasio's recent book which comes at these issues – that is to say, questions concerning the relationship between cognitive and passional components of the human psyche – from a neurophysical and cognitive-psychological angle but which none the less adopts a broadly 'analytic' rather than 'continental' approach.[17] My point is that Spinoza's thinking resists any adequate classification in terms of the standard, textbook account of how philosophy has developed over the past four centuries. For if Spinoza is undoubtedly a thoroughgoing rationalist who maintains that true wisdom can only be achieved through a reasoned critique of common-sense notions or intuitive, self-evident ideas then he is just as much a radical empiricist (more aptly: a radical naturalist and materialist) according to whom such wisdom consists in a due recognition of the various physical, causal, and socio-political factors that bear upon human knowers in their quest for more adequate self-understanding.

Of course the mere fact that he cannot be placed on either side of these deep-laid philosophic rifts doesn't mean that he manages to bridge them effectively or achieve the ultimate reconciliation between subject and object, mind and world, reasons and causes, or free will and determinism that has eluded philosophers from Descartes down and continues to preoccupy analytic and continental thinkers alike. However, it does provide a telling reminder of just how anomalous a figure Spinoza must appear by the lights of any orthodox historiography or any attempt to assimilate his thought to this or that certified line of descent. Where responses do tend to divide in fairly predictable ways is by reacting to the 'scandal' that Spinoza represents either in downright celebratory terms – as a salutary challenge to the norms and pieties of orthodox philosophic thought – or with various degrees of suspicion, mistrust, or hostility. Thus Bennett, as we have seen, has a high opinion of the *Ethics* just so long as it remains on analytically respectable ground, i.e., just so long as Spinoza is concerned with the corrective capacity of 'adequate ideas' when applied to the various confusions thrown up by the realm of sensory appearances or 'ideas of imagination'. However, this attitude switches very sharply to one of disappointment or shocked incredulity when Spinoza moves on, in Book V, to expounding the third kind of knowledge, that which involves a direct apprehension of the nature or essence of things somehow conceived as present to thought without any form of conceptual mediation. Such claims can only strike Bennett as amounting to a quasi-mystical doctrine whereby the mind is taken to possess something very much like the power of 'intellectual intuition' that Kant likewise denounced, that is, a capacity to pass beyond the realm of phenomenal appearances where sense-data are brought under adequate concepts and thus lay claim to an immediate knowledge of ultimate, noumenal reality.[18]

Yet it is hard to see the point of any 'rational reconstruction' in the analytic mode that adopts so partial or selective a view of those elements in Spinoza's thought that are deemed to merit serious attention by present-day analytic standards. For what drops out of sight in this process is also what constitutes the singular challenge of a thinking that runs directly counter to the whole tradition of epistemological enquiry which began with Descartes, found its systematic high-point in Kant, and is still very much a part of the present-day analytic agenda. That is to say, it is the radically monistic approach that typifies not only Spinoza's claims with regard to *scientia intuitiva* but also his entire conception of knowledge or, more precisely, his entire ontology of mind and nature conceived as twin aspects or attributes of a single, indivisible substance.

II

It seems to me that analytic philosophy has long been striving to escape or overcome this Kantian legacy while in fact coming up with nothing more than a series of minor variations upon it.[19] Spinoza alone, among the great thinkers of philosophical modernity, goes so far in his rejection of the dualist epistemological paradigm and his embrace of a radically monist ontological alternative as to provoke bewilderment not only among his good-willed exegetes but also among those analytic types who are themselves in quest of some such (albeit less radical) alternative. As I have said, this contrasts with the positive, even celebratory response to Spinoza's thinking in the recent continental – mostly French – reception-history where he has been recruited to a range of philosophical causes whose main (in some cases even sole) point of contact is the link they propose between issues of ontology and issues of an ethical or socio-political nature.[20]

Not that this dimension is altogether ignored by analytic commentators, forming as it does a crucial component of Spinoza's case for the role of philosophy in achieving a clearer, more distinct idea of the various factors (causal and social) that operate either to expand or to contract our scope for the exercise of human creative and emancipatory powers.[21] They have also shown some interest in pursuing the relation between Spinoza's more formal or logically articulated procedures of argument in the *Ethics* and the kinds of concern that animate those other portions of that work where he discusses the affective or passional aspects of human knowledge and experience, along with more overtly *engagé* writings such as the *Tractatus Theologico-Politicus*.[22] After all, any serious attempt to make sense of Spinoza's project as a whole will have to find some plausible way of explaining how the exercise of reason may contribute to a better, more enlightened understanding of the factors that make for psychological, social, and political well-being through a wise acceptance of our place in the natural order of things. More than that, it will have to offer an account of this process that ties in convincingly with Spinoza's critique of religious revelation and his arguments concerning the complex background of historical and cultural conditions that alone provide an adequate contextual basis for reading the scriptures in a critically informed and non-dogmatic way.

So of course the broadly analytic reception has included some work on this aspect of Spinoza's thought and on relevant details of his own socio-political background as one much involved in the various debates – as well as the front-line struggles for power and influence – within the Dutch Republic of his time.[23] However, it has not shown anything like the commitment of thinkers like Althusser, Balibar, Deleuze, or (most strikingly) Antonio Negri to produce a reading of the 'new Spinoza' that brings these multiple aspects together in a strictly inseparable fusion of politics, life-history and work.[24] What unites the various 'continental' approaches – despite their otherwise large divergences of aim – is a shared conviction that Spinoza's thought cannot be understood except through a reading that takes due account of both its immanent ('purely' philosophical) modes of argument and its close imbrication with the various historical, social, and political events that made up its formative background. That is, they start out by rejecting the analytic principle that requires a clear distinction between 'context of discovery' and 'context of justification', or the kind of strictly second-order research that has to do with matters of cultural-historical or psycho-biographical interest, and the kind of first-order investigation that pertains to the assessment of philosophic claims in accordance with distinctly philosophical criteria of truth and validity.[25] For an *echt*-analytic commentator like Bennett this distinction is so very basic – so definitive of what properly counts as philosophy rather than intellectual or cultural history – that the worth of Spinoza's intellectual achievement is to be judged solely with reference to the context of justification, which for him means in keeping with present-best ideas of conceptual rigour and precision.[26] For others of a broadly similar but somewhat less hard-line analytic persuasion (among them Alan Donagan) the distinction holds in matters of conceptual exegesis or strictly philosophical content but doesn't prevent such 'extraneous' interests from making some (albeit very limited) contribution to our better understanding of Spinoza's thought.[27] However, this allowance doesn't go so far as to invoke a contingent, i.e., historical, geographically specific and socio-politically emergent context for his central philosophic concerns, that is to say, his monist ontology and metaphysics along with whatever implications they might hold for current debates in epistemology, philosophy of mind, or cognitive psychology.

Nor is it surprising that this should be the case, given both the analytic premise that issues in philosophy cannot be reduced to second-order questions of history, politics, psychology, etc., and also – reinforcing that belief among his analytic commentators – Spinoza's commitment to the idea of philosophy as aimed toward an order of truth transcending any mere particularities of time and place. Yet of course there is another whole dimension of Spinoza's thought that is inescapably rooted in the social conditions and political events of his time, and which cannot be understood without reference to those same conditions and events.[28] Moreover, it is one that touches so directly on his chief metaphysical concerns – especially the issue of free will versus determinism which lies at their very heart – that any attempt to apply the two-contexts principle and distinguish clearly between 'life' and 'work' is sure to end up by offering a highly partial, not to say distorted view of those concerns. This is where his 'continental' readers have an edge since they reject that principle – at least in its more doctrinaire form – and make a point of relating life to work not just as a matter of more-or-less relevant psycho-biographical or socio-historical 'background' but as offering the only adequate means to grasp what is most distinctive and uniquely challenging about Spinoza's project. For of course it is a main part of that project to explain how we can think of human beings *both* as belonging to an order of causal necessity that allows no appeal to some imaginary realm of purely autonomous agency or choice, *and yet* as possessing the capacity to transform 'passive' into 'active' modes of experience. This capacity comes about – so he maintains – through the achievement of 'adequate ideas' which in turn make possible some measure of freedom from the realm of unknown and hence blindly operative causal forces.

Of course this way of putting Spinoza's case – like his own formulations in the *Ethics* and elsewhere – is very far from resolving the free-will/determinism issue and might well be seen as merely re-stating it in a sharpened or more intransigent form. Yet it is the merit of readings like those of Balibar, Deleuze, Macherey and Negri to insist that he alone among the great thinkers of early philosophical modernity faced up to that issue without taking refuge – like Descartes before and Kant after him – in a dualist metaphysics of subject and object, mind and body, or a noumenal domain wherein reason gives the rule for its own autonomous exercise and a phenomenal realm wherein everything is subject to the dictates of causal necessity. Moreover, they do so most often with specific reference to that complex background of historical, political, religious, and socio-cultural events that exerted such a crucial formative influence on Spinoza's thinking about issues of free will and determinism. Of course this may be said to beg the question yet again since, after all, there is a prima facie contradiction – or at any rate a sharp clash of priorities – between the claim for Spinoza as one who possessed a sufficient degree of intellectual autonomy to think those issues through in a novel, creative, and independent-minded way and the claim that his ideas were crucially affected by the distinctive pressures and specific challenges of the time. Indeed, these commentators might be seen as of going out of their way to emphasize the problem and ensure that Spinoza's readers have to face it fair and square, rather than seeking a convenient escape-route or evasive compromise solution that would purport to bring him out as a moderate determinist and upholder of free will in some likewise moderate, qualified or 'compatibilist' form.

Thus the main thrust of interpretations like those just mentioned is to insist – *contra* such face-saving or emollient accounts – that Spinoza's was an outlook radically opposed to any notion that the problem might be assuaged by adopting a sensible line of least resistance midway between those strictly unthinkable extremes. For instance, they stress that he took time off from composing the *Ethics* and before proceeding to those parts of Book V where, if anywhere, his doctrine of freedom might attain its definitive statement in order to write the *Tractactus Theologico-Politicus* as an urgent and topical contribution to debates about politics, religion and the future of the Dutch Republic.[29] That work was primarily concerned to explain how the supposed timeless truths of scriptural revelation should rather be understood as products of their own historical, cultural and socio-political locale along with the motives of those various, far-from-disinterested parties who first wrote them down and then engaged in the process of editing, transmission, and selective deployment to overt or covert manipulative ends. That is to say, the *Tractatus* was a thoroughgoing exercise in the mode of materialist, causal-explanatory, socio-diagnostic, proto-secular and demythologizing critique that would emerge to full view only after two centuries of an afterlife and impact that were largely underground since often forcefully repressed.[30] Of course one reason for this, quite apart from its explosive theologico-political content, was the fact that Spinoza here seemed to adopt a thoroughly determinist approach that purported to demonstrate the false and illusory character not only of religious truth-claims but also of our cherished, theologically sanctioned self-image as believers whose faith – or lack of it – could properly be ascribed to our own God-given capacity for autonomous belief-formation. What the *Tractatus* drives home to painful effect, for anyone who wishes to retain such faith, is both the logical impossibility of squaring this latter pair of requirements and the extent to which that entire belief-system along with its various doctrinal, scriptural, and institutional props can be seen to rest upon a basis of merely contingent historical events. Thus it leaves no room for such 'imaginary ideas' as those of revelation, divine intervention, or miracles, all of which Spinoza treats (like Hume after him) as resulting from a mixture in various proportion of natural, historical, and psychological causes joined to the effects of ignorance, fear, and predisposed or passive credulity.

In short, as these commentators acknowledge, the free-will/determinism issue is by no means resolved or quietly laid aside but in some ways rendered all the more intractable by Spinoza's decision to interrupt work on the *Ethics* and devote several years of intensive research under often very difficult personal and social circumstances to composing the *Tractatus*. Their point, like his, is to wean us away from any idea that thinking might achieve a genuine – as distinct from merely notional – margin of autonomy or freedom by claiming to rise above the conditions of its physical or causally constrained, as well as its historically situated place and time. Yet their commentaries would surely miss something crucial if they didn't all the same make allowance for the strong countervailing tendency in Spinoza's thought, that is, his commitment to a doctrine of 'adequate ideas' that affirms the power of intellect to criticize false beliefs and pass beyond them to a knowledge no longer in the grip of illusory 'common-sense' or ideological notions. This is what lends a degree of credibility to the sorts of analytical approach, like Bennett's, that pretty much ignore any 'background' matters of historical, cultural, or socio-political context, or again, the attitude summed up by Donagan when he remarks that '[g]enerally [Spinoza's] life was of a piece with what he wrote: discoveries about its details – apart from facts about his intellectual exchanges – bear dubiously on disputed questions about what he thought'.[31] It is also the aspect of his thinking that most captivated Althusser and the early Macherey when they recruited certain pregnant formulations from Spinoza as a prime exhibit in their structural-Marxist case against Hegelian, Lukacsian or other such 'expressive' ways of figuring the link between socio-economic base and political-cultural superstructure.[32] Rather we should try to conceive it as a complex, decentred, and over-determined mode of relationship wherein there exist certain 'structures in dominance' but where economic forces should be taken to predominate only 'in the last instance', or just in so far as they are assigned that role by the entire existing conjuncture.

This is not the place for anything like a critical exposition of Althusserian Marxism. Sufficient to say – in the present context – that Spinoza's influence is often plain to see in its emphasis on structural (as opposed to expressive or 'totalizing') modes of explanation and on the crucial role of philosophy as a form of 'theoretical practice' aimed toward resisting or breaking the hold of intuitive, self-evident, or common-sense (for which read: ideological) beliefs. Thus, according to Althusser and Balibar,

> [e]ffects are not outside the structure, are not a pre-existing object, element, or space in which the structure arrives to imprint its mark; on the contrary, the structure is immanent in its effects, a cause immanent in its effects in the Spinozist sense of the term, that the *whole existence of the structure consists in its effects,* in short that the structure which is merely a specific combination of its peculiar elements, is nothing outside its effects.[33]

Moreover, Spinoza should also take credit for having pioneered the mode of critical or 'symptomatic' reading that enabled commentary to go beyond its traditional, fideist attitude in matters of textual (especially scriptural) warrant, and thereby reveal those moments of unresolved aporia, strain or contradiction that signalled the effect of some repressed yet disruptive ideological content. Clearly what Althusser and Balibar have in mind is the *Tractatus Theologico-Politicus* and its precocious combination of textual exegesis with a range of approaches – hermeneutic, source-critical, historical-reconstructive, 'cultural-materialist' in no very stretched sense of the term – that would have to wait a good two centuries before they were taken up and developed. Hence their very striking claim in *Reading Capital*: that

> [t]he first man ever to have posed the problem of *reading*, and in consequence, of writing, was Spinoza, and he was also the first to have proposed both a theory of history and a philosophy of he opacity of the immediate.

> With him, for the first time ever, a man linked together in this way the essence of reading and the essence of history in a theory of the difference between the imaginary and the true.[34]

So there is another side to the recent reception-history, one that has more to do with 'adequate ideas' and with the power of thought to criticize and thereby transcend its own formative context or background conditions than with the need to recall theory to a sense of its own inescapable enmeshment in those same conditions.

However, it is worth recalling once again how urgent were the social and political circumstances that bore upon Spinoza when his main concern – and a chief motive for writing the *Tractatus* – was to stave off the threat of religious dogmatism and the warring factions whose claim to exclusive possession of scriptural truth looked set to destroy the Dutch republic. Nor is the *Ethics* by any means free of such turbulence, since – as must strike any attentive reader – there is a notable contrast between the order of numbered axioms, propositions and corollaries with their appearance of impassive, (quasi-)geometrical precision and the various interpolated scholia where Spinoza finds room for some powerful expressions of positive and negative affect. This why, as Deleuze puts it, there is need for

> a double reading of Spinoza: on the one hand, a systematic reading in pursuit of the general idea and the meaning of the parts, but on the other hand and at the same time, the affective reading, without an idea of the whole, where one is carried along or set down, put in motion or at rest, shaken or calmed, according to the velocity of this or that part.[35]

Thus it was always very much on the cards that the high theoreticist 'moment' epitomized by Althusserian Marxism would at length give way to a reactive trend – in Spinoza scholarship and also in the wider context of post-1980 French philosophical debate – which mounted a vigorous challenge to it. That challenge took shape among thinkers like Deleuze in the name of difference, intensity, 'desiring-production', 'molecular' versus 'molar' forces, 'de-territorializing' lines of flight versus 're-territorializing' modes of control, and other such attempts to evoke or connote what lay intrinsically beyond the grasp of adequate conceptualization.[36]

Along with this went a drastically changed estimate of Spinoza's significance, one that located the potentially transformative and liberating power of his thought not so much in the process of conceptual critique whereby 'confused' or 'imaginary' ideas yielded place to their 'clear' or 'adequate' counterparts, but rather in those passages from the *Ethics* that affirmed the priority of positive over negative or 'joyous' over 'sad' affects and emotions. Thus the image of Spinoza that predominates in Althusser's work – that of an elective precursor to Marx who somehow manages to construct (or discover) in advance the entire theoretical apparatus of Marxist *Ideologiekritik* – now yields place to the image of one who adopted a simulacrum of rationalist method so as to impart a certain order to his otherwise unmanageably prolix and tumultuous thoughts.[37] What is so remarkable is that both these conflicting accounts find warrant not only in a few, carefully selected passages from Spinoza but on the basis of readings that adduce large amounts of highly relevant textual evidence, and which do so moreover with consistent and well-defined interpretative ends in view.

III

I should perhaps make it clear that I am not for one moment presenting Spinoza as some kind of textual Rorschach-blot into whom various parties can read – or onto whom they can readily project – whatever meanings or messages they choose. On the contrary, as I have said, these variant readings each have a claim to exegetical rigour and fidelity that redeems them from any such charge. More to the point is Derrida's remark that certain thinkers – maybe all great thinkers, or those who have given rise to a significant reception-history – tend to generate sharply opposed interpretations which cannot be reconciled or subject to settlement one way or the other since they can both cite chapter and verse in their own support and can both very plausibly assert their credentials as the authorized version.[38] Very often these debates fall out between 'left' and 'right' lines of intellectual descent, as can be seen in different ways – so Derrida observes – with philosophers from Aristotle to Kant, Hegel, Nietzsche and (not least) Marx. He also provides a useful metaphor by which to think about this curious feature of intellectual history, namely that of the tape-recording machine with multiple playback heads, such that any given segment of tape (or passage of text) may be decoded in different ways yet without this necessarily entailing any drop-outs or distortions (interpretative oversights, errors, or symptoms of gross ideological bias).

Thus philosophers and literary theorists tend to distort the issue by constructing a false *tertium non datur*, that is, by supposing it to fall out between defenders of a strict intentionalist or 'single-right-reading' position and those who adopt an attitude of 'anything goes' or total hermeneutic licence. However this is merely to sidetrack attention from the more challenging question as to *just what it is* about certain passages in certain authors that somehow engenders such instances of deep-laid scholarly-critical dispute, given that there do exist certain constraints upon the range of admissible readings. For if one thing is clear from Derrida's work on thinkers from Plato and Aristotle to Rousseau, Kant, Nietzsche, Husserl and J. L. Austin it is the fact (as he puts it) that interpretation cannot develop in just 'any direction at all' or 'authorize itself to say almost anything', but rather requires 'all the instruments of traditional criticism' – of philology, textual scholarship, and a due regard for authorial intent – as an 'indispensable guardrail' in the process of critical exegesis.[39] On the other hand – crucially – this guardrail 'has only ever *protected*, it has never *opened* a reading', so that criticism has to go beyond 'the effaced and respectful doubling of commentary' in order to reveal how 'the writer writes *in* a language and *in* a logic whose proper system, laws, and life his discourse cannot dominate absolutely'. And again: 'the reading must always aim at a certain relationship, unperceived by the writer, between what he commands and what he does not command of the patterns of the language that he uses'.[40]

It is Nietzsche who provides one of Derrida's most striking examples since in no other case have the divergent 'left' and 'right' interpretations run to such extremes and been able to cite such a range of good (or at any rate highly plausible) supporting evidence from the texts in question. But there is also a sense in which Nietzsche lends himself too easily to Derrida's purpose since his writings contain such a mass of provocative, wilfully extreme and often downright contradictory remarks assembled with such scant regard for all the normal (to his way of thinking: inertly conformist) protocols of rational discourse. Spinoza offers a more interesting test-case in so far as his thinking manifests a high degree of logical consistency, even if his style of reasoning *more geometrico* in the *Ethics* is apt to make it seem more rigorously argued and tightly structured than it is. So where 'left' and 'right' Nietzscheans can always point to different, often conflicting passages in the text which provide support for their likewise divergent interpretations it is not so easy to explain how Spinoza could have spawned such a multifarious reception-history. This challenge becomes yet more acute when one considers that his was the most resolutely monist and hence – one might expect – the most unambiguous, clearly stated and multiple-reading-proof philosophic system to have appeared in Western philosophy since the great monists of antiquity such as (at opposite extremes) a metaphysician such as Parmenides and radical materialists such as Epicurus or Democritus. And yet, as I have said, it is the utterly unqualified or uncompromising character of Spinoza's monist ontology that has given rise to this likewise extreme pattern of contrasting interpretations. Thus his reputation has always been a battle-ground between those who considered him a pantheist, a mystic, a well-nigh saintly figure, or (in the famous words of Novalis) a 'god-intoxicated' thinker and, on the other hand, those – including the vast majority of his contemporaries – who deemed him an out-and-out materialist, atheist, and wicked subverter of every last moral value.

Of course the terms of this controversy have changed and one is nowadays unlikely to find Spinoza either praised or vilified for any such reasons. All the same it is not too hard to discern the legacy of those old battles in the more restrained and philosophically specialized yet none the less sharp divergences of view that continue to attend his present-day reception-history. For there is just as great a difference between, say, Althusser's high-structuralist or rationalist reading and Deleuze's take on the prophet of unbridled desiring-production as any that arose in his own time or during the subsequent two centuries when 'Spinozism' was a watchword – and a dangerous charge to bring or to face – in various philosophical, theological and socio-political disputes. Or again, there is just as much at stake in doctrinal terms between those who take Spinoza to be offering intimations of a new and radically distinct mode of cognition in Book V of the *Ethics* (where he talks about the 'third kind of knowledge', or *scientia intuitiva*) and those, like Bennett, who come at it from a strongly analytic or 'rational-reconstructive' angle and who tend to throw up their hands in despair at just this point.[41] What unites them, all the same, is a strong sense that Spinoza is venturing into strange seas of thought where established philosophical distinctions break down, among them most obviously those between mind and body, subject and object, or self and world. Hence no doubt, his renewed appeal to philosophers of otherwise diverse persuasion who see in Spinoza's radical monism – or something very like it – the hope of achieving a clean, conceptually unencumbered break with the whole bad legacy of Cartesian dualism and its various, e.g., Kantian and present-day (whether 'analytic' or 'continental') successor movements.

Thus, for instance, Donald Davidson cites Spinoza as the one early-modern philosopher who managed to think his way through and beyond the Cartesian impasse and who thereby prefigured the current turn toward a thoroughly naturalized or non-'metaphysical' yet also (just as crucially) non-reductionist account of the mind/body relationship.[42] What Davidson puts forward, briefly stated, is a theory of 'anomalous monism' which seeks to maintain the following principles (1) that for any mental event there is a corresponding physical (brain) event in the absence of which the mental event would not have occurred, (2) that this correlation, though strict, is non-law-governed or 'anomalous' since construed as holding only between event-tokens rather than event-types, and (3) that although the mental supervenes upon the physical there is no prospect of reducing the former to the latter – of pushing right through with any radical behaviourist, eliminativist, or central-state-materialist programme – since they involve altogether different descriptive or conceptual registers. Where the one requires a discourse (like those of neurophysiology or cognitive science) adequately stocked with causal-explanatory terms and predicates the latter requires a language equipped with just the sorts of vocabulary that figure in our everyday as well as philosophical talk of human meanings, motives, reasons, intentions, beliefs, desires, attitudes, and so forth. On Davidson's account – and on Spinoza's as Davidson reads him – we can enjoy the full benefits of both while preventing any possible conflict between them by keeping in mind the central tenet of anomalous monism, that is, the lack of any nomic (i.e. any fixed, invariant, or type-type) connection that might lend credence to the 'strong'-reductionist or physicalist case. This theory Davidson takes to be implicit in Spinoza's idea of mind and body as two different 'attributes' of the self-same substance, and likewise in his cautiously coded talk of *deus sive natura* – 'god or nature' – as two distinct yet compatible ways of referring to a single ultimate reality whose various 'modes' (or particular instantiations) could just as well be conceived under one or the other attribute.[43]

It is obvious enough why Davidson should look to Spinoza not only as a thinker who arrived at some strikingly similar ideas about the mind/body relationship but also as one who went a long way toward Davidson's closely connected views on the nature of actions and events. Thus he cites a well-known passage from the *Ethics* where Spinoza sets out his cardinal distinction between active and passive powers: '[o]ur mind does certain things and undergoes other things, viz. Insofar as it has adequate ideas, it necessarily does certain things, and insofar as it has inadequate ideas, it necessarily undergoes other things'.[44] He might just as well have cited a range of other passages to similar effect, as when Spinoza states that '[w]hatsoever increases or diminishes, helps or hinders the power of activity in our body, the idea thereof increases or diminishes, helps or hinders the power of thought in our mind', or – by the same token – that '[t]he mind, as far as it can, endeavours to conceive those things, which increase or help the power of activity in the body'.[45] Davidson is clearly attracted to this as a way of re-conceiving the free-will/determinism issue that would seem to overcome that old and (on its own terms) strictly insoluble antinomy by locating the distinction between active and passive affects in the degree to which the self-realising powers of this or that particular mind/body have attained their fullest scope of expression or exercise.

Thus: '[o]n the positive side, Spinoza gives analyses of volition, perception, and the emotions consistent with his thoroughgoing naturalism and determinism. Perhaps the most striking feature of his concept of action is that it differs from being acted on just to the extent that its causes and effects lie within rather than outside us, and that this in turn is a matter of the extent and character of our knowledge'.[46] However he goes straight on to remark that this notion 'follow[s] directly from [Spinoza's] objective view of human beings as integral parts of the causal chain of events', and that 'even to us to whom this attitude comes perhaps more naturally than to [his] contemporaries, it is a sobering perspective'.[47] Yet, after all, that perspective is one that Davidson must surely be taken to share if his monism is not to appear just a notional concession to the science-led intellectual climate of the time, and his use of the qualifying term 'anomalous' (i.e., his rejection of psycho-physical laws) a somewhat shifty device for avoiding the nemesis of a hard-line physicalist, central-state materialist, or downright determinist creed.

In fact it seems to me that Davidson's approach is not so much an answer or a working solution to the problem of Cartesian dualism but more a conceptual sleight of hand or finessing of the issue which leaves that problem very firmly in place. This emerges in the following passage where his usual, briskly problem-solving tone gives way to a flat re-statement of the old puzzle and what sounds very much like an outlook of resigned acceptance that it cannot be resolved. 'I confess', Davidson writes,

> that I do not see how even the most complete understanding of human psychology can avoid essential reference to the material forces that impinge on us.

> Nor do I see how psychology, as long as it deals with concepts such as those of action, intention, belief, and desire, can either be reduced to the natural sciences or made as exact and self-contained as physics. As I suggested, we may even take Spinoza as having shown why such a psychology is impossible; the nomological irreducibility of the mental to the physical can be taken to point in this direction.[48]

So it seems that Davidson's theory of anomalous monism is not so much clarified or rendered more plausible as pushed even further out on a limb by this analogy with Spinoza on the dual, self-subsistent and causally isolated 'attributes' of mind and body. That is to say, it becomes even harder to conceive how the theory could amount to a genuine or working monism, as distinct from a self-defeating attempt to reconcile two contradictory or at any rate mutually exclusive claims. Moreover, although his appeal to Spinoza has a certain prima facie plausibility, it gives a similar impression of evading the radical implications of Spinoza's thought and reducing what is truly exorbitant about it – not only in the view of his more orthodox-minded contemporaries but also, as we have seen, in that of many present-day admirers – to little more than a convenient *façon de parler* adopted by way of heading off certain otherwise intractable philosophic problems. Thus one gets little sense from Davidson's account of what opponents at the time – and for a good two centuries afterwards – were wont to denounce as Spinoza's crypto-atheism, or his promotion of a thoroughgoing determinist and materialist creed under cover of a vaguely mystical or pantheist rhetoric.

That sense comes across far more powerfully in Deleuze, Balibar and other voices from the recent 'continental' reception-history who emphasize both the uncompromising nature of Spinoza's ontological monism and the fact that it cannot be understood – at least without massive distortion – unless one interprets it in equally uncompromising naturalistic and materialist terms. This is no doubt one reason, as Althusser and Balibar remark, why '[t]he history of philosophy's repressed Spinozism unfolded as a subterranean history *at other sites*, in political and religious ideology (deism) and in the sciences, but not on the illuminated stage of visible philosophy'.[49] For despite all philosophy's periodic claims, more than ever during the past few decades, to have moved decisively beyond any form of Cartesian mind/body dualism that idea still exerts so strong a residual hold that any truly radical break with it – such as Spinoza represents – is apt to strike many as un- , pre- or downright anti-philosophical.[50] This applies especially to his doctrine of the positive and negative (or 'joyful' and 'sad') affects, one that finds room for freedom of will and the notion of autonomous agency only in so far as these can be squared with the single most basic Spinozist precept, that is, the indivisible unity of mind and body or the claim that they must always be conceived as two attributes of the one, self-identical substance. From which it follows that the process whereby passive dispositions take on an active character or sad affects are converted into their joyful, affirmative, life-enhancing counterparts cannot consist in some conscious effort of will or deliberate exercise of 'mind over matter' such that the one (mental) attribute would assert its claim to freedom and self-fulfilment independently of – or in isolation from – the other. What we have to envisage, rather, is the activity of jointly mental and physical striving (Spinoza's *conatus*) through which every sentient creature endeavours to persist in its own proper and distinct mode of being, this latter conceived as a development toward its state of maximum possible flourishing. What we are *not* to suppose – at least if convinced by Spinoza's radically monist ontology – is that the process might occur (as philosophers have often thought) through the bending of our will or the summoning of mental, intellectual, or moral resources which somehow belong to a realm quite apart from that of our physical or bodily movements and affects.

So there are certain caveats that need to be entered before going along with Davidson's case for viewing Spinoza as a champion of 'anomalous monism' *avant la lêttre.* On this account his strongest elective affinity in present-day terms is with the broadly compatibilist or 'no problem' thesis that physicalism and mentalism or brain-talk and mind-talk can perfectly well be reconciled just so long as one assigns them to different languages or conceptual-descriptive registers and thus heads off any conflict between them.[51] From the same distinctly modern standpoint there is nothing to be lost – and a good deal gained – by translating Spinoza's ontological claims out of their original, overly scholastic and metaphysically otiose idiom into one that sheds all that surplus baggage through a careful re-statement in more up-to-date, i.e., analytic and linguistic terms. Thus it looks as if the Spinozist doctrine of substance and attributes can be rendered acceptable by present-day standards of intelligibility and yet retain everything distinctive about it through a straightforward process of assimilation to something like Davidson's theory of anomalous monism. Yet this does entail certain significant losses or – from a standpoint not so much in sympathy with the governing interests of analytic philosophy after the 'linguistic turn' – a tendency to opt for the line of least resistance with regard to Spinoza's most challenging claims. That is, it has recourse to a 'linguistified' version of the double-aspect theory which no doubt succeeds in outfacing the threat of hard-line physicalist or central-state-materialist arguments but only at the cost of so far re-interpreting Spinoza's doctrine as to render it consistent with Davidson's studiously non-committal (not to say somewhat evasive or shuffling) view of these matters.

Of course there is a sense in which Spinoza's writings lend themselves to just this kind of revisionist treatment to the extent that they adopt a systematically equivocal language or a trick of constantly playing on phrases – such as *deus sive natura* ('God or nature') or the mind as an 'idea of the body' – which leave commentators hard-pressed (or perhaps gratefully unobliged) to decide between rival interpretations. All the same, this ignores several points that bear directly on the question as to what counts as a valid understanding of crucial Spinozist concepts and categories. One is the fact that Spinoza would have seen absolutely no merit in the notion that those concepts and categories made sense only to the extent that they played an accepted role in some existing (whether everyday or specialized) linguistic register.[52] On the contrary, he thinks of natural language and its various (supposed) imperfections – vagueness, ambiguity, metaphor, referential opacity, and so forth – as among the chief obstacles to philosophic progress and hence as standing in need of treatment through an 'emendation of the intellect' that would lead from confused or 'imaginary' notions to clear and distinct concepts.[53]

In this respect Spinoza agrees with his contemporaries – rationalists and empiricists alike – that language is at best an efficient (i.e., transparent or non-distorting) means for the conveyance of ideas from one mind to another, and at worst a continual source of error and mutual misunderstanding.[54] Thus: '[s]ince words are a part of the imagination – that is, since we form many conceptions in accordance with confused arrangements of words in the memory, dependent on particular bodily conditions – there is no doubt that words may, equally with the imagination, be the cause of many and great errors, unless we keep strictly on our guard'.[55] David Savan spells out the implications of this in a passage that leaves no doubt as to the distance that separates Spinoza from many of his present-day (including his best-willed or most sympathetic) exegetes. According to Spinoza, he writes,

> [a]n idea is not an image and does not consist of words. A true idea can neither arise from experience of words and images nor can it be verified through such experience, for experience can give no knowledge of essences . . . . Whereas an idea is certain, words are uncertain . . . . And whereas it is of the nature of reason to consider things as necessary and under a certain form of eternity, words are connected with contingency and time.[56]

It is only from our own ensconced position on this side of the post-1950 and nowadays near-ubiquitous 'linguistic turn' that such a notion might seem philosophically naive, and then only in so far as we adopt a Wittgensteinian rather than a Frege-Russell view of such matters.[57] At any rate, Spinoza would have had little time for any argument that construed his central doctrine – the mind/body identity thesis – in such a way as to empty it of all substantive content by lopping off its ontological claims, espousing a linguistified version of the double-aspect theory, and thereby allowing us to carry on talking in the same old dualist terms.

That is to say, if Spinoza is indeed to be recruited in the name of 'anomalous monism' without resorting to a strong-revisionist (or grossly inaccurate) account then the monism needs to be taken at full strength and the 'anomaly' interpreted *not* as an absence of real, objectively existent nomic regularities or causal laws but rather as a frank admission that they are – and may forever remain – beyond our utmost powers of comprehension. Although Davidson's position sometimes lends itself to glossing along these lines he is more often to be found insisting that, since mind-events and brain-events are 'under different descriptions', therefore it is pointless (or strictly nonsensical) to advance any strong, ontologically committed version of the mind-brain identity thesis.[58] And again, since causes and reasons must likewise be thought of as falling under different descriptions – since they make sense only as playing a role in (respectively) the physical or natural and the human or social sciences – therefore, on Davidson's account, we shall go badly wrong and risk summoning the twin spectres of reductionism and determinism if we allow ourselves to mix them up. Yet of course it was precisely Spinoza's willingness to raise those spectres and to do so, moreover, without having recourse to any such strategic or crypto-dualist fallback stance that prompted his expulsion from the Jewish community of Amsterdam and thereafter drew the wrath of orthodox religionists against himself and anyone brave or incautious enough to earn the label 'Spinozist'.[59] Nor can Davidson's reading gain much in the way of added plausibility from Spinoza's 'other' reputation as a mystically-inclined pantheist thinker or – its close equivalent in psycho-physical terms – his notion of mind as 'an idea of the body', thus seeming to reverse the order of priority and suggest some kind of mentalist (even panpsychist) view of how their ultimate identity is best understood.[60] What disappears in Davidson's account is once again the extremity – by any 'normal' philosophic standard – of Spinoza's monist doctrine. For those alternative idioms don't so much soften or defuse its determinist impact as insist (like the phrase *deus sive natura*) on the need for all divine, supernatural, dualist, and suchlike 'imaginary' terms to be read metaphorically or allegorically and thereby rendered consistent with a thoroughly naturalized world view.

IV

It is this aspect of the 'radical' Spinoza – the philosophically exorbitant character of his thought – that comes across to more singular and striking effect in the recent 'continental' (mainly French) reception than among his analytically-minded commentators. One reason, no doubt, is that the continentals have fewer inhibitions about pursuing metaphysical themes, or pursuing them (so to speak) in a full-bloodedly metaphysical way, rather than in the scaled-down descriptivist or semantic mode that has been the hallmark of most recent work in the analytic tradition. Another is the way that commentators such as Balibar, Macherey, Deleuze, and (most conspicuously) Negri have shown a far greater willingness to integrate their often highly speculative theses concerning Spinoza's philosophic thought with a detailed interest in the various historical and political events that left a visible, sometimes decisive mark on the course of his life and work.[61]

The great exception here is Althusser who himself claimed Spinozist (as well as Marxist) warrant for insisting on a sharp conceptual distinction between ideology and science. On the one hand is the realm of common sense, confused, or 'imaginary' notions and on the other hand that of 'adequate ideas' arrived at through a process of rigorously theorized rectification and critique. By the same token – and taking a lead from Gaston Bachelard's method of 'applied rationalism' in philosophy of science – Althusser assumes that any interest attaching to Spinoza's life and its socio-political context may well be legitimate for the purposes of intellectual biography or the history of ideas but must always give way to quite different, altogether more demanding standards of accountability when it comes to issues of knowledge and truth.[62] Thus Althusser's approach, premised on Spinoza's distinction between the first ('confused' or 'imaginary') and the second ('adequate') kinds of knowledge, to this extent has more in common with those analytic modes of thought which likewise make a cardinal point of distinguishing 'context of discovery' from 'context of justification', or the background conditions of emergence for this or that scientific hypothesis and the validity-conditions that properly apply when it is subject to the rigours of empirical testing and further theoretical analysis. In Althusser this went along with the notion of a Marxist 'theoretical practice' – or labour of conceptual critique – whereby we might achieve an adequately theorized grasp of how ideology exerted its otherwise ubiquitous power to interpellate subjects and their various modes of 'common sense' knowledge and experience.[63]

As I have said, there was good Spinozist warrant for Althusser's claims in this regard, whatever the well-known problems with them from a practical-political as well as philosophical standpoint. It is also fairly clear how Spinoza could be read as offering support for Althusser's case – in opposition to the humanist, Hegel-influenced readings of Marx dominant in France at the time – that a chief prerequisite for the better understanding of Marx's thought was to break with such anthropocentric residues through a form of 'theoretical anti-humanism' that sought to reveal its underlying conceptual or epistemo-critical structures. Yet just as this involved a highly partial and selective reading of Spinoza – one premised exclusively on the doctrine of adequate versus inadequate or 'imaginary' ideas – so Althusser's larger project ran aground (or so it has seemed to many commentators) on its failure to explain how this ascetic imperative could possibly provide any motivating impulse for political agency and change.[64] Indeed, it was largely in reaction against that high-structuralist or theoreticist project that thinkers like Deleuze, Balibar, and Negri – albeit in different ways – set out to make their case for the 'other' Spinoza, that is, the thinker whose undoubted commitment to the rationalist way of ideas went along with an equal (and by no means opposite) commitment to the maximization of active over passive or 'joyful' over 'sad' affects. This is why the recent continental reception has typically managed to find room for a range of Spinozist themes and concerns that tend to be excluded – or to figure only marginally – in most analytic treatments. Or rather, it manages to take simultaneous account of those various aspects of Spinoza's work which more often receive separate attention from specialists in different quarters of the Anglophone academic community.

Thus, for instance, there has been some ground-breaking work by intellectual and cultural historians who have uncovered far more than was previously known about the socio-theologico-political contexts of Spinoza's thought and later reception. Most notable is Jonathan Israel's study of Spinoza as the central figure in that 'radical enlightenment' that emerged among certain dissident factions in the seventeenth-century Dutch republic and whose critical-emancipatory character was tamed or repressed by the mainstream (bourgeois) enlightenment that flourished in France and then Germany a century later.[65] Israel's work belongs very much to a tradition of Spinoza commentary where detailed scholarship has often coexisted with a strong sense of political partisanship or a marked sympathy for just those motivating values – secularism, tolerance, republican beliefs, freedom of thought and expression, cosmopolitanism in the legal and juridical spheres – for which he stood as an emblematic figure. It is a tradition that goes back, via champions like Samuel S. Feuer, to the centuries-long history of his underground reputation when 'Spinozist' (for which read: atheist, materialist, and politically subversive) ideas were subject to vigorous church-and-state repression in many parts of Europe and beyond.[66] Then again, recent developments at the interface between neuroscience and cognitive psychology have led some admirers – like Damasio – to hail Spinoza as a thinker far ahead of his time in his outright rejection of Cartesian dualism, his recognition of the strictly indissoluble tie between mind and body or thought and emotion, and moreover his having set forth these prescient ideas in way that falls square with some of the most advanced current thinking in the field.[67] Yet there is a strong sense throughout Damasio's book that he sees this approach as one that holds the promise of rescuing Spinoza from the dead hand of philosophy and granting him the favour of a reading informed by the latest scientific knowledge. In Israel's narrative likewise, the richness and sophistication of historical scholarship goes along with a fairly routine treatment of Spinoza's philosophical ideas and a relative lack of concern with their specific rather than general bearing on that ferment of radical ideas which his book brings so vividly to life. In fact one has to look elsewhere – to the French reception and its complex interweaving of metaphysical or speculative with historical, political, and socio-cultural themes – in order to gain a more adequate (i.e., philosophically informed and integrated) view of Spinoza's galvanizing role.

My point is not to suggest that we should reconcile or integrate these diverse claims on Spinoza's behalf in some kind of grand synthesis. Such a prospect is neither realistic nor even desirable, given what we surely should have learned from Spinoza and his best exegetes, that is, the fact that his thinking stubbornly resists assimilation to any of the ready-made concepts or categories by which philosophers, no less than intellectual historians, try to keep things safely under control. This is no doubt why the many attempts to recruit Spinoza to this or that philosophic cause – for free will or determinism, pantheism or atheism, compatibilism vis-à-vis the mind-body issue or a hard-line physicalist position – have always and inevitably run into problems when striving to square these doctrines with a properly attentive, impartial reading of his work. Indeed, a chief lesson of the modern French reception, and one that could only have resulted from its greater degree of speculative licence, is the strange co-presence in Spinoza's thought of a rationalism aimed toward removing or correcting all the causes and effects of the mind's enslavement to the passions and a countervailing will to intensify those passions to the point where they are no longer merely suffered but actively promoted and enjoyed. Thus the compact history of visions and revisions that leads from Althusser to Deleuze and which takes in the shift of focus exhibited by thinkers like Balibar and Macherey is one firmly grounded *at every stage* in a detailed engagement with Spinoza's texts and *in no way* contravening the letter or the spirit of those texts. Rather it is a faithful reflection of the fact that Spinoza remains the great exception – in Negri's phrase, the 'savage anomaly' – whose thinking is able to solicit and support (if not fully accommodate or reconcile) such a range of conflicting interpretations.

All the same, there is something distinctly unsatisfactory about the way that historians, philosophers, critical theorists, and lately cognitive psychologists have tended each to go their own way and stake their claim to the 'new' or 'radical' Spinoza without much sign of either knowing or caring what the others have to say. This is not, I think, just because they come at his ideas from such divergent and specialized angles of interest but also because the philosophers – who might be expected to mediate these various concerns – are themselves so deeply at odds with regard to the nature, import, and present-day significance of Spinoza's work. That is to say, there exists such a gulf between the two philosophical cultures and their respective ideas of what constitutes a valid or worthwhile contribution to debate that it is hard for people in other disciplines to get any confident purchase on the issues involved. Thus it is reasonable to suggest that Jonathan Israel might usefully have looked to Deleuze and Balibar by way of philosophic support for his conception of Spinoza as a seminal thinker of the radical enlightenment. Moreover, Damasio might likewise have drawn upon Deleuze – on his detailed working-out of the Spinozist idea of active versus passive affects – as offering a line of approach very much to his own neuroscientific and cognitive-psychological purposes. What has so far got in the way of such productive exchange is the fact that most workers in those other fields tend to take their philosophic bearings from a received, often analytically filtered account of Spinoza's thought that finds no room for such suspect – hermeneutically venturesome – approaches. This is not so much a matter of tough-minded analytic commentaries in the mode of 'rational reconstruction' descending from Bertrand Russell and taken up with comparable vigour by Jonathan Bennett.[68] In their case the treatment is so frankly selective and the ideas they can't accept so forcefully denounced that readers are left in no doubt as regards the dominant agenda. Rather it is the more moderate 'descriptivist' programme – of which Stuart Hampshire is the best-known exponent – that has exerted such a deep and pervasive influence on the way that Spinoza is typically read by Anglophone philosophers and also by those (intellectual historians among them) who take their cue from that same dominant tradition.[69] Here the main object – with Spinoza as likewise with Leibniz, Kant and others – is to coax these thinkers down from the heights of 'metaphysical' or speculative thought and lead them back to a sensible acceptance that in truth philosophy can do no more than describe or map the conceptual geography of our various, whether everyday or scientific modes of knowledge and experience.[70]

Yet if there is one thing that Spinoza stands up for *contra* this whole way of thinking it is the need for philosophy to challenge, criticize, and – where necessary – amend or rectify such taken-for-granted notions. Moreover, as again emerges very strongly from the recent French reception, this applies just as much to Spinoza's radically heterodox thinking about ethics and politics as to his thinking on matters of ontology, epistemology, and philosophy of mind. In my book, I put the case for Spinoza as the single most important, albeit under-recognized source for many then current or lately emergent ideas in critical and literary theory. What I have sought to do here is make a similar case for the potential of a close engagement with Spinoza's thought to revitalize large areas of present-day philosophic discourse by cutting across their more restrictive and narrowly specialized concerns. Above all, it brings out the impossibility of addressing such core topics – even after the much-heralded linguistic turn – without raising substantive metaphysical issues which go well beyond the scaled-down descriptivist conception that marks the outer bound of acceptability for most analytic thinkers. In this respect Spinoza poses a greater challenge than Kant, despite the latter's role in a good deal of recent analytic philosophy as representing either the kinds of metaphysical seduction that we have nowadays thankfully laid to rest or else (more in keeping with Kant's self-estimate) a means of putting philosophy back on its feet by precisely defining its operative scope and limits.[71] For that role has much to do with the clear continuity – pointed out by recent commentators – between Kant's epistemological project and the analytic enterprise which took shape largely as a linguistic (or logico-semantic) transposition of central Kantian themes.[72]

Thus one way of writing the history of analytic philosophy is as a series of attempted and loudly proclaimed breaks with Kant – especially with his subject-centred epistemology and his doctrine of synthetic a priori knowledge – that have then periodically given way to various attempts at a partial reconciliation along descriptivist, naturalized, or other such revisionist lines.[73] What has led to this oscillating pattern of rejection and qualified acceptance is the presence within analytic philosophy of just the same chronically unstable dualisms – subject and object, mind and world, concept and intuition – that Kant claimed to resolve in his First *Critique* but which have none the less continued to vex thinkers from the German post-Kantian idealists to John McDowell. It is the same with those recent efforts to 'dismount from the see-saw' (McDowell's phrase), or at any rate to damp down its movements, whether by following Sellars and Davidson in their attack on the empiricist 'myth of the given' along with its attendant scheme/content dichotomy, or else by invoking some naturalized (or de-transcendentalized) version of Kant.[74] That these efforts have failed – that the dualism always crops up again in a more-or-less covert, displaced, or surrogate form – is a case borne out (as I have argued elsewhere) by a good many episodes in recent analytic debate.[75] My point now, by way of conclusion, is that Spinoza's radically monist understanding of mind and body or mind and world offers by far the most effective counter-instance to this whole way of thinking and its hold over philosophers from Descartes down. Moreover, it constitutes a standing reproof to that other tenacious dualism that has had such a damaging effect on recent philosophical debate, namely the split – however one perceives it – between the analytic and continental traditions of thought. For there is no hope that philosophy will rise to the Spinozist challenge unless it puts away some of the fixed preconceptions that have so far acted as a strong disincentive for taking that challenge at its full philosophic, i.e., onto-epistemological as well as ethico-political force.

V

Of course, there is a third highly simplified or typecast dichotomy that has long exerted its distorting hold on perceptions of the history of philosophic thought over the past four centuries and more. This is the standard textbook idea of thinkers falling into one or other camp on a field – very often a field of battle though sometimes the site of an uneasy truce – where the opposing parties are 'rationalism' and 'empiricism', and where any hope of a somewhat more durable peace is likely to be hedged around with all manner of exemptions, fallback clauses, and carefully drafted territorial guarantees. The dividing-line is all the more difficult to draw with any assurance in so far as the effect of those various treaty provisions, like their equivalents in the geo-politico-historical realm, is to create large problems for any attempt to fix borders in a rigorous, precise, or even a widely agreed-upon way.

In part, this has to do with the suggestive term-for-term alignment or homology between the various other dualisms instanced here (analytic/synthetic, scheme/content, analytical/continental, etc.) and its failure to apply – or to make much philosophic sense – when extended to the rationalist/empiricist pair. Where that attempt breaks down most conspicuously is on the fact that much analytical philosophy has been deeply empiricist in its methodological (especially its science-led epistemological) commitments, even while – as with manifold developments in the wake of logical positivism – retaining the basic rationalist idea of empirical data as subject to certain prior (though perhaps empirically challengeable) forms of conceptual and logical regimentation.[76] Moreover – as appears plainly in the original (mainly seventeenth-century) disputes and in the wrangling of numerous partisan commentators since – the very problem of defining 'rationalism' or 'empiricism' with anything like an adequate degree of conceptual precision has often meant that they are epitomized through a process of contrastive characterization each against the other, which in turn tends to generate further such conflicts and boundary-disputes. Nor are the tensions noticeably eased by the hallmark analytic practice, from Bertrand Russell to Jonathan Bennett, of interpreting the Great Originals (Locke, Berkeley and Hume on the one hand; Descartes, Leibniz and Spinoza on the other) through a highly selective mode of 'rational reconstruction' which brings them as far as possible into line with current best practice by currently prevailing standards, and which consigns whatever cannot be so reconstructed to the scrapheap of dead 'metaphysical' ideas.[77] Despite its frequent acuity and occasional power to generate insights that mitigate the charge of gross anachronism this approach has done little to bring about a better, less fraught or conflict-prone understanding of the rationalist/empiricist dichotomy.

What that would require, it seems to me, is a radical re-thinking of issues in ontology, epistemology and philosophy of mind such as Spinoza essayed to remarkable effect and such as finds its currently most eminent protagonist in the French philosopher Alain Badiou.[78] It is for this reason chiefly – along with the fact that his work stands out for its sheer range and depth of philosophical concerns, in contrast to the narrowly specialist character of so much present-day analytic thought – that I have focused on Badiou in my next chapter. More specifically he is a thinker who, like Spinoza, points a way through and beyond the dead-end predicament which some would now diagnose as having been the end-product of that conceptual-epistemological 'turn' in philosophy that started out with Descartes, received its most elaborate treatment in Kant, and was then taken up in 'linguistified' form by a great variety of latter-day thinkers in the Russell-Frege or Wittgensteinian lines of descent.[79] In this regard he agrees with a sturdy upholder of the analytic virtues like Timothy Williamson who none the less considers that enterprise to have suffered a highly damaging restriction of its scope and its capacity to establish truths about the world rather than truths about the sense, purport, semantic structure or logical form of various statements concerning it.[80] The former Williamson deems properly 'metaphysical' in the sense that they exceed any purely empirical means of discovery or verification, that they involve certain distinctively modal (i.e., necessity- or possibility-related) considerations, and also that they are capable of being arrived at through an exercise of 'armchair' or a priori thought which despite that delivers knowledge of what must be the case with regard to certain matters of physical or natural as well as conceptual warrant. Thus, '[e]ven if one does not fully understand *how* thinking can provide new knowledge, the cases of logic and mathematics constitute overwhelming evidence that it does so'.[81]

Again like Badiou, Williamson envisages the philosophic project as 'disciplined' or 'conditioned' by other kinds of knowledge – pre-eminently, in his case, mathematics and the natural sciences – but also as needing to maintain a due sense of what sets it apart in terms of both its proper working methods and the nature of those truths that philosophers can reasonably hope to attain through application of them. 'Indeed', Williamson writes, 'philosophy subject to only one of those disciplines is liable to become severely distorted; several are needed simultaneously.' At the same time – a further point of agreement with Badiou in principle and practice – 'to be "disciplined" by X here is not simply to pay lip-service to X; it is to make a systematic conscious effort to conform to the deliverances of X'.[82] That is, these latter have to be regarded both as providing certain needful constraints on the exercise of philosophic thought and as themselves on occasion standing in need of the at once constraining and intellectually enabling or disinhibiting effect of such thought. Although Badiou makes the case with what might appear a far greater degree of speculative licence – crossing disciplinary boundaries in a way that would surely strike Williamson as downright reckless or profligate – his thinking has a rigour and a power of conceptual analysis as well as a scope of conceptual creativity unrivalled by any philosopher of our time.

I shall therefore focus on those aspects of his work that hold out the promise of a passage beyond all the vexing antinomies referred to above, that is to say, which render such typecast distinctions wholly off-the-point since demonstrably lacking any purchase on Badiou's own practice as a thinker who conforms to none of them. Here for once it is justifiable – indeed, only fitting – to make use of a grossly overworked adjective and assert that Badiou has made groundbreaking advances (and not mere shifts in the range of currently available linguistic-conceptual resources) with regard to some of the most longstanding and intractable problems in ontology, epistemology, ethics, political theory, and – pre-eminently – philosophy of mathematics. All the more unfortunate, and symptomatic of prevailing attitudes, that his work has so far received minimal attention from philosophers in the mainstream-analytic line of descent. Here I try to help along the process of reception by describing Badiou's remarkably original and ambitious approach to issues of mathematical (more specifically: of set-theoretical) ontology, and by explaining just where his project stands in relation to some major points of debate within that 'other' tradition. Chief among them are the dispute between realists and anti-realists – along with various avowed middle-ground or compromise solutions – and those oddly tenacious problems-from-Wittgenstein (e.g., concerning what it means to follow a rule) that have so preoccupied philosophers over the past decade.[83]

In particular I stress the remarkable combination in Badiou's thought of inventive brilliance with logical precision and depth as well as acuity of philosophic insight, qualities which tend to make those other discussions often appear self-absorbed to the point of wilful parochialism. Most importantly, Badiou engages these issues at a level of technical grasp which puts his thinking closely in touch with the way that set theory has itself evolved through a constant process of – in his own phrase – 'turning paradoxes into concepts'. Along with this claim for the absolute priority of reasoning *more mathematico* in matters of formal and ontological concern goes an equally firm rejection of the 'linguistic turn' in its manifold analytic and continental variants, all of which (but especially those of a Wittgensteinian provenance) he regards as little more than a come-lately version of ancient sophistical and cultural-relativist themes. On the other hand, that whole set of precepts and commitments is itself strongly offset by Badiou's signature idea of the 'event' as that which inherently eludes or surpasses the conceptual resources of any received ontology, whether in mathematics and the physical sciences or in the case of those genuinely epochal changes – albeit few and far between – that have punctuated the record of political and world-historical events. Just as his work in philosophy of mathematics (a description he would emphatically reject) transforms all the normal, acculturated terms in which these matters are nowadays discussed so likewise Badiou's conception of the event is such as to entail – if his arguments go through – a far-reaching challenge to our customary grasp of how such strictly exceptional episodes may be thought to come about.

Even in mathematics, logic and the formal sciences it is question of grasping how thought is sometimes – if rarely – able to overleap the boundaries of a deeply entrenched ontological scheme and achieve the kind of conceptual breakthrough (like Cantor's set-theoretical insight concerning the multiple orders of infinity) which marks an event in Badiou's qualitative sense of the term.[84] Thus a mathematical discovery of that magnitude is one that, like a revolution in politics or a major shock to the political system (such as the 'failed' or 'abortive' Paris Commune of 1871), opens up a tract of as yet unexplored conceptual or political terrain wherein to pursue its more radically disruptive or transfigurative implications. Moreover, in keeping with a central thesis of my book, this involves an equally drastic revision to received, quasi-Cartesian ideas about the privileged role of conscious as opposed to unconscious or preconscious thought, at least when it comes to the more advanced or innovative modes of human intellectual endeavour. Such notions may have been subject to a fair amount of critical questioning over the past half-century – from followers of Wittgenstein not to mention Spinoza's prescient challenge as discussed in this chapter - but they still retain a strong hold in many quarters of philosophical debate. What Badiou brings out through his detailed rehearsal of those 'truth-procedures' that can now be seen to have typified the conduct of enquiry in the run-up to some epochal event (whether in mathematics, the natural sciences, or political history) is the sheer *inconceivability* that any breakthrough of that order could ever be achieved in or through an exercise of lucid conscious thought. It is at just this point, I suggest, that we can start to see a way beyond the conflict between naturalism and rationalism that has had such an unfortunate, progress-blocking effect on attempts to overcome those various tenacious dualisms. For if there is one thing on which both parties can agree it is the need to throw off that Cartesian legacy that has weighed so heavily on both as well as on the state of relations between them.

Chapter 5

Alain Badiou: Mathematics, Politics, and the Venture of Thought

I

At the close of Chapter 4 I described Badiou as a thinker whose work combines remarkable powers of intellectual creativity and inventiveness with the highest degree of analytic rigour and logical-conceptual grasp. The idea that these attributes cannot go together – that creativity and rigour just don't mix – has widespread currency not only among analytic philosophers who take a dim view of everything 'continental' (i.e., mainland-European) from Hegel to Derrida but also among some of their 'continental' counterparts who are no more impressed by what they regard as the narrowly technical, pedantic, self-occupied, and trivial character of much that bears the name 'analytic' philosophy. In this chapter I set out to challenge that drastically separatist conception of how things have gone with the two main lines of philosophical descent since Kant. I do so by engaging principally with Badiou's work in philosophy of mathematics, a description that he would no doubt strongly contest (for reasons that should become clear later on) but which may serve for want of any better, analytically received alternative.[1] Badiou makes the case that philosophers should (indeed must) take their lead from the most advanced developments within mathematics if they wish to contribute anything of value as regards the various regional ontologies that mark out their fields of interest. Thus they need to go to school with the mathematicians rather than presuming to regulate or dominate a whole range of other fields – mathematics among them – from a standpoint possessed of its own self-validating truth-conditions or criteria. For very often this results in a reactive swing of the kind that has typified so many developments in recent (post-1960) analytic thought. That is to say, it produces a swing away from the prescriptive high ground whence philosophy seeks to adjudicate in matters of truth and falsehood to the strictly subaltern, e.g., late-Wittgensteinian role whereby it cedes all authority to the wisdom enshrined in common-sense usage or 'ordinary language'.[2]

So there are several reasons for my choice of mathematics as a focal topic by which to introduce Badiou's work to Anglophone readers who will probably come at it from a background in analytic philosophy. First, this approach takes Badiou at his word when he affirms the pre-eminent role of mathematics – more specifically, of post-Cantorian set theory – as the basis of any ontology worth considering and therefore as the indispensable point of reference for a wide range of philosophical issues including (crucially) the scope and limits of any such ontologically-based approach. Second, to repeat, it brings out the fallacy involved in that commonplace (chiefly analytic) assumption concerning the downright incompatibility of intellectual rigour or conceptual precision with any thinking that ventures onto speculative ground unlicensed by existing methodological decree. Third, it is the best way of doing justice to a quite extraordinary body of work that should engage the interest of analytic philosophers since it addresses some of the core issues on their own priority list. These include the debate between realists and anti-realists – posed with particular clarity and force in the case of mathematics – and the question whether such issues (traditionally counted as belonging to the province of ontology or epistemology) might find their quietus through some version of the turn toward language as a means of resolving all philosophical disputes.[3] Fourth, it allows me to point up the contrast between his way of thinking through a range of issues in the philosophy of mathematics – that is, an immensely detailed, rigorous, historically informed yet creatively engaged treatment of set-theoretical themes – and the kinds of discussion that typically hold sway in the mainstream analytic line of descent.

Thus when Badiou treats these topics in relation to other, on the face of it strikingly remote areas of thought one is driven to reflect on how tightly boxed-in and how utterly devoid of such creative-exploratory flair is a great deal of work in the broadly analytic (whether post-Frege/Russell or post-Wittgensteinian) tradition. Nowhere is this more evident than in the endless Wittgenstein-inspired debate around rule-following, private language, and the problem (if such it is) of specifying standards or decision-procedures for what should count as correctly or properly following a rule.[4] Here again there is much to be learned from Badiou with regard to the distorting effect upon philosophy of allowing its own specialist concerns – in this case a preoccupation with scepticism and the problem of knowledge – to overwhelm what should be its guiding principle, namely the commitment to *thinking mathematically* rather than using mathematics as a springboard for its own pet interests or obsessions. Above all, what he shows (both by way of explicit diagnosis and implicitly by contrast with his own work) is the kind of danger that philosophy courts by giving in to one or other of its twin present-day temptations: either that of hubristically claiming to lay down rules for the better conduct of mathematical thought or that of reducing itself and mathematics alike to a matter of conformity with communally sanctioned customs and practices.

Indeed, Badiou is among the keenest critics of that standard appeal to language as an ultimate horizon of meaning, truth, or intelligibility which has been so prominent on both sides of the (supposed) analytic/continental split.[5] In its place, he proposes a full-scale return to those central themes of Western philosophy from the pre-Socratics down – the one and the many, the finite and the infinite, stasis and flux, the eternal and the temporal – which can now be addressed in a different, immensely more productive way since the advent of modern set theory with its new-found conceptual and ontological resources. What lies within reach, once philosophy has taken this decisive turn to mathematics (rather than language) as its point of departure, is the power to distinguish clearly between the disparate orders of being and event. That is, it makes possible a sharp demarcation between that which falls within the proper domain of a set-theoretically based ontology and that which entirely eludes or exceeds ontological specification. For it is here that we encounter an order of occurrence – a mathematical discovery, a scientific breakthrough, a genuine (revolutionary) transformation in the prospects for political change – that could never have been predicted with assurance on the strength of preceding developments.

So philosophy still has its work cut out, though neither as a discourse with delusions of epistemological grandeur nor again – at the opposite though likewise doctrinaire extreme – as one that follows Wittgenstein in offering a course of ordinary-language therapy for philosophy's self-induced ills. That work consists partly in pointing up the wider, i.e., the extra-mathematical implications of various signal events that have marked the progress of set-theoretical enquiry from Cantor to the present. At the same time its task is to maintain contact with the primary 'conditions' that between them provide philosophy's most important subject-matter and its chief spurs to reflection. It is at this point that the gulf opens up between Badiou's highly speculative thinking and the kinds of work that conventionally count as serious, responsible, or disciplined philosophy of mathematics in the mainstream analytic mode. For the conditions that Badiou has chiefly in mind – and that figure centrally in his major work *Being and Event* – are those of science, politics, art and love, the coupling of which (and moreover their treatment in set-theoretical terms) is such as to conjure bafflement or downright disbelief from thinkers trained up in that other tradition. Suffice it to say, for the moment, that he sees them as the four most basic modalities of human thought, knowledge and experience, and also as the strictly indispensable means by which philosophy can set about its task of linking the truths of mathematics, logic and the formal sciences with those of our more or less commonplace or specialized being in the world.

However, if it forms *too* close or exclusive a bond with any one of them – to the point of yielding up its own autonomy and also its obligation to the others – then philosophy is gravely at risk of undergoing various kinds of ethical and political as well as intellectual deformation. What helps to prevent such abuses from occurring is the willingness to reflect on its own scope and limits through a better, mathematically informed understanding of the relationship between being and event. For instance, the particular abuse represented by Heidegger's way of subjugating philosophy to poetry is one that finds voice – with dire political effect – in his strongly nationalist conception of language and his idea of certain German poets as uniquely authentic conduits of a truth inaccessible to other, less privileged mortals or cultural-linguistic communities.[6] So likewise with science, politics and love, each of which – the latter as brought within range of conceptual, albeit highly speculative grasp by psychoanalysis – has the greatest capacity to challenge and energize but also, if granted such unilateral determinative power, the maximum potential to distort and subvert the character of philosophic thought. Examples are not hard to come by and would include its uncritically positivist recruitment to the programme of physical science, its sometimes willing annexation to the purposes of totalitarian politics, and the hostile or at best condescending attitude shown toward philosophy by many psychoanalysts from Freud to Lacan.[7]

Thus, one main reason for Badiou's placing mathematics squarely at the centre of his remarkably wide-ranging project is that it serves to remind us – as against other, more partisan views such as Heidegger's – that if ancient Greek thinking was indeed the *fons et origo* of all subsequent philosophy then this instauration wasn't confined to just one natural language. Still less were its benefits restricted to certain poet-thinkers fortunate enough to have that language as their native tongue or – like Hölderlin, Rilke or Trakl – to inherit them by way of cultural birthright. For ancient Greece was just as crucially the setting for that other, more decisive (since universally valid) instauration that witnessed the emergence of mathematics as the chief point of reference for philosophy's dealing with issues of knowledge and truth. This is why Badiou makes his claim for the epochal nature of Cantor's discovery that it was possible not only to conceive of the infinite without running into all manner of paradox or conceptual affront but also to reckon with multiple orders or 'sizes' of infinity and still avoid any such inherently absurd or stultifying upshot.[8] Moreover, the prospect of advancement thus achieved was by no means solely a matter of increased conceptual grasp in mathematics, logic, and the formal sciences. On the contrary – he staunchly maintains – it extends far beyond those specialized disciplines to regions of ethico-political as well as scientific and philosophical endeavour that likewise involve a well-developed capacity for turning obstacles, limits, or setbacks into a source of renewed intellectual or creative impetus. As regards set theory this process finds its perfect exemplification in the leap of thought by which Cantor was able to think his way through and beyond the aporias or dead-end paradoxes of earlier thought about the infinite and use them – so to speak, in slingshot fashion – as a means of overcoming precisely those erstwhile sources of conceptual blockage.

This pattern of advancement *despite and against* the kinds of resistance repeatedly encountered at the limits of thought is a main topic of Badiou's philosophy and one that he pursues with the utmost degree of analytic rigour and precision. Its high point, he suggests, is the set-theoretical concept of 'forcing' proposed by Paul Cohen and designed to explain how thought could possibly achieve such advances through the speculative positing of theorems or hypotheses that are well-formed and possessed of an objective truth-value even if it lies beyond our present-best power to prove or ascertain.[9] Thus the process or procedure of 'forcing' is one that can retrospectively be seen to have enabled a great range of discoveries not only in mathematics but also – according to Badiou – across the entire range of human intellectual or practical-purposive activities wherever it is a question of staking some claim to truth that cannot be verified or falsified by any method currently to hand. His treatment of these issues is detailed, rigorous, and lucid in a way that bears ample witness to Badiou's command of the relevant mathematical issues. This contrasts with a good deal of analytic work in the philosophy of mathematics which often seems more concerned with a range of fairly standard philosophical problems – such as scepticism and the realism/anti-realism debate – than with anything that might be expected to engage the interest of working mathematicians.[10] It also explains why Badiou is so strongly opposed to the various forms of intuitionist, constructivist, anti-realist, or fictionalist thinking that have lately captured the high ground – or at any rate set the agenda for discussion – within the analytic community. What these approaches exclude a priori is any question of so-far unproven conjectures being thought of as objectively true or false in virtue of the way things stand in mathematical reality quite apart from the way things stand with regard to our present-best or even future-best-attainable state of mathematical knowledge. By taking the anti-realist position that this bivalent (true-or-false) axiom of classical logic simply doesn't apply to statements of the so-called 'disputed class' these thinkers effectively restrict truth to the compass of human knowledge and knowledge in turn to the range of statements that happen to lie within our human, all-too-human powers of proof or ascertainment.[11]

Hence Badiou's steadfast opposition to all such (as he sees them) latter-day variants on an age-old sophistical theme, whether those that have taken the linguistic, hermeneutic or textualist 'turns' or those – very often with their home-ground in philosophy of mathematics and logic – that have embraced an anti-realist conception of truth as epistemically constrained. Hence also his insistence on the need for philosophers to put in the necessary homework (that is, to acquire at least a basic grasp of set-theoretical concepts and methods) if they are to have any chance of engaging these issues at an adequate level of debate. Yet despite his caution with regard to their habit of putting the philosophic cart before the mathematical cart, Badiou makes it clear that philosophers do have the vital task – one mostly shunned by mathematicians – of drawing out the ontological consequences of set theory and explaining just how these relate to the other main spheres of involvement which jointly constitute philosophy's element or source of sustaining insights. In what follows, I shall look in more detail at Badiou's approach to these issues in philosophy of mathematics and also – by way of introduction – at the way they have been treated by thinkers in the mainstream analytic camp. We may then be better placed to appreciate just how distinctive is Badiou's contribution to this and to other, seemingly disparate fields of enquiry where he has brought mathematical concepts to bear. On the other hand, we should have gained a sharper sense of just where this project runs up against its limits as signalled in the title of *Being and Event.* That is to say, even the most resourceful set-theoretically based ontology will fail to account for the irruptive, anomalous, exorbitant, or strictly inassimilable character of those singular occurrences that spring as if from nowhere and exert a profoundly transformative impact on the subsequent course of developments in this or that field. If Badiou is not averse to claiming such status for his own rethinking of these issues then I hope that my discussion will help readers to assess that claim on its merits.

II

Philosophers have often puzzled about the relation between mathematics and reality, or just how it is that a discourse dealing in such abstract entities as numbers, sets and classes can possibly have played so central a role in the advancement of scientific knowledge from Galileo to Einstein and beyond. The quantum physicist Eugene Wigner expressed this puzzlement most aptly in the title of his essay 'The Unreasonable Effectiveness of Mathematics in the Physical Sciences'.[12] Where thinkers divide is on the question how best to cope with the problems that arise as soon as one asks two basic questions. First is the ontological question (What exactly is the nature of those entities? Where can they be thought to reside?), and second, the epistemological question (How can we achieve knowledge of them? Whence our assurance with regard to such matters, from the elementary rules of arithmetic to the existence of proofs or truth-values for the most advanced mathematical theorems and conjectures?). They are combined in Wigner's somewhat plaintive query as to why – in Galileo's famous words – the book of nature should turn out to have been written in the language of mathematics. If it is hard to explain how we could ever make epistemic contact with such abstract entities as numbers, sets, and classes then it is even harder to conceive how those entities should somehow mesh so closely with the structures of physical reality. For otherwise we are faced with the even more difficult task of explaining – *sans* recourse to miracles or massive cosmic coincidence – how mathematically-informed science has managed to come up with so impressive a range of applied technological advances.[13] At this stage the realist has a vast amount of highly persuasive empirical and documentary evidence on her side but also the problem of bridging that troublesome gap between *knowledge* in so far as it lies within our powers of epistemic grasp and *truth* in mathematics or the physical sciences as a matter of objective, i.e., epistemically unconstrained or verification-transcendent validity. Indeed some philosophers have thought this dilemma to be strictly insoluble and declared that we can *either* have truth *or* knowledge but surely not both unless by fudging the issue.

One alternative response – that favoured by anti-realists – is to take the epistemological line of least resistance by redefining truth in terms of 'warranted assertibility' and thus bringing it back within the scope and limits of human ascertainment. For how could we ever acquire knowledge of objects – whether in the physical or (more especially) the formal sciences – that are thought of, on the realist (or Platonist) account, as always potentially transcending the utmost capacities of human perceptual or epistemic grasp? To which the realist will typically respond with some version of the 'no-miracles' or 'cosmic coincidence' arguments, that is, some further variation on the theme that we cannot make sense of the history of science to date (not to mention our other, more everyday processes of knowledge-acquisition) unless we accept her three most basic priorities and principles. These are (1) that the world contains a whole vast range of objects on various scales from the subatomic to the astrophysical that exist and exert their causal powers, properties, dispositions, etc., for the most part quite independently of whatever we happen to know or believe concerning them, (2) that the truth-content of our scientific statements, hypotheses, theories, or predictions is determined by the way things stand in reality, rather than confined to the scope and limits of our current-best knowledge, proof-procedures, or methods of verification, and (3) that this case is most strongly borne out – and best defended against any form of sceptical attack – as a matter of inference to the best explanation, i.e., as by far the most plausible account of how knowledge has advanced in various fields through a process of by-and-large successful or truth-conducive investigation. A standard anti-realist move at this point is to bring up the 'argument from error', that is, the observation that a very large amount of what was once taken as established scientific truth has now been rejected or falsified, and hence – by a form of 'sceptical meta-induction' – that we must be mistaken (hubristically deceived) if we suppose ourselves to be any better placed in that regard. But this in turn lies open to the realist rejoinder: how can the sceptic or the anti-realist think to rest his case on evidence of this negative kind unless by presupposing the existence of truths (whether known or yet to be discovered) in the absence of which it simply *cannot make sense* to speak of past or present error?

Such, in brief, is the way that this debate has typically gone in recent mainstream analytic philosophy of language, logic, mathematics, and the physical sciences. It has tended to produce the sharpest disagreements in connection with mathematics since here, more than anywhere, the problem arises as to how – if at all – truth can be conceived in objectivist (recognition-transcendent) terms and yet as somehow accessible to human knowers or as falling within the epistemic ken of those equipped with the requisite conceptual, logical, or calculative powers. To some, as I have said, this dilemma has appeared simply insoluble, or at least to present a *tertium non datur* wherein one has to plump either for objectivist truth or for humanly attainable knowledge with no possibility of managing to reconcile these two (on the face of it) equally compelling desiderata. For others it has seemed to require the adoption of an anti-realist approach that grasps the latter horn – 'truth' as epistemically constrained and hence ipso facto as humanly knowable – and which thereby resolves (or dissolves) the problem but only at the cost of creating others that will strike the realist as equally or more worrisome. Thus it seems to entail that any well-formed but as-yet unproven or maybe unprovable conjecture – such as Goldbach's that every even number is the sum of two primes – must be thought of not merely as neither-true-nor-false so far as we can ascertain but as neither-true-nor-false *sans phrase* since (according to Michael Dummett's influential statement of this case) any gap in our powers of mathematical proof is also a gap in mathematical reality.[14] Also, it would have the yet more absurd (to the realist's way of thinking) consequence that, in the case of a long unproven hypothesis such as Fermat's Last Theorem, the conjecture was neither-true-nor-false during the three centuries or so after Fermat first jotted it down but acquired the value 'true' as soon as Andrew Wiles made good certain flaws in his celebrated proof.

Still the problem remains as to how mathematical truths can be both objective – the very paradigm of objectivity, according to many philosophers – yet also (some of them) accessible to knowers whose cognitive or intellectual capacities are subject to certain limits. Those limits may be thought of as pertaining to this or that temporally indexed state of knowledge (and hence as capable of future change and amelioration), while others might perhaps have to do – so far as we can tell – with innate restrictions on our powers of formal or logico-mathematical reasoning. For the realist this simply goes to show that truth in such matters can always in principle surpass the current or even the future-best-possible scope of human ascertainment. For the anti-realist, conversely, it shows that we had much better scale down our talk of objectivist (recognition-transcendent) truth to talk of warranted assertibility or 'truth' as epistemically constrained. For we shall otherwise be courting the realist nemesis of an argument that falls plump into the sceptic's sights since it places truth inherently beyond reach of human apprehension. Such has nevertheless been the preferred option with many analytic philosophers since Frege launched his well-known assault on what he regarded as J. S. Mill's vulgar-empiricist idea that mathematics even in its higher, more abstract or 'purely' theoretical dimensions might come down to a construction out of our everyday procedures of counting or grouping together.[15]

Some thinkers, Philip Kitcher among them, have taken a more sympathetic view of Mill's position – not least because it seems to offer some hope of answering Wigner's query with regard to the singular effectiveness of mathematics as applied to the physical sciences – whilst also seeking to strengthen it against the Fregean line of attack by building in a greater measure of objectivity.[16] Others have made the case for a 'response-dependent' or middle-way approach that would allow for some measure of epistemic constraint but prevent that allowance from going too far in an anti-realist direction by tying it to certain, carefully specified conditions on what should count as a competent, expert, or properly qualified response.[17] However this attempted compromise solution can be shown to run aground on one or other of two decisive counter-arguments. Thus either it equates the conditions for warranted assertibility with best judgement conceived in the ideal limit (thus amounting to objectivist or verification-transcendent truth for all practical purposes), or else it falls back to some hedged-about version of anti-realism which strives to set the justificatory standards as high as possible but still within the bounds of human knowability or epistemic constraint. That is to say, such approaches cannot make good their claim to offer a deliverance from all our philosophic woes by establishing a sensible middle-ground position between the twin problematical extremes of hard-line objectivist realism on the one hand and anti-realism on the other. Rather, this attempted splitting of the difference ends up by posing the same dilemma in terms that are no less stark and intransigent for its appearance of judicious even-handedness.

I have revisted these issues in recent analytical philosophy of mathematics by way of pointing out the way that they contrast with the role played by mathematical concepts, arguments and analogies in Badiou's *Being and Event*. Let me say straight off – so as to pre-empt certain stereotypical responses – that it really is a matter of well-defined concepts and clearly articulated arguments rather than vaguely suggestive analogies whereby to gain a hearing for some highly speculative theses across an otherwise large and disparate range of subject-areas. After all, it is hard to imagine any analytic philosopher taking mathematics (more specifically, post-Cantorian set theory) as their basis for a radically heterodox ontology that claims to have large – indeed transformative – implications for our thinking about matters of science, politics, art, and love. Still less would one expect such a claim to enjoy credence among those for whom the parameters of serious, mathematically informed and philosophically pertinent debate are such as I have described in outline above, that is, concerned chiefly with the issue between realist (or objectivist) and anti-realist (epistemic or response-dependent) conceptions of truth in the formal sciences. Yet these doubters will be wide of the mark – and will miss the chance of some constructive exchange across the analytic/continental rift – if they conclude from the sheer ambitiousness of Badiou's project that he must be just another of those typecast builders of metaphysical castles in the air with nothing in the least worthwhile to say about matters of genuine philosophic import. On the contrary, Badiou has long been deeply engaged with just the kinds of issue (in ontology, epistemology, philosophy of mathematics and science, ethics, politics, and aesthetics) that preoccupy his counterparts in the broadly analytic line of descent, even if his approach to those issues diverges rather sharply from theirs at various points. More than that, he is about as far as could be from the stereotyped notion of the 'continental' – especially the French – philosopher as one who ignores the most basic standards of logic, rationality or plain good sense so as to indulge a deplorable penchant for pursuing the wildest flights of metaphysical or speculative fancy. Indeed it is fair to say that Badiou is among the most rigorous and logically exacting, as well as the most creative, original and idiosyncratic of present-day philosophers.

Most likely this last combination of epithets will strike many analytic philosophers as a *contradictio in adiectis*, or a yoking-together of wildly discrepant terms. For one very marked feature of most (if not all) work in the mainstream analytic tradition has been its ethos of patient, meticulous devotion to various well-defined problem-areas or tasks in hand, these latter assumed to repay such treatment through a gradual and progressive clarification of the issues involved. Hence its deliberate eschewal of the two main traits that are taken to typify thinkers in the 'other' tradition, namely their proneness to sweeping or grandiose metaphysical claims and their tendency – or so it is maintained – to substitute idiosyncrasies of style or inventive 'literary' turns of thought for the more serious (philosophically reputable) business of conceptual, linguistic, or logico-semantic problem-solving. This is why analytic philosophers are apt to look askance at claims like that of Gilles Deleuze, in his late work *What Is Philosophy?*, which reject any such self-denying ordinance and affirm what they see as philosophy's right – indeed, its primary vocation – to create new concepts rather than analyse old ones.[18] Such is at any rate a chief source of the friction that has often kindled into flare-ups over the past half-century and more, from A. J. Ayer's mordant animadversions on the supposed logical blunders of existentialism to Gilbert Ryle's influential change of mind with regard to the philosophic interest (as he had once thought) of Husserlian phenomenology and John Searle's brusquely dismissive treatment of Derrida on the topic of Austinian speech-act theory.[19] In each case a major cause of this communicative breakdown has been the deep-laid analytic suspicion of any approach that goes beyond the limits of conceptual exegesis so as to explore – some would say exploit – the various expressive or performative (including 'literary') aspects of language that cannot be fully captured in constative, i.e., theoretical or purely conceptual terms. It is unfortunate that Derrida's contretemps with Searle has become something of a locus classicus for partisans on both sides of this dispute. The result has been to polarize opinion all the more sharply between those in the analytic camp who take it to have shown Derrida up as a latter-day sophist with nothing of philosophic value to impart and those on the 'continental' wing who rejoice in his having run deconstructive rings around Searle's earnestly truth-seeking approach.[20]

Now there is no doubt that Badiou (like Derrida) is a highly original, creative and inventive thinker, one who has a strong title – in these respects – to pre-eminence among contemporary philosophers of whatever school or persuasion. Thus he claims nothing less than to have radically transformed our conceptions of, inter alia, the relationship between truth and knowledge; the issue between realism and anti-realism in logic and mathematics; the role of mathematics as fundamental ontology vis-à-vis every aspect of the physical, social and human sciences; the nature of commitment or fidelity in its various (principally ethical, political, and amorous-erotic) dimensions; and the question whether language should be thought of – as it is by a great many present-day philosophers – as the ultimate horizon of truth or intelligibility. It is safe to say that no analytic philosopher would regard this extraordinary range of topics as making up a viable or integrated project of enquiry, rather than a loose assortment of themes that might just conceivably engage the interest of a single thinker at various times or when wearing different philosophical hats. On the other hand, it is also safe to say – on the strength of his published work – that Badiou's creativity and range of interests have not been achieved at the cost of intellectual rigour or through a cheerfully eclectic habit of mind for which everything serves as grist to the philosophic mill. Indeed it is only in the narrowest, most parochial sense of the term 'analytic' that one could properly or justifiably reject its application to Badiou's work. Nor does he conform in the least to that stereotype – put about by proponents and detractors alike – which would have it that 'continental' philosophy has pressed much further than its analytic counterpart with the currently widespread linguistic turn, so that any appeal to standards of logic, reason and truth gives way to a depth-hermeneutic or strong-descriptivist approach. Indeed few philosophers have argued so forcefully against this 'language-first' idea, whether in its moderate (including its prevalent analytic) forms or the more extreme form that it takes among thinkers such as Heidegger, Gadamer, and the later Wittgenstein. Thus Badiou sides very firmly with Plato in drawing the sharpest possible contrast between philosophy as a truth-seeking discipline of thought and sophistics as an exercise in rhetorical persuasion which assumes, like Protagoras or Wittgenstein, the non-existence of truths beyond whatever finds a place in some particular language-game or cultural form of life. Hence the various 'truth-procedures' that Badiou conceives in terms very different from those of mainstream analytic philosophy, but which he takes to be the sine qua non of any thinking that would break with the currency of handed-down (common sense or ideological) beliefs.

However, by far the most strikingly original aspect of his thought – and the point at which it touches most closely on a range of core analytical concerns – is Badiou's claim for the role of mathematics (and set theory in particular) as the basis for any adequate conception of truth in those various domains. Set theory has to do with relationships of membership, inclusion, and exclusion among numbers or other entities that are taken as forming a unit of assessment for some given purpose.[21] Thus sets are defined – in Badiou's terminology – as products of the 'count-as-one', that is to say, the classificatory procedure that consists in grouping together a certain range of such entities and treating them as co-members of a single assemblage whatever their otherwise diverse natures or properties. The great benefit of set theory as envisaged by Cantor, Frege, Russell and its other early proponents was the prospect it offered of reducing mathematics to a purely logical or axiomatic-deductive structure of entailment-relations that would leave no room for anomaly or paradox. That claim encountered its first major setback when Russell showed – by purely logical means – that set theory was intrinsically prone to generate just such problems, namely the kinds of self-reflexive, self-predicative or auto-referential paradox that resulted from its dealing with formulas such as 'the set of all sets that are not members of themselves' or 'he who shaves the barber in a town where the barber shaves every man except those who shave themselves'. Russell's answer was to make it a stipulative rule that statements in formal languages such as those of mathematics or the logical calculus should not be self-referring in a way that gave rise to difficulties of this sort. Rather they could best be averted by a 'Theory of Types' which distinguished clearly between various orders or levels of statement, i.e., those belonging to the first-order language of direct or material-mode assertion, those that referred to such first-order statements from a higher logical level, and so on up through successive stages of increasingly abstract formal (i.e., meta-linguistic) specification. Only thus, Russell thought, could set theory – as a crucial component of present-day developments in logic and mathematics – be kept on its path toward an ever more secure, since ever more precisely codified conception of validity or truth.

Still his purported 'solution' to these problems struck many, then and now, as objectionably ad hoc and as having more to do with interests of pragmatic or methodological convenience than with principles self-evident to reason. Indeed the set-theoretical paradoxes have remained a spur to philosophic thought and a potent source of speculative ideas both within mathematics and across a range of other disciplines ever since Russell first discovered them. Their impact was intensified by various related developments, including – most notably – Gődel's undecidability-proof to the effect that any formal system sufficiently complex to generate the axioms of elementary arithmetic or first-order logic would necessarily include at least one theorem the truth or validity of which could not be proved within the system itself.[22] In other words, one could have either truth as matter of rigorous logical procedure or consistency ('completeness') as a matter of intra-systemic coherence but surely not both unless by some manoeuvre, like Russell's, that looked suspiciously like a mere device for saving logico-mathematical appearances. Nevertheless set theory survived these and other challenges through the effort of various thinkers to provide some method of formal re-statement in axiomatic terms that would keep the paradoxes safely out of view or at least prevent them from doing any real harm. During the past century it has become absolutely central to every branch of pure and applied mathematics, as well as to every mathematics-based development in the physical and even (in certain contexts) the social and human sciences. Philosophically speaking it has figured chiefly as a focus for reflection on the scope and limits of formal (i.e., deductive) logic as applied to various regions of enquiry and, in particular, to problems in epistemology and philosophy of mind.

Badiou's work is notable for *not* losing sight of the set-theoretical paradoxes – indeed, for placing them squarely and expressly at the centre of its philosophic interests – while none the less maintaining a staunchly realist outlook as regards the objective (or verification-transcendent) character of mathematical truth. Thus he views the history of advances in mathematical knowledge as having most often come about through a dialectical process whereby various sorts of problem or paradox eventually gave rise to some new concept or agreed-upon way of proceeding which in turn – when its consequences became clear – could be seen to involve a further, deeper, and yet more thought-provoking challenge to the project in hand. If this sounds decidedly Hegelian then the impression is not entirely wide of the mark although it does demand qualification in ways that will emerge later on. At any rate, Badiou is absolutely firm in his realist belief that although knowledge must be held distinct from truth – since truth might always transcend the utmost limits of human knowability – nevertheless knowledge is attainable, albeit with the strict fallibilist proviso that all present claims in that regard might conceivably be subject to future revision or outright disconfirmation. Indeed, one of the philosophic traits that lifts his work well clear of post-structuralist, postmodernist and other recent Francophile movements of thought is Badiou's insistence that belief in and commitment to the existence of objective truth-values is not in the least a sign of dogmatism or entrenched doctrinal adherence. On the contrary, it is only by accepting this realist premise and conceding the possibility of error in even our most deeply-held theories, truth-claims, or items of belief that we are saved from equating truth *tout court* with what counts as such for ourselves and fellow-members of our own (whether specialized or culture-wide) belief community.

III

Hence, as I have said, Badiou's unremitting polemic against all attempts – whether Wittgensteinian, neo-pragmatist, anti-realist, or postmodernist – to redefine truth as warranted assertibility or knowledge as that which holds good to (what else?) the best of our knowledge. This sets him decidedly at odds with many – indeed, most – of the movements of thought that have dominated recent (post-1980) developments in cultural criticism and theory. It also goes some way towards explaining why his work has only lately begun to enjoy the kind of adequate, i.e., philosophically informed attention and detailed analytic commentary that it clearly requires, despite having been taken up much earlier – a full decade back – by cultural and literary theorists. These exegetes naturally tended to recoil from the set-theoretical and scientific (not to mention the properly philosophical) aspects of his thought and focused rather on its various connections with post-structuralism and other, to them more familiar realms of theoretical enquiry.[23] Thus they very often approached his work with the preconception that any thinking which espoused a 'radical' position in political or socio-cultural terms must ipso facto endorse at least the main tenets of post-structuralism, among them the language-dependent character of all human knowledge or cognition and the non-availability (indeed non-existence) of any truth that would claim to transcend the scope and limits of linguistic or discursive representation.[24] For Badiou, on the contrary, philosophy is defined as that singular exercise of thought which takes truth as its object and its constant point of reference vis-à-vis those various other disciplines – or 'conditions' – that constitute its field of engagement. Thus he has absolutely no time for the idea that truth-conditions are relative to some given state of knowledge and knowledge-claims relative to this or that 'language-game', 'discourse', or representational schema.

To this extent at least, post-structuralism has much in common with developments within analytic philosophy such as the strain of anti-realism espoused by thinkers such as Michael Dummett or again, more broadly, the kinds of linguistic-communitarian approach spawned by Wittgenstein's later writing.[25] What these notions exclude – if taken at anything like their full philosophical force – is the very idea that thinking might attain a critical perspective outside or beyond the horizon imposed by its own cultural-linguistic terms of reference. To this extent they go clean against Badiou's most basic commitment in philosophy as likewise in matters of science, politics, and ethics: namely, the principle that truth with respect to whichever of these subject-domains must be thought of as always potentially at odds with our present-best state of belief, knowledge, or judgement. Here again Badiou's work defies classification according to the standard (analytically received) view which typecasts all 'continental' thinkers as endorsing some form of linguistic or cultural constructivism, that is, some presumptively far-out version of the anti-realist doctrine whereby any talk of 'reality' and 'truth' must always be treated as relative to – or 'constructed' by – some particular language or discourse. Indeed he belongs very firmly in the company of those analytical philosophers who have taken a squarely opposed view of the matter. Thus, for instance, some have argued that Gődel's undecidability-proof and its various cognates – such as the Russellian paradoxes about set theory – are evidence that the human mind is capable of grasping, perceiving, or discovering truths that must be thought to exceed the utmost bounds of any purely formal (including even the most powerful computer-based) methods of demonstrative proof.[26] After all, how could we possibly be in a position to follow the logic of Gödel's theorem and evaluate its truth-content if indeed – as it purports to show – any proof of that theorem on demonstrative, i.e., on axiomatic-deductive grounds will necessarily be incomplete or incapable of rigorous formalization as regards at least one of its component parts? In which case, so the argument runs, there must be certain truths that the mind is able to grasp through a mode of knowledge that transcends or surpasses the powers of formal reasoning.

Moreover, as is sometimes claimed, those truths may exceed the furthest bounds of artificial, computer-based 'intelligent' systems in so far as the latter are constrained by their reliance on digital (binary) hardware circuits and formally specified software programmes. That Gödel himself espoused a Platonist view of mathematics, logic and the formal sciences would seem to support this inference.[27] Thus he conceived of mathematical entities – numbers, sets, classes, functions, and so forth – as existing in a realm of absolute ideal objectivity to which the human mind has access although not through the commonly-accepted modes of empirical (sensory-perceptual) acquaintance or logical (deductive) thought. Rather, it involves something like what many philosophers following Kant have ruled strictly off bounds, that is to say, a form of intellectual intuition that would close or negate (rather than mediate) the gap between sensuous-intuitive knowledge and concepts of understanding.[28] That gap has been installed within the discourse of philosophy and subject to various attempted solutions, evasions or compromise settlements in both the main lines of post-Kantian philosophical descent, i.e., the 'continental' (or mainland-European) and 'analytic' (chiefly Anglophone) traditions of thought. Indeed it could well be seen as the central problem of post-Kantian epistemology and philosophy of science, one that has pitted realists against anti-realists – or objectivists against constructivists – in various disciplines and contexts of debate. It has also spawned various attempts to stake out a midway position that would somehow combine an adequately robust conception of truth – at least something stronger than can possibly be had from current anti-realist accounts – with a due regard for those philosophic problems that make it unwise (or downright nonsensical, as the anti-realist thinks) to conceive truth as potentially transcending the scope and limits of human discovery, recognition, or ascertainment.

I have shown elsewhere that such attempts always fail either by leaning so far in an anti-realist direction as to let the whole issue go by default or – conversely – by making so many crucial concessions to the realist (objectivist) case that they end up by pretty much endorsing that case with just a token gesture to the standard range of anti-realist counter-arguments.[29] What has led to this situation is the twofold 'turn' in philosophical thinking over the past century and more, namely that which re-directed attention from ontology to epistemology as the main focus of interest and then – pressing yet further along the same path – that which placed language squarely at the centre of epistemological debate. The first, to repeat, had its most direct source in Kant's *Critique of Pure Reason* and his claim that scepticism could be defeated only through a theory of knowledge that played down the concern with ontological problems (or with issues of objective, mind-independent truth) so as to prevent any gap from opening between subject and object or mind and world. This Kantian gambit can now be seen to have had just the opposite effect, producing as it did a multiplicity of problematic doctrines – from Fichte's subjective versus Schelling's objective versions of idealism to sundry analytic variants on the old logical-positivist disjunction between sense-data and conceptual scheme – that have plagued the discourse of epistemology ever since. The second major turn was that which substituted language for the rationalist or empiricist (or indeed Kantian) 'way of ideas' as purportedly the surest, most reliable means of rescuing epistemology from this sorry predicament. For instance, according to Wittgenstein, it is the fact of our belonging to a certain community with its own language-games, customs, practices, rule-following procedures and other such linguistically mediated forms of life that constitutes 'reality' so far as we can possibly know, describe or comprehend it.[30] Other, more strictly 'analytical' approaches in the Frege-Russell line of descent reject this recourse to ordinary language as an ultimate court of appeal and insist on digging down to the logical structures that are often obscured – and sometimes distorted – by our everyday habits of speech.[31] However what they share is the basic conviction that philosophy of language is first philosophy, along with the belief that all those old, intractable issues about knowledge and the subject/object or mind/world 'interface' can be resolved simply by re-phrasing them as issues concerning the scope and limits of language. Since language is inherently a 'public' affair – since, as Wittgenstein is commonly taken to have shown, there could not be any such thing as a 'private language' – therefore the shift to a linguistic-communicative register is sufficient to dispel all those needless qualms of epistemological doubt.

Yet nobody who has read very far into the recent analytic literature can have failed to notice how wide of the mark is any such upbeat prognosis. On the contrary, if one thing has emerged very clearly from this present-day linguistic turn it is the fact that all the issues bequeathed by Kant crop up yet again, and to just as problematical effect, when language is taken as the 'limit of my world' (in Wittgenstein's famous phrase) or as the ultimate horizon of intelligibility. If this is the case, then we shall have to conclude, as anti-realists readily do, that it cannot make sense to suppose the existence of objective or verification-transcendent truths – mathematical, scientific, historical, or whatever – that exceed the bounds of our present-best knowledge and hence our best powers of articulate statement. More precisely, if knowledge is indeed co-extensive with the scope and limits of linguistic expression and if truth is itself epistemically constrained (i.e., co-extensive with the range of humanly attainable knowledge) then anti-realism becomes the default option since truth must then be thought of as likewise subject to the various constraints placed upon us by our no doubt limited or partial modes of linguistic representation. And again, even more to the anti-realist point: we are in no position to speak like this about the limits of our language or its various supposed epistemic, cognitive, or representational shortcomings since – after all – how could we possibly get to know of them, given the thesis as set forth above?

So we should perhaps not take Wittgenstein too much at his word when he claims that his philosophy 'leaves everything as it is' as concerns our various acculturated language-games or communal forms of life, including – presumably – those that involve an objectivist or realist outlook (e.g., with respect to mathematics, science, or historical events) as a part of their basic orientation. Some exegetes, among them Cora Diamond, have seconded this claim as accordant with Wittgenstein's non-interventionist stance and denied that his thinking has any implications for the realism/anti-realism debate, apart from disabusing us of certain 'metaphysical' notions that we should anyway be keen to offload.[32] However 'metaphysics' is here defined in such a way that these very often turn out to include a large proportion of the various, chiefly ontological commitments that characterize realism across the whole range of disciplines or fields – philosophy included – where its claims are at issue. Thus an anti-realist such as Dummett is more closely attuned to the character of Wittgenstein's thought when he takes it to entail that 'warranted assertibility' (rather than objectivist or verification-transcendent truth) is the furthest we can get or the most we should hope for by way of epistemic assurance once Wittgenstein's lesson has truly sunk in.[33] This conclusion is unavoidable, he thinks, since truth cannot exceed the bounds of attainable human knowledge and knowledge cannot be acquired, manifested, or recognized – the three main prongs of Dummett's anti-realist argument – except through possession of the various linguistic means whereby those capacities are realized. So when Dummett says that the turn towards language as opposed to the previously dominant 'way of ideas' is the single most significant development in modern analytic philosophy he is also implicitly staking his claim for an anti-realist approach that would accept these jointly epistemic and linguistic constraints. If the concept of truth is thus scaled down to the notion of epistemic warrant, and this in turn to the range of descriptive, theoretical, or explanatory means whereby to express it, then clearly there is no conceiving of truth in objectivist (i.e., recognition-transcendent and language-independent) terms. From which it follows – *contra* those thinkers like Diamond who would interpret Wittgenstein's quietist stance as strictly neutral in this regard – that in fact the linguistic turn leans strongly in an anti-realist, cultural-relativist, or social-constructivist direction.

IV

We have seen already how emphatically Badiou rejects any version of this thesis which presents it explicitly as a matter of treating language as the end-point or bottom line of philosophical enquiry. However his opposition to it extends well beyond that obvious source of present-day cultural-relativist thinking to encompass a range of developments, especially (but not exclusively) in philosophy of mathematics, which he takes to entail a similar restriction of truth to knowledge and of knowledge, in turn, to the scope and limits of current investigative thought. It is primarily against this whole recent slew of, as he sees them, highly retrograde developments that Badiou brings to bear the conceptual resources developed by set theorists from Cantor to Cohen. Moreover their discoveries have large, indeed momentous and as yet unrecognized implications far beyond the purely mathematical domain. Thus: '[t]he fact that a generic procedure progresses to infinity entails a reworking of the situation; one that, while conserving all of the old situation's multiples, presents other multiples'.[34] Such is the 'evental caesura' or decisive break whose effect would be 'that the truth of a situation, with this caesura as its principle, *forces the situation to accommodate it*: to extend itself to the point at which this truth – primitively no more than a part, a representation – attains belonging, thereby becoming a presentation'.[35]

No doubt there is much here that goes clean against the grain of any common sense, even common-sense-realist idea of what properly counts as truth or knowledge. Indeed it is a conception that could find no room in those various constructivist, intuitionist, or anti-realist approaches to issues in mathematics and the formal sciences that include among their central tenets the conceptual impossibility that truth should be thought of as somehow exerting such pressure – such potentially transformative force – on the limits of present-best knowledge.

This point is well made in *Being and Event* when Badiou stresses the counter-intuitive, indeed on the face of it the downright absurd character of certain set-theoretical axioms. Thus, for instance, it seems a matter of rational self-evidence to treat the void – or the empty set – as 'by very definition' containing nothing and hence as having no subsets. However that assumption takes no account of the difference between 'belonging' and 'inclusion', that is to say, what follows in strict logic from the twin set-theoretical axioms (1) that 'the void is a subset of any set: it is universally included', and (2) that 'the void possesses a subset, which is the void itself'.[36] This leads on to a passage of detailed and complex but by no means impenetrable logico-mathematical reasoning whose explication in simpler terms would require far more space than is available here and which, besides, will bring great benefit to the reader if carefully followed through. Sufficient to say that Badiou demonstrates by strictly logical means that it makes no sense (and is indeed self-refuting) to deny that the void is *included* in every multiple even if it fails to register as *belonging* thereto in accordance with the membership conditions laid down by this or that instance of the dominant count-as-one.

His proof starts from the basic Aristotelian axiom *ex falso sequitur quodlibet*, or, as usually interpreted, 'from the assertion of any two contradictory statements it follows that every assertion whatsoever is both true and false'. In brief, the argument runs (1) 'there exists a negation (there exists a set for which "to not belong to it" is a universal attribute, an attribute of every multiple)'. But in that case, given this true (set-theoretically valid) statement, it also necessarily follows (2) that should the statement be denied – should we suppose that some multiple does or might conceivably belong to the void – then on this basis we can, with perfect logical consistency, assert (or deny) both the truth and the falsehood of any other statement, even when (as Aristotle was careful to stipulate) that statement is taken to refer to the same thing at the same time and in the same respect. Moreover, (3) 'the absurd chimera – or idea without being – of an "element of the void" implies that this element – radically non-presented of course – would, if it were presented, be an element of any set whatsoever', and would therefore have to be conceived as 'that ultra-nothing, that ultra-void with regard to which no existence-multiple could oppose it being presented by itself'.[37] From which one is logically obliged to conclude (4) that this mode of reasoning *ex falso* – or by reductio ad absurdum – leads back to the truth of that basic premise, the excess of inclusion over belonging, that Badiou takes to have such momentous consequences for our thinking with regard to issues in and beyond mathematics and the formal sciences.

Chief among them, to repeat, is the standing possibility that truth may always exceed the compass of present-best knowledge and also – more difficult to take on board from most philosophical viewpoints – that certain as-yet undiscovered or unknown truths may already be latent in the gaps, anomalies, or contradictions (even if themselves so far undiscovered) that characterize some given stage of advance. Such is Badiou's claim with regard to the 'future-anterior' modality of truth-oriented thought, i.e., its dependence on the standing possibility of eventually being brought to the point of proof or verification through a procedure that quite simply could not have been discovered in the absence of just that prerequisite stage in its often distinctly uneven or (on the face of it) wholly unpredictable sequence of developments to date. I should explain that this claim involves an appeal to the formal technique of diagonalization whereby Cantor established the existence – that is to say, the preordained place within the range of conceivable abstract realia – of multiple infinite or transfinite sets that could actually be reckoned with rather than treated (like the old theological or more recent romantic conceptions of infinity) as belonging to some realm of quasi-mystical transcendent being. Thus,

> [f]or an infinite faithful procedure to generate as its positive result-multiple – as the post-evental truth – a total of (+)s connected to the name of the event which 'diagonalize' a determinant of the encyclopedia, it is sufficient that within that procedure there be at least *one* enquiry which avoids this determinant. The presence of this particular finite enquiry is enough to ensure that the infinite faithful procedure does not coincide with the determinant in question.[38]

In other words, if indeed it is the case (*contra* anti-realists and intuitionists) that truth always in principle surpasses knowledge – that there exist verification-transcendent truths such that a conjecture may be warranted by them even though they cannot as yet be known through any available method of proof – then this requires that there should also exist some potential but so far unutilized analogue to the procedure of 'diagonalization' that would, if applied, be found to yield just that sought-after yet so far elusive result. To the question 'Is this a reasonable requisite?' Badiou responds that in truth it is since 'the faithful procedure is . . . in no way determined by knowledge', and since '[i]ts origin is the event, of which knowledge knows nothing'.[39] Hence his stress on the indiscernible, on whatever in the situation cannot be clearly and distinctly made out, or – as he more paradoxically puts it – whatever may be 'indiscerned' through the problems, contradictions, anomalies, discrepancies, excesses of inclusion over belonging, of inconsistent over consistent multiplicity, and so forth, which constitute an integral and fully objective (though so far unrecognized) feature of that same situation.

So it is indeed possible, pace Dummett, to affirm without absurdity or self-refutation that there do exist unknown and at present unknowable truths along with various likewise unknown procedures for eventually bringing them to light. In addition, there is Badiou's claim – even more unacceptable to the intuitionist or anti-realist – that despite being inaccessible as yet to any means of formal proof or demonstrative warrant these truths are none the less (if 'indiscernibly') present as potential indices of a 'faithful' procedure that would point decisively beyond the current state of knowledge or stage of advance. Nor are such truths by any means restricted to the domain of mathematics and the formal sciences, as often emerges unexpectedly at just those moments where he seems most deeply and intently occupied with just such 'technical' issues. On the contrary, it is basic to Badiou's idea of the relationship between mathematics and philosophy, as likewise of that between philosophy and its fourfold enabling 'conditions', that the most significant questions should be seen to arise at the point where a rigorously formal approach turns out – as in Roland Barthes's well-known saying – to bring us all the more pointedly back to history rather than leading us away from it.[40] Thus:

> [i]n collective situations . . . politics (if it exists as *generic politics*: what was called, for a long time, revolutionary politics, and for which another word must be found today) is also a procedure of fidelity. Its events are the historical caesurae in which the void of the social is summoned in default of the State; its operators are variable; its infinite productions are indiscernible (in particular, they do not coincide with *any part nameable according to the State*), being nothing more than 'changes' of political subjectivity within the situation; and finally its enquiries consist of militant organized activity.[41]

I should stress that when Badiou refers to 'subjectivity' or 'the subject' – and to epochal events as involving 'nothing more' than a transformation in the forms and modalities of subjective engagement with the political process – he is very far from proposing any kind of irrationalist, decisionist, or (in the commonplace sense of that term) subjectivist approach to these issues. Rather he takes it, at this fairly late stage in *Being and Event*, that readers will by now have thought their way beyond the crudely dichotomous conception of subject and object that he sees as having captured the philosophic mindset from Descartes to Kant and nowadays in just about every main quarter of debate, whether 'continental' or 'analytic'.

What this involves is a capacity to grasp those as-yet strictly 'indiscernible' elements that reveal the constant (on occasion critical) excess of inconsistent over consistent multiplicity, subsets over sets, parts over members, inclusion over belonging, or the 'state of the situation' over the situation as currently construed according to the dominant count-as-one. This is why, as he says,

> [a] subject alone possesses the capacity of indiscernment . . . [and] also why it forces the undecidable to exhibit itself as such, on the substructure of being

of an indiscernible part. It is thus assured that the impasse of being is the point at which a Subject convokes itself to a decision, because at least one multiple, subtracted from the language, proposes to fidelity and to the names induced by a supernumerary nomination the possibility of a decision without concept.[42]

It would therefore be wrong to locate Badiou exclusively on either side in the quarrel between 'humanism' and 'anti-humanism' that has periodically occupied stage-centre in so much French philosophical debate from Sartre down. To be sure, he is very far from endorsing the idea – often more than hinted at by mathematical intuitionists – that our conception of 'truth' must be tailored not only to the scope and limits of human knowledge but even to the contours of aesthetic sensibility among more discerning mathematicians.[43] Nor is he at all disposed to resurrect that idea of the thinking, acting, willing, and judging subject whose various successive incarnations from Descartes to Kant and Husserl have rung so many changes on the theme of transcendental constitution. To this extent Badiou is very much aligned with the strong anti-humanist current of thought that can be seen to have gained its chief impetus from the structuralist turn against Sartrean existentialism announced by such otherwise diverse thinkers as Lévi-Strauss, Althusser, Lacan, Foucault, and Barthes.[44] On the other hand his debt to Sartre is often and handsomely acknowledged, not least with respect to precisely this question – so central to Sartre's later writings – of how one can wrest sufficient space for the exercise of human discovery, invention, creativity, and ethical as well as socio-political praxis from the strongly determinist implications of a full-scale structuralist approach.[45]

As concerns politics Badiou's working premise (like Sartre's before him) is, I think, captured well enough by the famous passage from the *German Ideology* where Marx and Engels write that human beings 'make their own history, but do not make it as they please; they do not make it under self-selected circumstances, but under circumstances existing already, given and transmitted from the past'.[46] In mathematical terms it is perhaps best characterized by the following statement from *Being and Event* which, despite its cryptic style, does much to clarify Badiou's thinking on a question that goes to the philosophic heart of his project. 'A subject', he writes,

> is at the intersection, via its language, of knowledge and truth. Local configuration of a generic procedure, it is suspended from the indiscernible. Capable of conditionally forcing the veracity of a statement of its language for a situation to-come (the one in which the truth exists) it is the savant of itself. A subject is a knowledge suspended by a truth whose finite moment it is.[47]

This may help to explain how Badiou is able to assert the double and, as it might seem, contradictory claim that truth is (1) 'forced' to emerge through a procedure that develops according to its own, strictly formal and hence recognition-transcendent logic, yet also (2) discovered by subjects whose fidelity, commitment or *truth to* the singular demand of some given inaugural event enables them to press that procedure through to its ultimate conclusion. I should add, lest my phrasing of (2) leave room for any misapprehension: Badiou is absolutely clear on the point (much emphasized by philosophers of science in the mainstream analytic tradition) that socio-cultural or psycho-biographical 'context of discovery' and intra-scientific 'context of justification' involve very different standards of relevance or criteria for what counts as a valid mode of argument.[48] Where he does implicitly claim to go beyond that fairly standard philosophical distinction is through his cardinal idea, compactly expressed in the passage just cited, that the 'subject' as defined in relation to the formal, physical, and even (certain branches of) the social and human sciences is by no means synonymous with 'the subject' as conceived in psychological or – conventionally speaking – in subjectivist terms.

One can find a good number of passages where Badiou's formulation of the case runs close to denying – or seeming to deny – any role for the subject in a more than purely nominal or place-holder role. Such is, for instance, his statement that '[a] term forces a statement if its positive connection to the event forces the statement to be veridical in the new situation (the situation supplemented by an indiscernible truth)'.[49] Yet it is clear by this stage that such statements are by no means incompatible with the ascription of a genuine, indeed a decisive role for the subject just so long as the latter is construed in such a way as to respect Badiou's crucial point about the strictly indissoluble tie *in this context* between the subject and the 'post-evental' project of enquiry to which that subject is committed. Thus if '[t]he opening of a generic procedure founds, on its horizon, the assemblage of a truth', nevertheless 'subjectivization is that through which a truth is possible', since it 'turns the event towards the truth of the situation for which the event is an event'.[50] It is on this condition – and this condition only – that one can make sense of his peremptory statement 'I will term *forcing* the relation implied in the fundamental law of the subject', and his yet more rigorously formalist or ultra-objectivist-sounding dictum that '[g]rasped in its being, the subject is solely the finitude of the generic procedure, the local effects of an evental fidelity'.[51] We can now look more closely at Badiou's antagonistic relation to those various currents of thought – 'analytic' and 'continental' alike – that have conspired, as he sees it, to conceal the extent to which truth outruns the state of knowledge at any given time yet also, and quite compatibly with that, the extent to which thought may have a purchase on truths that lie beyond its present-best capacities of conscious or knowing grasp.

V

As I have said, Badiou stands out for his absolute and principled refusal to have any truck with the linguistic turn in its continental (mainland-European) as well as its analytic (chiefly Wittgensteinian) variants. This refusal has a long and philosophically distinguished pedigree, going back to Socrates' engagement with the sophists and his claim that these professional teachers of rhetoric – or instructors in the art of linguistic persuasion – were purveyors of a false wisdom or pseudo-knowledge that disguised their ignorance of the various topics about which they held forth with such eloquent assurance.

Nowadays the sophists have their defenders who are apt to point out that Socrates himself was not above exploiting certain tricky turns of argument and also that the modern, strongly pejorative sense of 'rhetoric' (especially in political contexts) is one that makes insufficient allowance for the role of persuasive language in promoting good or desirable as well as tendentious or downright pernicious beliefs.[52] However it is not just this kind of modest or broadly pragmatist rehabilitation exercise that Badiou has squarely in his sights. Rather, it is the philosophic turn toward language as the *ne plus ultra* or putative end-point of all investigation, whether this takes the form of a hermeneutic, post-structuralist, Heideggerian, Wittgensteinian, neo-pragmatist or other such (in his view) philosophically inadequate and ethically evasive doctrine. What these approaches have in common – despite their otherwise large differences of view – is the idea that it can ultimately make no sense to conceive of our various statements, hypotheses, theories, conjectures, and so forth as possessing an objective truth-value beyond their acceptance or non-acceptance within some existing language-game, discourse, or acculturated form of life. What sets Badiou firmly, even fiercely at odds with all of them is his conviction that truth is *not* just a matter of whether or not some candidate item happens to fall square with the currency of in-place, linguistically expressible, and hence communally warranted belief. On the contrary: truth is the ultimate standard by which to assess our various theories, scientific hypotheses, ethical values, political commitments, and whatever aspects of our personal (including our affective or erotic) lives involve a fidelity to past events that possessed a life-changing or radically transformative character.

Moreover this is *truth* as distinct from *truthfulness*, that is to say, truth conceived as a matter of duly specifiable truth-conditions and not in some vaguely defined sense having to do with acting in good faith or with sincerely and genuinely meaning what one says.[53] To be sure, these latter kinds of consideration have a prominent place in Badiou's thought since it is the mark of a 'militant' fidelity to truth – in whichever of those disciplines or spheres of life – that the subject should acknowledge and seek to live up to the demand placed upon them by the truth-event in question. This applies more obviously in the case of events such as amorous encounters or moments of decisive personal or ethico-political choice when the question of fidelity – of somehow continuing to act, think and feel accordingly – is posed with particular urgency and force for the individual concerned. Yet it also applies in those other dimensions of human intellectual, scientific and moral endeavour where the traditional philosophic view would have it that issues of objective truth and falsehood have to be treated as belonging to a realm quite apart from issues of personal psychology or private-individual motivation. Nor would Badiou for one moment wish to deny this latter claim, given his unremitting emphasis on truth as that which transcends or surpasses any merely subjective realm of meanings, beliefs, values, or intentions. His point is most emphatically *not* to recommend that the primary concept of truth be revised or scaled down in this direction but rather to insist that the operative standards for what counts as truth with regard to our lives as ethical or political agents be conceived more in accordance with that same, rigorously specified and highly demanding concept. For the criterion in both cases is whether or not the subject comes to recognize the truth-event in question and, if so, to what extent and how resourcefully they keep faith with its requirements. Moreover, any failures in this regard – any fallings-short of fidelity thus conceived – will be defined not merely in subjective or personal terms but in terms of their evading the truth-conditions or requirements for an adequate or faithful pursuit of the logically, scientifically, or historically emergent implications laid down by that specific event.

Thus Badiou yields to no-one in the analytic tradition as concerns his high valuation of truth, its central role in his thinking across a wide range of topic-areas, and also his realist insistence that truth be conceived as wholly independent of our language-games or cultural life-forms and therefore as always potentially verification-transcendent. This is why he is so markedly at odds with that nowadays widespread anti-realist or constructivist trend that has characterized a good deal of analytic philosophy in the wake of logical empiricism. Where his work is most likely to conjure a hostile response from thinkers in that 'other' tradition is on account of Badiou's blatant disregard, as they will see it, for certain crucial distinctions between the different kinds of truth-claim that characterize different areas of thought. These are taken to range from mathematics and the physical sciences, through the various (more-or-less rigorous or truth-apt) branches of philosophy itself, to the sort of 'truth' that is in question when it comes (say) to personal life-choices, relationships, and other such subjective or inter-subjective matters. In its hard-line or resolutely top-down form this conception goes back to the heyday of logical positivism and the so-called 'Unity of Science' programme. Here the unity concerned was very much that of a science-led ranking order with physics at the top and a place for the lower-order human or social 'sciences' only in so far as they espoused the methods of empirically based observational, inductive, or hypothetico-deductive enquiry.[54] Analytic philosophers over the past half-century have pretty much abandoned that programme under pressure from various quarters, some of them – like Quine's all-out attack on the two last 'dogmas' of logical empiricism – coming from thinkers securely ensconced within the analytic tradition and also committed to a thoroughly physicalist or 'naturalized' epistemology.[55] Hence the emergence of what is sometimes referred to as 'post-analytic' philosophy, a hybrid development whose chief common feature is its way of hanging onto the typical idioms (along with a good few central concerns) of old-style analytic discourse while expressly renouncing most of its technical and 'metaphysical' baggage. At the same time these thinkers often look further afield, i.e., to mainland-European developments such as phenomenology or hermeneutics – albeit suitably 'naturalized' – in the quest for alternative resources.[56]

My point is that Badiou stands at an odd angle to each and every one of these developments. Thus he insists – as against proponents of a positivist or science-led approach – that there are more things properly deserving of a truth-based, rigorous, or justificatory account than are dreamt of in any such narrowly partisan outlook. Yet he also insists – as against adepts of the present-day linguistic, cultural, sociological, pragmatic, hermeneutic, and other such 'turns' – that this doesn't mean letting go of those standards by stretching them around subject-areas that some would regard as beyond the pale of a truly philosophic, let alone a mathematically based approach. After all, it is hard to imagine an analytically-trained philosopher of logic or mathematics finding anything remotely plausible in Badiou's claim for the absolute centrality of set-theoretical concepts and methods to the discourses of politics, ethics and aesthetics, not to mention psychoanalysis and love. For despite unsubscribing from the stark dichotomies of 1930s-style logical positivism – as between cognitive (truth-apt) and emotive (empirically vacuous) kinds of utterance – still the analytic tradition is heir to that deep mistrust of any thinking that would venture into speculative regions too far beyond the limits thus laid down. So Badiou is a decidedly anomalous thinker whether viewed in the context of an analytic mindset that would share his demand for conceptual precision but find something suspect about his sheer range of philosophical involvements, or again from a 'continental' standpoint that would happily endorse this latter aspect of his work but look awry at his claims concerning the primacy of truth conceived *more mathematico* and, more specifically, of set theory as the basis for all such enquiry. Yet his work also offers a needful reminder that this typecast distinction is one that encounters anomalies, exceptions, or complicating cases on both sides. Among them are analytic philosophers whose thought reveals striking points of contact with phenomenological concerns and also a number of eminent French thinkers of an earlier period – such as Gaston Bachelard, Georges Canguilhem, and (most importantly in this context) the mathematicians Jean Cavaillès and Albert Lautman – whose distinctive contributions to philosophy of science and mathematics have their source in a critical reading of Descartes and the legacy of seventeenth-century rationalism.[57]

It is against this more complex and internally differentiated background – as opposed to the stereotypical bi-polar view – that Badiou's work begins to reveal both its true originality and (quite compatibly with that) its various, densely interwoven lines of affiliation. What also emerges to striking effect is his heterodox view of the relationship between mathematics (conceived in Platonist terms) and those areas of human praxis, commitment or shared endeavour that would normally be thought of as belonging to a separate realm where the relevant standards are those of inter-subjective meaning and value rather than objective truth. For Badiou, the idea of a truth-procedure – a rigorously consequent mode of investigative thought or practical-exploratory conduct in the wake of some decisive or epochal event – is just as valid when applied in these latter kinds of context. Thus, for instance, his high regard for Cavaillès has to do not only with the latter's ground-breaking work in mathematics but also with his having been a highly active member of the French Resistance whose death before a German firing-squad bore witness to the likewise rigorous and consequential nature of his thinking about the wartime situation and his own responsibilities in that regard. This follows from Badiou's conception of truth in its four chief worldly or 'applied' domains (i.e., politics, science, art, and love) as also involving a dimension of *truthfulness-to* whatever requires such strict and unconditional fidelity. That is to say, it entails a uniquely personal or life-historical component which cannot be captured in terms of truth as classically defined, i.e., as a strictly impersonal matter of correspondence between veridical statements and factual states of affairs, or again, between valid mathematical or logical propositions and their truth-value conceived in objective (recognition-transcendent) terms. Indeed it is crucial to Badiou's whole conception of the subject as a 'militant' for truth that this be thought of as requiring just such actively partisan modes of commitment above and beyond what can presently be justified in terms of factual, empirical, evidential, probabilistic, or deductive warrant. Still, this should not be allowed to obscure the equally crucial point: that these various truth-procedures – say, political courses of action or ways of confronting the choices posed by some decisive personal (e.g., amorous) encounter – can have very definite, precise and exacting conditions of fulfilment or non-fulfilment.

To be sure, those conditions cannot be specified in advance since it is only later, that is, as a more or less long-term consequence of the inaugural event that one is able to discern what properly counts as an instance of fidelity or truth to that event and its hitherto concealed implications. Yet this is not for one moment to espouse some mystical, crypto-theological, or eschatological conception of truth that would place it in a realm of redemptive possibility beyond the scope and limits of rational thought. On the contrary, Badiou insists: transcendence of that sort can have no part in our lives as political, knowledge-seeking, or artistically creative subjects whose sights may be fixed on that which exceeds our present-best powers of perceptual or conceptual grasp, but for whom such truths are none the less a matter of what *might yet transpire* in terms of strictly intramundane or worldly possibility. This argument finds its paradigm case in mathematics – more specifically, in post-Cantorian set theory – since this is a discourse that poses ontological issues in such a way as constantly to drive thought beyond its current limits by turning up hitherto unperceived kinds of problem or aporia. Thus, to repeat, it was Russell's discovery of the paradoxes created by self-predication which gave rise to his putative solution, namely the Russellian 'theory of types' involving a strict hierarchical distinction between various levels of logico-linguistic analysis. That in turn led on to a normalizing phase in the development of classical set theory and the emergence of several internally consistent and (to this extent at least) logically well-founded systems, among them the Zermelo-Fraenkel system adopted by Badiou in *Being and Event*. However he is clear that the fact of his adopting it – and thereby implicitly lending credence to Russell's rather quick and peremptory way with the paradox of self-reference – should not for one moment be taken as a claim that set theory has now arrived at a stage of conceptual refinement and stability wherein all the major problems have either been resolved or brought within reach of solution. On his account the advancement of mathematical knowledge most often occurs through a process whereby some existing, apparently secure and problem-free framework turns out to harbour an unsuspected conflict or anomaly, one that might temporarily be laid to rest (perhaps through a stipulative rule, like Russell's, designed to avoid such trouble) but whose settlement eventually opens the way to further such conceptual provocations and developments.

VI

This is why Badiou makes the case for set theory as our surest route to the truths of fundamental ontology, that is to say, as a privileged mode of comprehension in so far as it allows us to think most directly about the nature and structure of humanly cognizable reality. It is also why he is careful to distinguish it from those other truth-procedures (of science, politics, art, and love) that are no less concerned with pursuing the consequences of some decisive change in our conditions of knowledge or experience, but which stand in an altogether different relationship to the truth-event which they affirm and bear out through the various kinds of fidelity involved. It is the difference between on the one hand a discourse, that of mathematics, whose 'evental' character – i.e., its history of discoveries, advances, setbacks, solutions, problems, etc. – must always have reference to an order of truth that potentially transcends any present state of knowledge, and on the other an order of contingent, i.e., radically evental truths that impose their conditions for investigative rigour only through a complex and open-ended series of strictly unforeseeable after-events.

Such is the thesis announced in the very title *Being and Event*, one that maintains this cardinal distinction even while yielding absolutely no ground to the various movements of thought – post-structuralist, postmodernist, neo-pragmatist, Wittgensteinian, and so forth – that would ignore it and simply press right through with this or that version of the cultural-linguistic-ontological-relativist creed. Nor are such ideas by any means confined to the so-called 'continental' line of descent since – as I have said – they are also common currency among the recent generation of analytic (more aptly: of 'post-analytic') philosophers who take their chief bearings from Wittgenstein, Quine, Dummett,

Nelson Goodman, or the later Hilary Putnam.[58] In this context Badiou stands out, not only for his firm insistence on a strongly realist (or anti-constructivist) conception of truth in mathematics, logic and the formal sciences, but also for his equally principled commitment to a philosophy of praxis – of ethical and socio-political engagement – that involves a quite different but none the less rigorous and exacting set of truth-conditions. For it is precisely in the rift between being and event, or ontology and that which inherently eludes any kind of ontological specification, that there emerges a space for the exercise of human choice, commitment, and fidelity.

Of course this is a space hemmed in by various pressures of socio-historical and personal circumstance, pressures whose nature and bearing may often not be evident unless and until the agent encounters some especially resistant block on the way to fulfilment of their project. All the same – and despite his uncommon emphasis (as in the case of Cavaillès) on the role of rigorously consequent logical reasoning in matters of ethical conscience – it is prerequisite to any adequate account of human action and its motives or justificatory grounds that this distinction be observed. That is to say, it is crucial that the issue of *truth to* some historically emergent event along with its largely unpredictable effects be conceived in terms categorically distinct from the issue of truth as that which obtains and exerts its conditions with a force of ontological necessity indifferent to any such strictly contingent or circumstantial factors. Where the latter has to do with objective truths that might always hold good despite and against what presently counts as best opinion or accredited knowledge the former has to do with an order of truths that reveal themselves in and through the temporal passage of events. This means that they are subject to possible shifts of social, political, or ethical perspective that may impose new criteria – that is, new conditions of fidelity and truth – in response to such unpredictable changes of circumstance. Yet it is still very much a matter of truth not just in the subjective sense 'truth by the lights of one's own moral conscience, personal convictions, strength of feeling', etc., but rather in the sense: 'truth to what imposes itself as the necessity of acting or thinking with maximum regard to the specific requirements of this situation at this time and (so far as possible) with a view to likely consequences'.

Thus the truth-value that attaches to commitments undertaken in response to events of a contingent or temporal character is not the kind of truth-value that pertains to mathematical (set-theoretical) advances or discoveries. On a realist (e.g., Fregean) account, these latter are like stages in the exploration of a so-far uncharted terrain that must be thought to pre-exist the explorer's arrival, along with all its various landmark features such as lakes, mountains, coastline, and so forth, whose location and scale can then be mapped with more or less accuracy.[59] The former, conversely, have to do with a range of open-ended possibilities that are not so much laid down in advance by the way things stand with regard to a certain (whether physical or abstract) object-domain as brought about through the various choices and commitments that make up an individual life-history, a political movement, an intellectual-scientific transformation, or a radical departure in the cultural-aesthetic sphere. Yet in these cases also – as Badiou insists – there are certain constraints or conditions for the exercise of a properly responsible freedom which can indeed be specified with some precision and even rendered more amenable to adequate conceptualisation through treatment in terms derived from the resources of post-Cantorian set theory. Such, after all, is Badiou's main argument in support of his distinctly heterodox claim for mathematics as the basis of all ontological enquiry whether in the formal, the natural or indeed certain aspects of the social and human sciences. It is crucially what underwrites his idea – otherwise merely a kind of extravagant pun – that there exists a close structural homology between, on the one hand, the *state* as a locus of juridico-socio-political order with specified membership-conditions and exclusionary powers, and on the other the basic operation of 'count-as-one' which constitutes the *state* of some given multiplicity as defined in set-theoretical terms.

For Badiou, this is not just a vaguely analogical way of thinking but a precise correspondence which enables us to grasp how structures of state power can exert a great measure of social or ideological control by deciding who should count as a citizen, a subject, or one whose identity is duly recognized and confirmed by their belonging to this or that self-authorized (e.g., national or party-political) collective. At the same time, that power is always threatened – exposed to a potentially dislocating force – by the lack of any fixed or stable correspondence between the two modes of subject-interpellation that Badiou describes in set-theoretical terms as the conditions of belonging and inclusion. It is precisely through the non-coincidence of these orders that political injustices arise, since it witnesses a failure (or refusal) of the state to accord a full measure of recognition to those whose proper rights and entitlements are thereby revoked or withheld.[60] Such anomalies may take a blatant form, as in places and periods of gross racial, ethnic, class, or gender oppression, or they may be passed off under cover of a more-or-less spurious rhetoric of democracy, equality, shared national values, education for citizenship, and so forth. Hence Badiou's firm and principled insistence on holding up a mirror to those Western 'democratic' states whose vaunted self-image as defenders of freedom and justice against the various threats of extremism, terrorism, religious fundamentalism, etc., is constantly belied by their patent double-standards or downright abuses of that humanitarian creed, both at home and abroad.[61] It is by way of bringing formal and demonstrative rigour to the discussion of social and ethico-political issues that set theory constitutes a prime point of reference for Badiou's entire project. This allows him to theorize the inbuilt deficits, shortcomings and distortions of past or present political systems – whether totalitarian or Western (pseudo) democratic – through a mode of analysis that focuses on the various ways in which those systems operate to reinforce dominant structures of power and representation. Thus it offers the best, most conceptually adequate resources for an immanent critique of that hegemonic discourse that plays on received (ideological) perceptions of Western or 'free-world' democratic values versus the forces of ignorance, bigotry, and cultural darkness.

What the set-theoretical approach brings out in this seemingly improbable context of application is the extent to which 'abstract' modes of reasoning in pure mathematics can be shown to bear upon real-world issues of an urgently practical or ethico-political nature. Nor should this be taken as any great cause for surprise, given what Wigner found so hard to explain yet also undeniable, namely (to repeat) 'the unreasonable effectiveness of mathematics in the physical sciences'. For that 'effectiveness' goes well beyond physics and other natural-scientific fields of enquiry to encompass the entirety of human dealings with the world of our shared knowledge and experience. This may seem flatly opposed to the claim for mathematics, logic and the formal sciences as belonging to an order of strictly a priori or necessary truth, and hence – as Platonists like Frege would assert – to a realm of absolute ideal objectivity beyond any question of empirical verification. However, as I have said, there are some realistically-minded philosophers, Philip Kitcher among them, who think that we can have a good measure of objectivity or verification-transcendence yet still accept the rootedness of even the 'purest', most abstract or speculative branches of mathematical thought in such basic operations as counting, sorting, assembling into sets, and so forth.[62] That is, they hold the view – *contra* the company of sceptics and anti-realists – that there is nothing inherently paradoxical or self-refuting about the claim that one can have *both* mathematical truth *and* mathematical knowledge. Moreover this desirable upshot can be had without truth being tailored to the scope and limits of attainable human knowledge or knowledge reduced to a matter of more-or-less firmly held communal belief on the grounds that (supposedly) by very definition it cannot make contact with the realm of objective, verification-transcendent truth.

It is precisely against this kind of pyrrhic argument – in brief, the case that quite simply 'nothing works' in philosophy of mathematics – that mathematical Platonists like Gődel have asserted the existence of truths and of corresponding modes of knowledge quite distinct from those that characterize our dealing with objects of sensory or phenomenal cognition. Badiou takes a somewhat different line of approach since he thinks that this whole debate has been skewed by a falsely polarized view of the issue between Platonists and anti-Platonists. After all, as he does well to remind us, Plato is very clear in the *Meno* and elsewhere that it is only through an actively involved or participant exercise of mind that the discoverer, learner or knower can genuinely come to grasp mathematical truths rather than merely reel them off as a matter of rote-like pseudo-knowledge.[63] Thus Badiou – like Kitcher in this respect at least – sees nothing but error and delusion in the spat between self-styled realists and anti-realists, as well as in the various attempts to come up with a 'sensible' middle-ground position that, so far from resolving the issue, gives us the worst of both worlds. Where he does very definitely part company with Kitcher and other, like-minded thinkers in the analytic camp is by taking mathematics – on this realist yet crucially mind-involving or conceptual-activist account – to possess such a range of decisive implications for our thinking about issues beyond the mathematical domain. That his arguments are likely to encounter resistance for just that reason, as well as on political grounds, is perhaps why most of his commentators to date have steered well clear of pursuing the relation between Badiou's work and debates within mainstream analytic philosophy. No doubt they would have a good few additional bones to pick – not least as concerns his reading of developments in post-Cantorian set theory – if they did make a serious effort to engage with that work. Even so, it might well have the strongly beneficial effect of lifting their discussion out of those ruts – those endless exchanges of minor variations on well-worn themes such as the rule-following 'dilemma' – into which the analytic ground has been trodden over the past few decades. For there is no doubting Badiou's stature as one of the most creative and ambitious philosophical thinkers of our time, whatever the problems he undoubtedly poses for anyone schooled in the standard analytical division of academic labour.

More than that, his work offers strong grounds for rejecting that deep-rooted habit of mind (deep-rooted at least among a good few academic philosophers) which takes for granted the existence and indeed the necessity of an outright conflict between naturalism and rationalism, in the broadest sense of those terms. Hence the idea, pushed hardest by followers of Wittgenstein, that in matters of will, motive, interest, desire, and (above all) ethical responsibility it is an error of the most grievous since humanly degrading kind to confuse the two wholly separate 'language-games' of *causal explanation* with reference to various features of a person's natural, cultural, or life-historical conditions and *rational-evaluative judgement* with regard to their imputed character as thinking, willing, and morally autonomous agents.[64] Here again the doctrine in its full-strength Kantian separatist guise is one that most philosophers nowadays would shy from embracing since it bears such a weight of metaphysical commitments and, besides that, seems so deeply in hock to a Cartesian conception of body and mind (or body and soul) as two substances with no point of contact between them. Kant himself tried and signally failed to deflect that charge by taking Descartes to task, in the 'Paralogisms' section of the First *Critique*, for supposing the subject (the transcendental subject of epistemological enquiry) to possess some kind of substantial unity or knowable nature, rather than figuring purely as a necessary place-holder – a transcendentally deducible 'condition of possibility' – for knowledge and experience in general. That, despite these efforts, he is seen as having fallen into yet another version of the same 'metaphysical' trap is one reason why so many present-day philosophers have striven to establish some viable middle-ground position that would allow them to avoid a similar fate. Such are the various quasi-naturalisms surveyed in this book, all of which turn out to re-enact Kant's signal failure, albeit in a different (less strenuously argued though often more heavily qualified or bet-hedging) key.

What those approaches have in common is also, as I sought to show in Chapter 2, what emerges very much against his overt intent in Harry Frankfurt's exposition of the case for second-order desires as constitutive of genuine or fully-achieved personhood. That is, we here witness a strong desire on the part of many philosophers to defend the pre-eminence of reasons over causes – or autonomous will over any admixture of 'external' or merely heteronomous prompting – in matters of human motivation while none the less making sizable concessions to the naturalistic Zeitgeist so as not to attract the charge of backsliding into Kantian, crypto-theological, or unscientific ways of thought. However, so far from deflecting that charge, the result is to invite it all the more pointedly through an opening up of yet further unlooked-for dualisms or conceptual tensions that the proposed remedy has failed to resolve despite its best efforts in this bridge-building mode. It seems to me that there remains only one viable alternative after so many major or minor variations on the rationalist versus naturalist quarrel that has been a main feature of Western philosophical debate from Plato and Aristotle down, and which has taken a particularly complex form (since so often disguised or disavowed) in recent attempts to draw up a new post-dualist or third-way agenda for thinkers in the wake of 'old-style' analytic philosophy. This is the alternative marked out by Parmenides when he asserts that the order of thought and the order of being are necessarily one and the same; by Spinoza in his similarly downright insistence on the unity of mind and nature; and by Kitcher in his well-advised refusal to accept the supposedly insoluble dilemma between objectivist and epistemically constrained ideas of mathematical truth. It gains further support from Badiou through his radical rethinking of philosophy's role in relation on the one hand to mathematics as our best (indeed only adequate) means of ontological orientation, and on the other to those discipline-specific 'conditions' – fields of scientific, political, or artistic endeavour – wherein some existing ontological order is subject to challenge and eventual overthrow by that which exceeds its utmost powers of foregone conceptual or even, at the limit, creative-imaginative grasp.[65]

In each case the point is not, most emphatically, to come up with a middle-way solution that avoids both the rationalist 'high priori road' and what is typically regarded as the opposite temptation, i.e., the appeal to an out-and-out naturalistic world view that would leave philosophy (as well as those other disciplines) devoid of normative or justificatory warrant.[66] On the contrary, what these thinkers envisage is a full-scale transformation in our thinking about such matters that would at last come out on the far side of all those vexing dualisms rather than striving to negotiate a path between their supposedly opposite, indeed starkly antithetical poles. That the order of ideas is also, and inseparably, the order of things – that the truths of mathematics and logic are there to be discovered both through an exercise of a priori thought and as a result of their 'applying' so directly to such a vast range of scientific and everyday activities – is a fact about the world and also about us as rational and real-world situated knowers that cannot be sceptically conjured away even though it has struck many as highly mysterious.[67] After all, were it not for this remarkable yet commonplace alignment or convergence between the way things stand on both sides of those various notional dualisms (mind/world, subject/object, reasons/causes, theory/observation, normative/descriptive, etc.) it would be nothing short of a massive and wholly implausible 'cosmic coincidence' that science should have made such progress to date and indeed that human beings should have managed to survive so long against some at times pretty heavy physical odds.[68]

No doubt there will continue to be large differences of emphasis among philosophers concerning the relative priority accorded on the one hand to strong-naturalistic modes of explanation (e.g., those offered by the more confident branches of cognitive science or the socio-biological 'new synthesis'), and on the other to issues concerning the 'context of justification' wherein such claims – and any arguments raised against them – are properly subject to an ongoing process of rational-evaluative testing and critique. However it is only through the ingrained human (more than that: the inbred philosophical) propensity for turning what should be perfectly amicable divisions of labour into ever renewable causes for dispute that this has become something more like an outright clash of metaphysical, moral, and scientific world views. Hence the firm conviction of many philosophers nowadays that we need to get beyond the whole tangle of perplexities bequeathed by that Cartesian-Kantian conception of knowledge as involving some kind of 'interface' or problematic point of conjunction between mind and world.[69] If so, then we shall have to accept the inextricable tie between thought in its jointly descriptive-explanatory-rational-normative dimension and those various everyday, natural-scientific, and abstract (mathematical or logical-conceptual) domains wherein thought finds a purchase for its equally various investigative methods and procedures. Nor is such deliverance to be had through a half-way naturalized return to Kant – to a Kantianism shorn of its more overtly metaphysical or transcendental aspects – which none the less, as with recent trips in that direction by John McDowell and others, ends up by reproducing all the same old dualisms in a slightly different key.[70] What it requires, and what these thinkers either reject or accept with sundry caveats and qualifications, is the sole adequate alternative: a full-strength naturalist approach with the courage of its primary conviction but with none of those resultant worries – like the sense of a chronic normative deficit – that come of its having so often been cast as the antithesis of an equally typecast purebred rationalist outlook.

VII

One way of helping this process along would be to question the frequent polemical usage of terms like 'reductive' or 'reductionist' to imply that naturalism in any but a weak or heavily qualified form must entail the treatment of epistemological and, more alarmingly, of ethical questions as if they could be tackled through a (quasi-)scientific mode of analysis wholly indifferent to their larger contexts of motivating interest and value. Here it needs saying – and precisely with reference to that larger context – that certain kinds of reductionism do have their perfectly legitimate (indeed indispensable) place at a certain stage in every project of enquiry that aims to understand both the workings or functional constitution and, quite compatibly with that, the purposive-agentive-communicative character of human thought and conduct.[71] In fact the main reason why that term has become such a bugbear in philosophy and the social sciences is that its use has been effectively co-opted or hijacked for the purpose of various, more or less overt or covert anti-naturalisms. Among them, as I have said, are a good many moderate versions of the kind whose advocates would strongly disavow that title – since they do make some non-trivial concessions to the naturalist case – yet whose arguments betray a constant uneasy sense, as with Frankfurt's regress-prone appeal to the notion of 'higher-order' desires, that a charge of reductionist tendencies might always await them just around the next corner. However that charge will lack all force or all power to provoke such defensive responses once the point gets through that naturalism and rationalism (or causal and reason-based modes of understanding) are by no means inherently at odds but should rather be seen as inseparably linked, at least in any plausible theory of knowledge or credible account of human social interaction. There would be more hope of significant advance in epistemology, ethics and (especially) philosophy of mind if that fact were impressed more firmly on philosophers still prone to various kinds of residual Cartesian or Kantian thinking.

'Postmodernism' is one name – albeit a name that is ill-defined and capacious to the point of near inanity – for an area of discourse (the phrase seems suitably vague) where that legacy has exerted a strong negative influence. For if postmodernist thinkers may be said to share any dominant preoccupation or distinctive angle of philosophic vision it is their resolute rejection of anything even remotely connected with the Cartesian quest for epistemic certitude grounded in the pure self-evidence of the *cogito*, or with Kant's attempt to recall that quest from its errant metaphysical path by a process of argument that aims to reveal the a priori conditions of possibility – the jointly enabling and limiting conditions – for human knowledge and experience in general. Here again, there is a strong suspicion that many of those thinkers have been driven to espouse a highly sceptical view of the prospects for achieving a workable conception of reason, knowledge and truth (or one that would be proof against familiar forms of sceptical attack) through their tendency to conflate two quite separate components of the rationalist approach to epistemology and philosophy of mind. That is, they presume that any such project must entail a commitment to that same old Cartesian-Kantian notion of the self-conscious subject-presumed-know as the last guarantee behind such values, or the sole means of securing them against an otherwise all-encompassing doubt. However this presumption amounts to nothing more than an inverted negative variant of

Descartes' original claim, i.e., a blanket rejection of that claim which takes rationalism – or the very capacity of human beings to exercise the power of truth-seeking rational thought – to stand or fall on the subject's foundational role pretty much as Descartes conceived it.

In Chapter 6 I shall argue that such attitudes have spread well beyond the confines of recent 'continental' (mainly French) thinking and are also to be found, if with various qualifications attached, in a good deal of present-day analytic – or 'post-analytic' – philosophy. What these thinkers share with the adepts of 'continental', i.e., predominantly French-influenced postmodernism is a clear sense of having come upon the scene at a time of intellectual or cultural exhaustion, at least as regards those main currents of thought that had hitherto been taken to define the character of serious, constructive philosophical endeavour. Postmodernists typically take their stand on a wholesale rejection of 'enlightenment' values and beliefs, these latter just as typically conceived in sweeping style as components of an ideological monolith that extends all the way from Kant to the advent of ambitious or systematizing movements of thought such as phenomenology and structuralism.[72] So likewise 'post-analytic' philosophers take their cue from the perceived breakdown or terminal impasse of that project which started out with the ground-breaking claims of Frege and Russell, and which finally came to grief with the Quine's famous demolition-job in his essay 'Two Dogmas of Empiricism'.[73] I shall therefore devote a large part of the next chapter to chipping away at the old analytic/continental distinction and putting the case that both traditions – more aptly, both elective genealogies or optative lines of descent – have undergone a similar crisis of confidence and arrived at a similar range of proposals (with, as I see them, similar problems and failures of intellectual nerve) by way of response. This will lead to the conclusion, once again, that a strong and principled form of philosophic naturalism – as distinct from the weak, inertly physicalist and normatively underspecified doctrine espoused by thinkers like Quine – is the only approach with any realistic hope of making good that crucial deficit.

Chapter 6

Deconstruction Naturalized: Beyond the 'Linguistic Turn'

I

According to what is nowadays a widespread account – the majority view among philosophers of a broadly analytic persuasion – 'postmodernism' and 'deconstruction' are good-as synonymous names for one particularly virulent strain of that larger enterprise known as 'continental philosophy' or sometimes simply as 'theory', the latter term marking it as yet more suspect owing to its likely provenance in such disreputable quarters as literary or cultural studies. Moreover, this antipathy has helped to shape the elective self-image of analytic philosophy over the past half-century and more by acting as a foil against which to measure its own self-ascribed virtues of logical precision, conceptual rigor, and a proper concern with well-defined philosophic problems rather than (as the contrast typically goes) pseudo-problems that are mostly the product of confusion, sophistry, or rhetorical hyper-inflation.[1]

In fact, I shall argue, the 'two traditions' have a lot more in common than these partisans might wish to concede, at least since they each took some version (or versions) of the turn toward language or its various surrogates (discourse, paradigm, conceptual scheme, framework, structure, and so forth) that has been a preeminent feature of both movements of thought during that period.[2] All the same – one likely rejoinder – such a claim is now of dubious relevance in so far as the linguistic turn has given way to a marked resurgence of interest in metaphysical topics that were pretty much off the agenda so long as its influence prevailed.[3] This has been accompanied by a striking renaissance of metaphysical-realist positions, often with a strong bearing on issues in ontology and epistemology, that had likewise suffered a prolonged period of eclipse – or had been so radically redefined as to render them devoid of substantive philosophic content – while language took pride of place.[4] Indeed those developments have been pressed yet further with the advent of a thoroughly naturalized approach to epistemology, philosophy of mind, cognitive psychology and other such disciplines where the effect has been to put language firmly back where, on this view, it properly belongs.[5] For it then figures as one, albeit highly significant item among the sundry human practices, activities, or ways of knowing that a fully fledged neuroscience – allied to a number of adjacent scientific disciplines, e.g., genetics or molecular biology – will take as falling wholly within its descriptive-explanatory scope.

In this case, clearly, issues such as that concerning the 'analytic' vis-à-vis 'continental' variants of the linguistic turn – let alone that concerning the precise relationship between postmodernism and deconstruction – will seem to be of no serious scientific or indeed philosophic import compared with questions concerning (say) the function, location, and connectivity of various cognitive-linguistic areas in the cerebral cortex. However, I propose to challenge both of these assumptions and show that they both take rise from a common misapprehension. To begin with I shall put the case that deconstruction – at least in the form best represented by Derrida's earlier work – should be thought of not only as distinct from but as standing squarely at odds with postmodernism in certain crucial respects. Chief among them is its qualified rather than wholesale subscription to the language-first doctrine that postmodernism takes to a strictly unsustainable extreme but which also constitutes the single most prominent trait of philosophy as practiced until very recently by thinkers across a whole range of otherwise disparate schools or technical persuasions. Then – in the final section of this chapter – I shall turn the received view on its head by arguing that it is just in virtue of those same distinctive aspects that deconstruction as exemplified in certain of Derrida's texts should be thought of not only as compatible with but as standing squarely on the side of a naturalized approach to epistemology, cognitive psychology, and related disciplines.

To some extent this agrees with the argument of a thinker like Stephen Stich who started out as a central-state materialist or hard-line eliminativist, i.e., one bent upon extirpating the entire set of 'folk-psychological' notions – thought, belief, attitude, intention, feeling, mind-state, or indeed 'mind' itself – but has since declared in favour of a more moderate approach that involves 'deconstructing' (his own choice of word) the very terms of this debate and seeking to show how both the mentalist conception and its materialist adversary run up against strictly intractable problems.[6] However Stich's usage alludes to one aspect only of Derrida's thought, that is, its regular practice of taking some value-laden conceptual opposition and locating the stress-points or aporetic moments where it undergoes a sudden reversal of priorities. At these points the lesser, derivative, deviant, or 'supplementary' concept turns out, in effect, to constitute a problematical ground or condition of [im]possibility for that which should in principle be self-sufficient since customarily thought of as the dominant term and hence as standing in no need of such adventitious support.[7] What drops out of Stich's later 'deconstructive' approach to issues in philosophy of mind is the further dimension – or strictly inseparable pair of dimensions – on which I shall focus throughout this chapter. In short, it is the singular (and utterly un-postmodernist) combination in Derrida's work of an extreme attentiveness to matters of logical implication – none the less so for their always emerging through a process of minutely close-focused textual exegesis – with an insistence on the sheer materiality of writing that very readily transposes into the register of a cognitive psychology premised on likewise materialist or radically naturalized working principles.

So there is, I maintain, a basic falsehood about the commonly-held idea of deconstruction as really just another, albeit more 'philosophically' geared-up variant on the standard postmodernist theme of out-and-out linguistic, rhetorical, discursive, or textual constructivism combined with a perfect indifference to issues of extra-linguistic reality and truth. In what follows, I attempt to dispel that idea through a running comparison between the two movements of thought and then – in order to emphasize the point – a discussion of Derrida's early reading of Freud's *Project for a Scientific Psychology* in relation to current developments in cognitive neuroscience.[8] In this context I shall also re-state the case for attaching less philosophical weight to the so-called 'hard problem' of consciousness and for acknowledging that thought very often runs ahead of anything that might plausibly be ascribed to a conscious (let alone self-conscious or reflective) state of awareness. Since I have contended throughout this book that naturalism is a sine qua non for any present-day philosophy of mind that lays claim to serious attention, and since Derrida's work has been central to my own thinking for many years now, it is clearly germane to my purposes here that deconstruction should indeed prove capable of recruitment to what might well seem – at least on the received 'analytic' view – a wildly improbable or downright perverse interpretation. However that response is merely another showing of the same misinformed root prejudice, as I shall seek to demonstrate here.

II

Let me start out by offering a few fairly basic distinctions without which any address to the topic of 'postmodernism' would amount to no more than a prolonged floundering in the same chaos of ideas that has witnessed more than its share of such floundering since the label achieved cultural prominence in the early-to-mid 1980s. Firstly, there is a need to distinguish full-strength from qualified or out-and-out from other, more moderate varieties of postmodernist thought. For the latter rationality is not so much an option as a strictly inescapable requirement since they take the point – traditionally pressed against cultural relativists by philosophers from Plato and Aristotle down – that anyone who claims to renounce the norms of rational discourse is sure to be stuck for a coherent or non-self-contradictory answer when challenged to put up a reasoned case in defence of their position. These moderate postmodernist types might well have all sorts of deep reservations concerning the 'philosophic discourse of modernity' or the kinds and modalities of rationalist thought that took root in the heyday of the European Enlightenment and thereafter – on this telling – came to exercise a baleful influence on the course of intellectual, cultural, and political events. Still, they tend to take the view that this was a massive aberration of reason – an oppressive, top-down, heavily bureaucratized, purely instrumental and hence grossly distorted version of rational thought – which could only be resisted and turned back through a reasoned (most often immanent-dialectical) critique of its motives, presuppositions and ideological investments.[9]

For the former, that is, advocates of postmodernism in its full-strength doctrinal guise any such resort to a middle-ground stance is most definitely out of the question since it is a sure sign of the lingering adherence to outworn, oppressive or coercive societal norms. From this point of view the very idea that there might be 'rational' as opposed to 'irrational' modes of thought – or that reason might somehow be equipped, through its own self-critical exercise, to reveal those various sources of error – shows up as just another product of that same epistemic will-to-power whose workings have been the regular target of antinomian sceptics from Nietzsche to Foucault and postmodernist gurus such as Lyotard and Baudrillard.[10] Full-strength postmodernism therefore constitutes an update on the old Protagorean relativist theme that 'man is the measure', albeit with the difference that Protagoras seems to have taken a transcultural or broadly uniformitarian view of human interests, values and concerns whereas the postmodernists typically reject what they see as the inbuilt cultural imperialism of any such view in favour of an outlook that insists on their sheer multiplicity and – in consequence of that – their incommensurable character.[11] Of course it has long been the standard refrain of those lined up on the opposite side, i.e., rationalists from Socrates down that this relativist claim must always self-deconstruct on an obvious (though presumably not to its proponents) self-contradiction. Thus there is something inherently absurd about the posture of offering arguments – or a show of rational justification – for the idea that all arguments should be treated as 'true', 'valid' or 'sound' only with the relativist rider attached: 'by the lights of this or that culture, community, life-form, belief-system, language-game, etc.'. Either the relativist/postmodernist claim goes through, in which case it is justified (persuasive) on its own terms but false for Socrates or anyone else who begs to differ, or else it fails to go through – indeed collapses at the first logical hurdle – on account of its being caught up in a strictly inescapable performative contradiction.

Of course the relativist may come back and say that persuasion, not logical rigor, is what they are aiming at and hence that they are simply not to be caught on the horns of this particular or any kindred (that is, logic-based) dilemma. Or again, they may say that the whole line of argument adopted by philosophers who have taken their cue from Socrates' supposed knock-down case against the sophists and other presumptive enemies of reason is one that banks on a narrow, overly formalized and hence context-insensitive idea of what should properly count as rational discourse. More specifically, they may fall back to the lower-strength or scaled-down version of postmodernism according to which it is not so much a question of announcing some radical break with the values of truth, reason, or other such pejoratively typecast 'Enlightenment' norms but rather of seeing that their rigid application – e.g., through insistence on the universal validity of classical or bivalent (true/false) logic – is unduly restrictive and hence a likely obstacle to other, more open or tolerant modes of thought. On this view 'postmodernist' becomes a descriptor that ranges far beyond the confines of recent, mainly French-influenced cultural and critical theory. It expands to take in a great variety of philosophic tendencies or movements of thought from the late-Wittgensteinian appeal to language-games or cultural life-forms as the ultimate horizon of truth, knowledge and intelligibility to Rortian neo-pragmatism, Quinean or Kuhnian paradigm-relativism, and the 'strong' programme in epistemology and sociology of science.[12] What these otherwise diverse projects all have in common is their contention that truth-claims are never straightforwardly cashable-out in objectivist, i.e., trans-culturally valid terms but are always in some degree context-relative or indexed to the current state of knowledge and belief in some particular epistemic community.

It is the first, full-strength version that analytic philosophers seem to have in mind when they dismiss postmodernism as just one of those curious aberrations to which 'continental' thinkers are unfortunately prone through their failure to take Socrates' point about the self-refuting or self-stultifying character of relativism when raised (as here) to a high point of radical doctrine. Otherwise the most frequent line of response is one that grants the limited remit of classical or bivalent logic – since, after all, in its formal guise it takes no stock of contextual factors and offers only a skeletal account of what constitutes a good, sound, cogent, or persuasive argument – but which views postmodernism (along with sundry other relativisms down through the ages) as merely the upshot of an over-reaction to that plain home truth.[13] On this view, postmodernists are all-or-nothing types who suppose that if formal logic turns out not to cover the field in certain well-understood respects then it must stand revealed as incapable of covering even a small part of it. Much better – so the diagnosis runs – to give up thinking in such drastically polarized terms and take the more sensible midway course of acknowledging that formal logic indeed has its place, albeit not the ultimately privileged place sometimes claimed, among the great range of discursive practices and justificatory procedures that make up our socio-cultural-communicative life-world. However there is ample warrant for concluding that arguments of this sort often lean right over into just the kind of relativist or postmodernist stance that they purport to show up as ill-founded and misconceived. This is notably the case when they are put forward by Wittgensteinian advocates of truth as internal to some given 'language-game' or communal 'form of life', or Rortian advocates of the 'cultural conversation of mankind' as our last, best hope for replacing truth-talk with talk of what's locally and contingently good in the way of belief.[14]

What these notions come down to is in fact something more like that low-strength or moderate version of postmodernism which stops well short of any frontal assault on truth, reason, and everything pertaining to the 'philosophic discourse of modernity' but none the less calls those values into question by treating them as strictly unintelligible outside or beyond the cultural-linguistic horizon shared by members of the relevant community. For on this account it is only within the perspective or the range of perspectives offered by that same horizon that they are able to make sense of the various sayings, doings, meanings, truth-claims, commitments, obligations, and so forth that constitute the very element of human thought, conduct, and action. It seems to me that the objections commonly brought against postmodernism by analytic philosophers committed to defending the values of reason and truth might just as aptly be brought against the sort of thinking – with its source in whatever localised breed of that protean beast, the 'linguistic turn' – that claims to have emerged at long last on the far side of all such epistemological or old-style analytic debates.[15] Not that the proponents of these various broadly 'post-analytic' modes of thought would for one moment endorse the claim that theirs is an approach that goes along readily enough with the kinds of position adopted by thinkers (mostly cultural or literary theorists) of a postmodernist persuasion. However, this would probably have more to do with the routine reluctance of thinkers brought up in the mainstream analytic tradition – however far they might have moved 'beyond' it – to acknowledge any kinship with fashionable trends in continental philosophy or, more emphatically, that particular variant that goes under the name of 'postmodernism' and is taken to constitute an affront to every last analytic virtue. Still, it is the case that this line of descent has produced some late-blooming yet still recognizably germane developments – those that can roughly be grouped under the rubric 'post-analytic' – which show a marked convergence with postmodernist ideas about the obsolescence of objectivist truth-talk and the need for a scaling-down of such talk to the measure of localized, contingent, or culture-specific modes of belief.[16]

Of course there would be nothing in the least objectionable about this convergence of aims from the viewpoint of a thoroughgoing pragmatist-relativist like Rorty, one for whom 'truth' equates without remainder to what's 'good in the way of belief' and reason cashes out wholly in terms of what counts as such by the lights of this or that cultural-linguistic community.[17] However, it would not go down anything like so well with those for whom there is still an important distinction to be drawn between knowledge and best opinion, or best opinion and majority belief, or that which holds as a matter of truth and that which holds as a matter of de facto (even if community-wide) consensus.[18] These thinkers – or some of them – hope to stake out a sensible middle-ground position that would allow them to avoid the twin perils of postmodern scepticism or out-and-out cultural relativism, on the one hand, and full-scale objectivist realism about truth on the other. This latter strikes them as most definitely not the right way to go since it entails the necessity that truth might always elude or transcend the utmost scope of present-best or even best-attainable human knowledge. In which case – so they argue – realism or objectivism about truth is the surest route to a sceptical upshot whereby truth is conceived as recognition-transcendent and hence, ex hypothesi, unknowable for beings (such as ourselves) whose cognitive capacities are subject to certain creaturely limits. Only by adopting the alternative, i.e., anti-realist or non-objectivist conception of truth – more aptly, of 'truth' redefined in terms of epistemic warrant or justification on the best evidence to hand – can thought escape this otherwise inevitable collapse into an outlook of ultimately self-defeating (since scepticism-inducing) objectivism.

Yet it is far from clear how this fallback position might be occupied without in the process giving up all claim to find a place for truth-values that transcend the currency of present-best or best-attainable belief. Once that possibility is renounced or treated (by analytic anti-realists and postmodernists alike) as a mere delusion or piece of metaphysical extravagance then the way is wide open to those various relativist doctrines that have lately captured large tracts of the high philosophical as well as cultural-theoretical ground. In the end it is no great distance – despite the claims to contrary effect by proponents of a 'moderate' anti-realist stance – from the idea that truth must somehow be indexed to the scope and limits of human comprehension to the idea that any notion of truth is just a product of the lingering (mainly philosophical) attachment to old and presumptively long since discredited modes of thought. The same applies to those sundry variants of the 'linguistic turn' across just about every discipline or branch of recent philosophical enquiry that likewise very often claim to point a way beyond all the vexing antinomies and unresolved aporias of earlier epistemo-metaphysical thought. Thus the idea is that with this widespread acceptance of language as the sine qua non of human understanding in general – as that which alone makes it possible for sapient creatures to think, reason, reflect, criticize, interpret, and (prerequisite to all such activities) communicate with others – there also comes the saving recognition that scepticism or flat-out relativism are simply not genuine options since language itself provides sufficient guarantee of our shared capacity for achieving cross-cultural, trans-paradigm, or inter-lingual communication.

This argument takes various forms, from those that provide little more than vague assurance (such as Wittgenstein's notions of 'family resemblance', language-games, or cultural life-forms) to Davidson's scaled-down version of a transcendental argument from the conditions of possibility for understanding in general.[19] Then there is Quine's radically holistic or context-relativist approach, one that sets out to demolish the distinction between 'truths of reason' and 'matters of fact' adopted in various guises by thinkers from Hume via Kant to the logical empiricists, but which offers in its place no anchor-points for reference and truth in the total 'fabric' of interrelated beliefs save those of pragmatic convenience plus a crude or grossly under-theorized appeal to behaviourist (i.e., stimulus-response) psychology.[20] Most talked-about at present is Robert Brandom's more elaborate proposal for an inferentialist-semantic approach that attempts to avoid all the well-known problems concerning realist epistemologies and theories of reference by looking to language as the social medium whereby we keep tabs on our own and other people's epistemic commitments and obligations.[21] Clearly this idea will possess a strong attraction for anyone who, like Brandom, leans toward a pragmatist conception of truth as what is good in the way of belief yet who thinks that if truth is to be thus redefined we had better supply the terms 'good' and 'belief' with more in the way of veridical content – or stronger epistemic credentials – than allowed by pragmatists like William James, let alone those like Rorty for whom pragmatism is really just a label of convenience for downright cultural relativism. That attraction has to do chiefly with its promising to hold truth very much in the picture – through Brandom's notion of discursive commitment and deontic-epistemic 'scorekeeping' – without what is seen as the deeply problematical appeal to a notion of reference that raises all the old realist/anti-realist issues about truth as objective and recognition-transcendent or truth as a matter of assertoric warrant and hence as epistemically constrained.

However, this approach brings its own share of problems, among them the fact that it still leaves a gap (and one that would appear unbridgeable on any purely semantic or discourse-based account) between truth *sans phrase* and what is held true by the standards of rational consensus, best judgment, or optimized epistemic warrant. To be sure, Brandom goes various elaborate ways around to mark his distance from a Jamesian, Rortian or postmodernist idea of truth as just the compliment we or other people routinely pay to those particular items of belief that happen to accord with our or their currently favoured values and priorities. Thus he makes a major point of distinguishing *de dicto* and *de re* ascriptions of epistemic warrant, the former denoting claims attributed in accordance with the subject's express (acknowledged or self-avowed) commitments, while the latter denotes what the listener – i.e., the epistemic/deontic score-keeper – takes to be the subject's actual, perhaps unwitting or as-yet unrecognized further commitments as a matter of consistency, logical entailment, or rational inference. Indeed Brandom goes out of his way to strengthen the referential, truth-functional and real-world-involving aspects of his semantically-based approach so as not to leave that approach wide open to the range of charges customarily levelled by realists or objectivists of various persuasion from Plato down. Still his programme places tight limits on what he is able to do in that direction since it counts valid inference as strictly a matter of semantic or intralinguistic entailment – of that which lays claim to truth through the differential warrant of speech-acts issued by parties to some given topic of debate – rather than a question as to whether or not some particular speech-act is justified or borne out by the way things stand in reality. This is why critics like Jerry Fodor can very plausibly put the case that Brandom's theory, despite its highly sophisticated character, simply misses the point when it comes to explaining how thought can hook up with certain aspects or features of reality via sentences that are capable of truth or falsehood precisely in virtue of their logical structure and their referential content or lack of it.[22]

Of course it is a main plank in Brandom's argument that we can have everything that's worth having in that regard – criteria of truth that pass muster for all practical as well as sensibly adjusted philosophic purposes – if we will just let go of the objectivist requirement that truth be recognition- or verification-transcendent and content ourselves with truth by the best standards of rational inference from speech-acts, subject to assessment by *de re* rather than purely *de dicto* criteria. Yet this still lies open to the realist rejoinder that it fails to capture the crucial difference between on the one hand statements, propositions, theories or hypotheses that are *taken as* or *believed to be* true and, on the other, those whose truth-value holds objectively despite whatever we or others may think or believe concerning them. Moreover the objection goes through despite Brandom's best efforts to deflect such criticism by beefing up the requisite standards of assertoric warrant as compared with any straightforward pragmatist account, that is to say, by distinguishing the score-keeper's tally from that of the informant (who may not have a fully perspicuous grasp of his or her own epistemic commitments) and using this distinction to motivate the contrast between *de re* and *de dicto* modes of assessment. For unless we are to take the purely thought-experimental route proposed by some thinkers, such as Donald Davidson, and postulate an 'omniscient interpreter' from whose epistemically exalted standpoint belief or holding-true are equivalent to truth *simpliciter* – a stance that Brandom, for obvious reasons, doesn't wish to adopt – we shall have to conclude that even the best (most expert, well-informed, rational, and lucid) score-keeper may sometimes get it wrong. Thus, no matter how scrupulous and far-sighted they will always at some point fall short of the mark and hence fail to achieve that ideal condition – that quasi-divine state of omniscience – wherein their beliefs would all be true and their reasoning upon those beliefs infallibly valid so that their epistemic commitments were, in the jargon, closed under logical implication. For of course we should then be in a realm altogether outside and beyond human possibility and involved not so much (as Brandom is) with issues of knowledge and communication but rather with the kinds of metaphysical debate that Plato addressed in his dialogue *Euthyphro*.

There it was a question of whether certain human acts were pious or impious because the gods deemed them so, i.e., because the gods had total jurisdiction over such matters and thus fixed the meaning or applicability of terms like 'pious' and 'impious', or – on the rival objectivist account – because the gods were infallibly truth-tracking and were therefore constrained to judge rightly since they lacked any sovereign or legislative power in such matters.[23] The Plato analogy is all the more relevant in this context since is one that has received a good deal of attention in recent years from thinkers, like Crispin Wright, for whom it has served as a handy device for setting up the realism/anti-realism issue along with that range of intermediate positions – among them variants on the Lockean response-dependence theme and Wright's own ideas of 'super-assertibility' and 'cognitive command' – that would claim to represent a viable alternative by staking out some specified region of the middle ground.[24] However, as I have said, those proposals always lean over in one or the other direction, that is, toward a realist/objectivist outlook that dares not quite speak its name or a relativist, constructivist, or anti-realist outlook which likewise seems anxious to avoid any too overt or explicit a statement of its own commitments in that regard. Thus the *Euthyphro* contrast mostly works out as a debate between shamefaced realists for whom 'best judgment' or 'optimized epistemic warrant' are conceived as infallibly truth-tracking and hence serve as just a roundabout means of avoiding the appeal to truth *tout court*, and those whose desire not to go too far along the relativist/anti-realist road is in the end outweighed by their failure to see how truth could possibly be conceived as epistemically unconstrained or recognition-transcendent. Either way it is clear that this debate is carried on between parties who, despite their specific points of difference, are playing very much by a common set of rules and a shared idea of what properly counts as worthwhile philosophic discourse, namely that which acknowledges the linguistic turn as a fait accompli and as setting the parameters for all such debate.

III

It is against this backdrop of a broad consensus on the priority of language or language-related issues – often taken as marking the arrival of philosophy at a stage where at last it could lay claim to constructive, progressive, and problem-solving power – that we can best view Brandom's inferentialist-semantic approach. What then becomes clear is that it occupies a position well out toward the rationalist end of a scale that runs all the way from full-strength versions of the turn, like Rorty's, according to which it is merely nonsensical to suppose that there might exist truths beyond the compass of presently attainable linguistic or discursive expression to the opposite pole at which thinkers such as Jürgen Habermas strive to explain how one can grant the priority of language while none the less preserving a robust sense of the distinction between truth and holding-true.[25]

However that scale stops crucially short of the point at which the language-first doctrine would give way to a realist or objectivist acceptance that truth might indeed surpass not only our best means of verbal articulation but also our best, most advanced or sophisticated powers of conceptual grasp. That is to say, there are certain well-defined limits to the scope for negotiation once thinkers have embraced the linguistic turn – in however strong or qualified and nuanced a form – as a kind of ultimate horizon that we cannot intelligibly claim to transcend, criticize, or call into question without thereby lapsing into nonsense or committing some logical absurdity. This type of response is something like a stock-in-trade among thinkers committed to one or another version of the turn, whether Wittgensteinians denouncing the presumption of rationalist opponents or Dummettian anti-realists routinely demanding how anyone could possibly be in a position to assert the existence of unknown (or to them unknowable) truths.[26] It is to be found issuing even from those, like Habermas, who seek above all to uphold and develop the critical discourse of modern post-enlightenment thought while none less acknowledging its need to undergo translation into terms more amenable to treatment in language-based, i.e., pragmatic and speech-act oriented terms.

Thus when Brandom and Habermas engage in debate over their respective ideas of how language enables the exercise of critical reason – for Brandom through an inferentialist theory of epistemic score-keeping, and for Habermas by way of the 'ideal speech-situation' and its normative or critical-evaluative bearing on our various modes of communicative discourse – they may disagree on detailed points of priority and emphasis but not as concerns that central precept.[27] What they share is the basic assumption that philosophy cannot get an adequate sense of its own operative scope and limits – and is therefore liable to fall back into all kinds of hitherto prevalent error and delusion – unless it acknowledges the role of language as our sole means of ensuring mutual intelligibility or communicative grasp across otherwise potentially unbridgeable divergences of sense, presupposition, or conceptual scheme. Moreover, both thinkers take it as pretty much axiomatic that the single greatest advance of modern philosophy was the decisive turn against any version of that old 'way of ideas' that had persisted in one form or another all the way from Descartes, Locke, Hume and Kant to its residual legatees among the logical positivists and logical empiricists. Such thinking could only be viewed as a massively protracted dead-end since it offered no means of rescuing the solitary mind from its state of self-locked private communion and leading it back to a public sphere wherein it might be subject to assessment on inter-subjectively valid or at any rate openly argued and accountable terms. That legacy had now at long last been left behind – so the story ran – with the ground-shifting interventions of (according to philosophic taste) Frege, Russell, Wittgenstein, Quine, Austin, Heidegger, Gadamer, Derrida, Rorty, or others who had pursued the linguistic turn to various, more or less radical conclusions. In this regard the issue between Habermas and Brandom, though crucial from their own respective points of view, is such as to leave them still sharing a large amount of common ground at the critical-reconstructive rather than the pragmatist or linguistic-communitarian (i.e., Wittgensteinian) end of the scale. That is, they still hold out for the value – indeed, the ethical imperative – of subjecting our beliefs, language-games or life-forms to a process of continuous assessment and critique that involves the capacity of thought to detach itself from those in-place societal or cultural norms that decide what is right, valid or correct by some given set of cultural lights.

However there is another sense, more germane to the present context of argument, in which Habermas and Brandon can both be seen as endorsing the turn toward a communal and language-based conception of those validating norms. This applies not only in the ethical and socio-political spheres where that approach – quintessentially Wittgensteinian – has at least a certain prima facie plausibility but also with respect to epistemological and even natural-scientific issues where it consorts most readily with cultural-relativist, linguistic-constructivist, and 'strong'-sociological modes of thought. As I have said, both thinkers have their own good reasons for resisting the currently widespread appeal of these and suchlike communitarian ideas whose effect, if carried through as a matter of principle and practice, would be to place a firm veto on independent thinking of any kind. They would foreclose the possibility of ever achieving that critical distance – that measure of detachment from currently prevailing doxastic norms – that is clearly required for the exercise of autonomous thought and which to that extent must be an inbuilt requirement of Brandom's inferentialist program with its stress on the critical-evaluative role of deontic score-keeping and likewise of Habermas's program for refurbishing (i.e., for bringing socio-politically as well as philosophically up to date) the 'unfinished project of modernity'.[28] So it is unsurprising that both thinkers make a regular point, although Brandom less forcefully, of staking their distance from a wholesale pragmatism in the Rortian mode as distinct from (so to speak) a retail pragmatism that derives some of its component parts from other, more critically-oriented sources.

Yet there is another angle from which those projects can be seen as further cases of the communitarian turn toward language as marking out the very horizon of intelligibility not only for those engaged in various kinds of ethical, social, political, scientific, or everyday exchange but also for that lingering Cartesian ghost, the solitary thinker, whose cogitations (so it is said) could make no sense even to herself outside or beyond that same horizon. Thus Brandom and Habermas both take for granted that there is no going back to the old 'way of ideas' whose empiricist and rationalist lines of descent had their joint culmination – along with their come-uppance – in Kant's critical philosophy, a project whose benefits can best be enjoyed (while avoiding its worst liabilities) by transposing its claims into an alternative linguistic-discursive register. Yet it is doubtful that this turn can be taken as far as they wish to take it without, in the process, retreating on so many fronts or yielding so many vital hostages to postmodern-pragmatist-communitarian fortune that the project seems doomed to failure. In each case the critical tribunal is subject to such drastic redefinition of its powers and remit – that is, such a marked switch of emphasis from rational autonomy to communally warranted and sanctioned modes of thought – that there is little room left for that exercise of independent judgment that is still supposed to play a significant role in the work of both thinkers. For it turns out to rest on very shaky foundation (even to be built on sand) if one accords full weight to their claims concerning the ubiquity of language as a means of social adjudication and the sole medium wherein, or whereby, such issues can achieve articulate or even intelligible form. 'Why this is postmodernism, nor are we out of it' as Marlowe's Mephistopheles might have quipped had he not had more urgent worries in mind.

Of course there is a basic truth about claims like this if one takes them as asserting nothing more than the obvious and trivial fact that ideas, beliefs, concepts, propositions, arguments, counter-arguments, and so forth cannot be subject to debate in the public sphere – cannot get a hearing and hence be worked out, developed, defended, or refined – unless and until they reach the stage of adequate linguistic formulation. So much one may readily concede without thereby being driven to endorse the further claim that is common currency among Wittgensteinians, Heideggerians, Dummettian anti-realists, neo-pragmatists, post-structuralists, 'strong' descriptivists, discourse-constructivists, and – as a veritable trademark feature – postmodernists of otherwise quite diverse philosophic and socio-political persuasion. This is the idea that thought depends on language not only in the sense just described, i.e., as a more-or-less effective and well-adapted means of communication but also in the more radical sense that it constitutes the very element of thought or – to repeat the well-worn metaphor that seems almost unavoidable in this context – the ultimate horizon within which we are constrained to think and reason. For beyond it there lies a linguistic-conceptual terra incognita whereof we can have no communicable knowledge and hence, so it is held, quite simply no knowledge at all. Indeed this line of argument is often pressed even harder to the point where, in Dummett's metaphysical and logico-semantic version of the case, any 'gaps in our knowledge' concerning (say) the record of historical events are also and quite literally 'gaps in reality'.[29] On this anti-realist conception of the matter – a conception halfway toward the kinds of far-gone linguistic-constructivist or cultural-relativist doctrine put about by the adepts of postmodernism – 'reality' *just is* coextensive with or restricted to the domain of known, verifiable, reliably vouchsafed, well-documented, adequately sourced, or at any rate sufficiently agreed-upon historical facts. Even if postmodernists would find no room for this latter range of epistemic terms and distinctions they could seize upon the basic anti-realist premise (the non-existence of objective or recognition-transcendent truths) as grist to their mill and all that is needed in order to advance more extreme or less qualified versions of the case.

So much is taken to follow necessarily from Dummett's three main precepts, i.e., that knowledge must be (1) *manifested* through a publicly demonstrable (i.e., linguistically articulated) grasp of the probative methods or criteria involved, (2) capable of shared *recognition* through some likewise verbally manifestable process of communal assessment and (3) able to meet the condition of having been *acquired* – whether by empirical, rational-deductive, historical-documentary, or other well-proven means – in such a way as once again to satisfy the standards of the relevant (whether specialist or broad-based cultural) community. Given these basic desiderata – in Dummettian shorthand the *manifestation, recognition,* and *acquisition* principles – we are obliged to conclude, so the argument goes, that truth cannot possibly transcend the scope and limits of best judgment or optimized epistemic warrant. In which case the anti-realist claim would extend to every branch or discipline of human enquiry, from history to geography, physics, biology, and also – very much Dummett's home-ground – mathematics, logic and the formal sciences.[30] For here it meshes perfectly with the intuitionist precept that any advance achieved in these fields is not so much a matter (as the realist or objectivist would have it) of discovering truths that were already 'there' awaiting discovery but rather one of inventing or creating new ways of thought which in turn open up new possibilities for future investigation. Thus Dummett's favoured type of imagery for describing this process tends to be drawn more often from the arts, especially the visual arts, and not from those metaphors of exploration – of striking out into uncharted terrain or strange seas of thought – which naturally appeal to those of a realist (that is, in this context, a Platonist) persuasion.

Again, it may be seen how close is the resemblance between anti-realism in this 'analytic' mode and the kinds of thinking that analytic philosophers are apt to lump together, most often for polemical or routinely dismissive purposes, under the title of 'postmodernism'. Indeed, one could draw up a detailed typology of post-1960 developments within the broadly analytic camp that would take in a whole range of 'weak'-to-'strong' positions on a scale that runs from slightly fretful realists anxious to defend their stance by making certain seemingly minor (though often in fact highly damaging) concessions, via defenders of a middle-ground approach like those that I have described and found wanting above, to fully fledged anti-realist arguments or out-and-out descriptivist/constructivist programs in the Rortian or Goodmanian 'ways of worldmaking' mode.[31] What these all have in common – and what throws them, willingly or not, into close contact with all those typecast 'postmodernist' thinkers – is the fact of their having taken the linguistic-discursive turn and moreover having carried it beyond that crucial line between versions of the thesis that amount to no more than a self-evident or trivial truth and versions that leave no room for truth in any meaningful sense of the term. That is to say, they are united in supposing that philosophy achieved an epochal breakthrough with the realization that language could and should be the basis of all constructive and genuinely problem-solving philosophical enquiry. This gave it a decisive advantage over those various other failed candidates – 'ideas', 'concepts', 'categories', 'intuitions', 'sense-data', and so forth – that had banked on the special privilege enjoyed by first-person epistemic access and had hence fallen prey to the objection (most influentially voiced by Wittgenstein) that their private character was such as to place them beyond the communal pale.[32] However, as I have said, that argument very easily leans over into the notion that language constitutes not only a more-or-less adequate medium or means whereby to articulate thoughts, concepts, propositions, etc., but also – a very different (even contradictory) claim – their very element or condition of possibility. From here the way is wide open to those sundry forms of cultural, conceptual, and (in Quine's case) ontological relativism that assume the priority of language over thought and of thought in its diverse, linguistically articulated modes over anything pertaining to the nature or structure of mind-independent reality.

Such is the line of argument adopted at its most logico-semantically refined but also its most metaphysically extravagant by those anti-realists who profess, like Dummett, a frank inability to comprehend how their realist or objectivist opponents could possibly conceive the existence of epistemically unconstrained, i.e., recognition-transcendent truths. Thus '[f]or the anti-realist', on Dummett's account, 'an understanding of [some given] statement consists in knowing what counts as adequate evidence for the assertion of the statement, and the truth of the statement can consist only in the existence of such evidence'. From which it follows necessarily that '[t]he notion of truth, when it is introduced, must be explained, in some manner, in terms of our capacity to recognize statements as true, and not in terms of a condition which transcends human capacities'.[33] And again, in a passage remarkable as much for the frankness with which it appeals to certain differences of personal-intellectual temperament as for the plain-speaking extremity of its philosophic view:

> [r]ealism about the past entails that there are numerous true propositions forever in principle unknowable. The effects of a past event may simply dissipate . . . . To the realist, this is just part of the human condition; the anti-realist feels unknowability in principle to be simply intolerable and prefers to view our evidence for and memory of the past as constitutive of it. For him, there cannot be a past fact no evidence of which exists to be discovered, because it is the existence of such evidence that would make it a fact, if it were one.[34]

This passage is all the more striking for the way that it falls back on a statement of undisguised personal preference, namely the anti-realist's feeling that unknowability in principle is a notion quite simply not to be borne, as against the realist's readiness to accept that certain truths will and must – for whatever reason – lie beyond the best efforts of human proof, knowledge, recovery, or ascertainment. It is as if that choice can be made only on the basis of a certain temperamental leaning since in the end it comes down to which prospect one finds more acceptable (or less disturbing), that of our statements having truth-values whose very objectivity must be thought to place them potentially beyond our epistemic reach or that of giving up any thought of objective truth, replacing it with standards of warranted assertibility, and thereby preventing any such worrisome gap from opening up in the first place.

It is plain enough from the statement just cited and others like it that Dummett inclines very firmly in the latter direction, despite his occasional claims to the effect that his is not so much a programmatic defence of anti-realism as a research-strategy designed to test it in various contexts or 'areas of discourse' and see whether it or a realist approach best suits our commonly-accepted intuitions. What this strategy most often involves is the trick of confusing two quite different but easily conflated issues, namely (1) whether we can have knowledge of truths that surpass our utmost knowledge, which self-evidently we cannot, and (2) whether we are able to conceive the existence of such unknown or even unknowable truths, quite aside from any issue with regard to this or that particular statement, hypothesis or theory and our state of knowledge regarding it (which realists say we self-evidently can while anti-realists take the opposite view). This may seem to lead us deep into the mist-laden bogs of

Donald Rumsfeld country where talk of 'things we know', 'things we don't know', 'things we don't even know we don't know' (and so forth) is apt to strike many people – perhaps with Rumsfeld's performances in mind – as just an exercise in pulling the wool over eyes too easily impressed by such semantic chicanery.[35] However, the realist/anti-realist debate does have important bearings beyond the 'purely' philosophical realm, not least with respect to precisely those issues of objective truth in relation to the scope and limits of attainable knowledge that constantly arise for historians, investigative journalists, and others seeking to sift the evidence and sort veridical from false or downright mendacious factual-historical claims.

To say that Rumsfeld's infamous utterance was more economical with the truth than with the number of words deployed is also to say that it raised verbal legerdemain to a high point of bafflement and deliberate mystification. I would not go so far in describing Dummettian anti-realism, although I think that this description would be close to the mark if applied to some of the ideas put forward by postmodernists – Baudrillard conspicuous among them – who press their own version of the anti-realist case to a terminal point where 'reality' becomes just the last forlorn refuge of all those hopelessly naïve and passé seekers-after-truth.[36] It is not hard to show how and where such arguments go wrong, as for instance by confusing epistemological with ontological questions and presuming on that basis to announce all manner of epochal shifts like the supersession of 'the real' by the simulacrum or the advent of a postmodern dispensation whereby any notion of truth is revealed as the merest of superannuated Platonist illusions. Nowhere can one find Dummett going in for such sweeping and (to say the least) philosophically dubious or under-argued theses, and nowhere does he manifest anything like the breezily offhand ultra-sceptical posture that Baudrillard constantly adopts. After all, Dummett's chief claim for anti-realism in the form given to it by his own work over the past four decades is that now, through the kinds of advance made possible by a selective application of Frege and Wittgenstein to issues in philosophy of logic and language, we are at last able to formulate terms for the realist/anti-realist dispute that will make it amenable to systematic treatment in an area-specific, case-by-case manner.

It is also worth noting Dummett's awareness – one not shared by most postmodernists – of the ethico-political problems involved in any full-scale endorsement of anti-realism with regard to historical events since in that highly sensitive context it can readily be recruited to the ends of some ideologically motivated revisionist programme.[37] Besides, he may well have certain positive reasons for thinking that an anti-realist approach to issues in the field of electoral politics is preferable, on grounds of democratic accountability, to a realist approach that would typically define 'representation' as a matter of achieving the optimal correspondence between that which does the representing (parliaments, governments, elected assemblies of various kinds) and that which finds itself represented (the electorate, here taken as a pre-existent or pre-constituted body of interests, opinions, voting intentions, social priorities, political commitments, and so forth). What Dummett is out to challenge – ethically as well as epistemologically – is the rightness of this latter presupposition, i.e., the defining realist axiom according to which the requirements of truth or justice always involve some such relationship of achieved correspondence with that which must in any case be thought to exist prior to, apart from, and irrespective of the particular mode or manner of its representing. The French word *justesse* captures these two senses ('truth = accuracy or truth-to-reality' and 'truth = conformity to the principles of just conduct, practice, or procedure') far better – with greater *justesse* – than any available English term.

The following passage from his book *Voting Procedures* brings out the connection clearly enough:

> Proportional representation has the object of making Parliament more representative: rival systems are to be judged, in the first instance, by how effectively they achieve this. It ought, then, to be the first task of any discussion of electoral reform to make precise what is meant by calling a parliament representative; yet the question is virtually never posed, Rather, it is taken for granted that everyone knows what 'representative' means; the assumption appears to be that a parliament is representative if its composition, by political party, corresponds closely to the composition of the electorate . . . . This assumption embodies a possible definition of the word 'representative', as used in this connection; it is far from obvious that it is the right definition to give.[38]

Anat Matar, who cites this passage in his book *Modernism and the Language of Philosophy*, makes the point that Dummett is here conceiving the political issue very much as he conceives the realism/anti-realism issue in philosophy of language and epistemology.[39] That is, he thinks of it as falling out principally between those (realists) who would regard truth or justice as always potentially transcending our best-present powers of recognition or achievement and those (anti-realists) who cannot make sense of that idea since, for them, truth – like justice – is a working concept that has no possible application beyond what lies within our scope of knowledge or capacities of moral grasp. One can well see why Dummett, given his strong commitment to the latter view, should think it socio-politically as well as epistemically advantageous that the term 'representative' should be so redefined as to bring it back into the space of human knowledge-seeking and justice-optimizing practices rather than cast adrift in some realm of objective, recognition-transcendent truth where it is apt to become wholly out-of-touch with any such sensibly scaled-down conception of the cognitive-communal good. Thus, as Matar puts it, Dummett's 'principal aim . . . is to prove that the reality that is to be represented is not given in advance of the voting procedure itself; that the "majority preference" does not have a "correspondent" definite composition prior to the procedure'.[40]

However the problem with this, once again, is that by tying justice so closely (or inseparably) to our best idea of what constitutes a just socio-political order it blocks the thought that existing arrangements *and also that our current-best existing idea* might fall drastically short of whatever 'representative' does or should entail as a matter of voting procedures and their adequate implementation. Indeed it seems to follow from Dummett's position that a term such as 'adequate' – at least on any strong (one is tempted to say 'adequate') understanding – has to drop out of for lack of any genuine critical-evaluative purchase. For it has no useful work to do once we take this turn toward the notion that truth is best redefined in terms of 'warranted assertibility', or whatever possesses a claim to that title according to whatever presently counts as expert opinion, optimized judgment, or plain best belief. In that case there would be little point in pressing, like Dummett, for a substantive, maybe radical and counter-consensual overhaul of current voting procedures or existing conceptions of the democratic good, any more than scientists would have any reason – pig-headedness aside – for pursuing research that went strongly against the dominant (professionally sanctioned) consensus view. More likely, his anti-realist approach would work out, if applied in this context, as a willingness to stay within the broad limits of that which would strike the majority of qualified (i.e., respected or communally esteemed) persons as a decent, just, and achievable set of aims.

IV

So – to repeat – I am not trying to sink the difference between anti-realism in the analytic mode where its claims of are of a broadly metaphysical character though grounded in certain highly formal, i.e., logico-semantic considerations and anti-realism of the postmodernist variety where any metaphysical assertions such as those to be found in Baudrillard's ideas about hyperreality are arrived at more through a process of free-associative musing on various cultural signs and symptoms than through anything like a logically constrained process of argumentation. All the same there are aspects of Dummettian anti-realism, especially when applied in the context of historical or factual-documentary discourse, that do run close (disturbingly close) to the wilder shores of postmodernist thinking. Thus he seems to intend that we should take it literally – not suspect the presence of metaphor, hyperbole, or over-statement for dramatic effect – when he asserts that any 'gaps in our knowledge' concerning historical events are also and for that very reason necessarily 'gaps in reality'. This follows straightforwardly enough from the conviction that Dummett attributes to his typecast anti-realist but which clearly reflects his own feelings on the matter, i.e., that 'unknowability in principle is simply intolerable', and therefore that one is left with no practical alternative except 'to view our evidence for and memory of the past as constitutive of it'.[41] For the realist, conversely, there may always be cases where we lack any direct, first-hand or conclusive warrant for affirming or denying some specific historical claim since – whether through accident of fortune, concealment, or wanton destruction – the vital evidence has now disappeared from the record of what actually, historically occurred. This is just what the anti-realist steadfastly refuses to accept, convinced as s/he is that talk of 'actual' occurrences or historical truth as opposed to falsehood simply cannot make sense – must perforce be counted a species of empty or pseudo-meaningful discourse – unless there is some evidence or validating protocol by which it can be backed up.

However, it must then be asked what considerations could stand in the way of a historical revisionist – perhaps with an ideological axe to grind – who argued from the circumstance of our possessing no decisive evidence for some hypothetical statement concerning this or that particular event to the conclusion that its truth or untruth as a matter of objective historical fact was simply not at issue since beyond our capacity to determine either way. After all, as the realist will be quick to point out, there exists a vast (indeed limitless) range of statements concerning the remote or even the comparatively recent past that are well-formed and truth-apt in the sense of specifying clearly what must have been the case in order for their truth-conditions to be met but which as it happens cannot be assigned a definite truth-value. Some concern events or circumstances that were merely too humdrum, trivial or inconsequential for anyone to have taken passing mental note of them let alone kept a written record, while others may be such that the issue of their having or not having occurred as purported by the unverifiable statement in question is a matter of the utmost ethical as well as factual-documentary significance. In other words, what we have here is just the kind of endemic limitation on our powers of perceptual, epistemic, recollective, reconstructive, and investigative scope that realists just accept as (in Dummett's words) 'a part of the human condition' in so far as 'the effects of a past event may simply dissipate', while anti-realists – himself included – find the thought of such unknowability downright 'intolerable'. However, so the realist may well respond, it is this latter (anti-realist) line of argument that should stretch our moral-intellectual tolerance to the limit and beyond since it is apt to provide much-needed succour to the hard-pressed revisionist seeking some handy philosophic back-up for the claim that there is really no choice but to view history as 'history of the present' or as a backward projection of our own ideological values and beliefs.

Still we should not think of this as a stand-off, stalemate or strictly unresolvable difference of views between parties who incline this way or that by reason of some deep-laid temperamental bias or cultural-intellectual world view. For just as postmodernism may be challenged on grounds of its confusing ontological and epistemological issues so anti-realism can be shown to involve a number of demonstrably false suppositions all of which concern that same basic failure to distinguish the two domains. Nowhere is this more evident than in Dummett's earlier-cited claim that 'there cannot be a past fact no evidence of which exists to be discovered, because it is the existence of such evidence that would make it a fact, if it were one'. To be sure, there is a sense – one much exploited by postmodernists, neo-pragmatists, post-structuralists, Wittgensteinians, and relativists of sundry persuasion – in which 'facts' are intrinsically linguistic or discursive in character. That is to say, they must be thought of not as somehow existing 'out there' in an objective, mind-and-language-independent realm but rather as products of human cultural activity which come into being only in and through the various practices (including that of verbal articulation) that make up their very condition of existence. However, like many claims in this area, one can take it either as a valid though trivial proposition (that facts must become known and be capable of adequate expression before they can possibly count as facts for members of any given knowledge-community) or else as a more extreme and dubious or at any rate contentious thesis (that facts are constituted through and through by the language, discourse, life-form, paradigm, conceptual scheme, etc., within which alone they make any kind of sense).

I have written elsewhere about the problems with anti-realism or, to speak plainly, the absurdities to which it gives rise if taken with less than the large pinch of salt that would render it amenable to a non-absurd (i.e. scientifically, historically and philosophically plausible) interpretation but would also, by the same token, remove the only intelligible motive for adopting such a viewpoint in the first place.[42] That motive was to defeat the sceptic by pre-emptively closing the troublesome gap – Kant's chief bequest to modern philosophy – between subject and object, mind and world, or truth and humanly attainable knowledge. However, this must seem a pyrrhic victory indeed if it means giving up the very idea that enquiry can sometimes take us from ignorance to knowledge and (a precondition for that) the idea that truth might always lie beyond our present-best attainable state of knowledge or rationally justified belief. It is the full-strength version of anti-realism – very often arrived at by way of the other, more moderate claim as if the two were strictly indivisible – that has characterized such a swathe of recent developments in philosophy and the human sciences. Moreover, as I have said, it is apt to crop up as a shared feature of programs or projects that would otherwise seem to have nothing in common and even to stand squarely opposed in terms of intellectual character, lineage, or cultural affiliation.

Thus, in Dummett's case, the anti-realist commitment stems from a certain confluence of philosophic sources – chiefly Frege and Wittgenstein – which between them produce the joint emphasis on language in its logico-semantic aspect as the key to resolving hitherto intractable disputes in metaphysics and language in its cultural-communal aspect as that which should resolve any problems of the sort created by attachment to misguided, e.g., realist or objectivist precepts. Among postmodernists the sources are likely to be different, most often giving prominence to Saussure as a systematic thinker whose concepts and categories can then be deconstructed through the sceptical rigors of a Nietzschean-Foucauldian genealogical critique.[43] However, these two lines of descent have one striking feature in common, namely their both having emerged through the encounter of a formal methodology wedded to certain (on its own terms) precise and rigorously defined concepts with a strongly reactive movement of thought which depends very heavily on those same concepts while avowedly casting them aside. Such, after all, is the famous ambiguity of the prefix 'post' as it figures in the lexicon of post-structuralism, postmodernism, and also – as I have suggested here – 'post-analytic' philosophy in so far as that label is nowadays applied to a range of ideas that would scarcely make sense if divorced from the context within which and against which they first took rise.

Indeed there is good reason to view this whole recent chapter of developments as a largely unwitting re-run (albeit in 'linguistified' terms) of the history of German idealism after Kant. That is to say, it looks very much like a suitably transposed or 'de-transcendentalized' repetition of the movement which led from Kant's ultra-systematized architectonic of the faculties to the subsequent reaction in favour of an outlook that stressed the historically shifting and culturally contingent character of human thought.[44] Along with this went an emphasis on the ubiquity of language as a formative or shaping factor in that process and the extent to which concepts such as truth and knowledge required a more context-sensitive treatment – an understanding more responsive to those shifts of cultural-historical perspective – than anything provided by Kant's universalist approach, that is to say, his transcendental mode of argument from the conditions of possibility for thought and judgment in general. One could press this analogy yet further and remark how some of the most prominent thinkers in the post-analytical line of descent – thinkers like Brandom and John McDowell – have quite expressly followed a path that led first back to Kant and then forward from Kant to Hegel, though not always (one suspects) with a full grasp of the problems that arose within that previous history of thought.[45] So it is scarcely surprising that those problems continue to vex this latest repeat episode, however scaled-down its metaphysical commitments through its having undergone the joint effects of the linguistic turn and the recent trend toward a strongly naturalized conception of epistemology and ethics.

This is not the place for a full-scale discussion of the way that such present-day 'rediscoveries' of the conceptual resources available from Kant and Hegel have tended not only to reinvent the wheel but to replicate some of the same old dilemmas. Among them is the Kantian veering-about between various unresolved dichotomies such as that between transcendental idealism and empirical realism, or again – at a more basic level – between concepts of understanding and sensuous intuitions. This latter is scarcely overcome by following McDowell's emollient advice that we simply switch terminologies and look instead to Kant's more promising talk of 'spontaneity' and 'receptivity' as faculties that are mutually inter-involved or 'only notionally separable' each from the other.[46] Moreover, as I have said, there are kindred problems – in this case a clearly-marked normative deficit as well as a lack of substantive epistemological content – with Brandom's more distinctly Hegelian take on the legacy of German idealism as one major source of his own proposal for an inferentialist-semantic account of how knowledge accrues through the various modes of social interaction and reciprocal communicative grasp. At any rate, there is good reason to think that these projects for moving beyond the linguistic 'turn' through a carefully monitored, i.e., semi-naturalized infusion of Kantian-Hegelian themes are thereby reverting – ironically enough – to just that stage in the pre-history of modern continental thought when philosophy set out on its course toward the various problems and dilemmas that are still being worked through by continental and analytic thinkers alike.

My point is that we will take an overly narrow view of this topic if we assume that postmodernism is a cultural phenomenon of exclusively continental (or mainland-European, chiefly French) provenance. In so far as there exists a 'postmodern condition' – or in so far as it has taken a distinctly philosophical as apart from a broadly socio-cultural form – this should rather be seen as having affected both main traditions of thought, not least through the well-nigh obligatory appeal to language, discourse or their various surrogates as the starting-point and end-point of all enquiry. One useful corrective to that standard view would be read some of Derrida's early essays where, contrary to presumed form, he takes a very strong and principled line against any version of the postmodernist (also the descriptivist, constructivist, neo-pragmatist, or cultural-linguistic-relativist) claim that language precedes thought or that our concepts and categories are contingent upon the particular semantic or syntactic resources supplied by some given natural language. In 'The Supplement of Copula' he focuses on Emile Benveniste's thesis to just this effect and shows it to self-deconstruct on the sheer impossibility of advancing any such claim without in the process deploying and imputing concepts and forms of categorial judgment which *cannot but* be thought of as pertaining to all languages.[47] For without them – so Derrida's argument runs – a language would lack even the most basic logico-syntactic resources (e.g., the copula or the 'is' of predication) that have to be a part of its conceptual apparatus in order for it to function or indeed properly to count as a language. What Derrida gives us here, although this will come as news (most probably unwelcome) to many disciples and detractors alike, is something much akin to Donald Davidson's case against various forms of linguistic and cultural relativism in his essay 'On the Very Idea of a Conceptual Scheme'.[48] In fact Derrida's is a stronger, more principled version of the case in so far as it rests not so much on a broadly pragmatic appeal to the conditions under which human beings from different cultures and speech-communities must be thought to make sense of each others' sayings and beliefs – the old Quinean problem (or pseudo-problem) of 'radical translation' – but rather on a full-scale transcendental argument from the conditions of possibility for thought and language.

Lest it be suspected that 'The Supplement of Copula' is an odd one out among Derrida's writings I would also instance his essay on Foucault – 'Cogito and the History of Madness' – which goes a similar way around in showing how

Foucault cannot consistently or logically maintain (that is, cannot assert without falling into a species of blatant performative contradiction) that he is writing a book about madness in the very language of madness and against all the protocols of rational discourse.[49] For of course, as Derrida is quick to remark, if Foucault had indeed – *per impossibile* – succeeded in writing such a text then the result would quite simply have made no sense at a basic or sentence-by-sentence level let alone with respect to its generic character as a long, immensely complex and elaborately argued discourse on the relationship between reason and madness at different stages in the history of European thought before and after the crucial moment of seventeenth-century rationalism. Thus Foucault's 'radical' gesture turns out in truth to be no such thing but rather something more like a 'Cartesian gesture for the twentieth century', a thought-experiment in hyperbolic doubt which cannot but confirm the security of reason within its own element. That is to say, it is obliged to argue its case and moreover to do so by deploying – what else? – the resources of full-stretch critical reason.

One could multiply examples from Derrida's work of this way that he has of entertaining various challenges to reason – or claims to have come out somehow on the far side of 'logocentric' discourse – only to turn that whole line of argument back against itself by showing how it must presuppose, incorporate and constantly avail itself of just those resources that it thinks to have consigned to the scrapheap of outworn ideas. In a different though closely related form the tactic reappears in his essay 'White Mythology' where Derrida's theme is the vexed relationship, from Plato and Aristotle down, between philosophy and literature, concept and metaphor, reason and rhetoric, dialectic and sophistics, and their various surrogate terms.[50] Here again, it is only on a superficial reading that this text can be taken straightforwardly to 'deconstruct' (i.e., to invert or undermine) the evaluative priority traditionally attached – by philosophers at least – to the first term in each of these pairs and to do so, like Nietzsche before him, through a mode of hard-pressed rhetorical analysis that brings out the sheer ubiquity of metaphor in the discourse of philosophy. To be sure, this is a part of Derrida's purpose but only as prelude to the more important business of showing how the various concepts and categories developed by philosopher-theorists of metaphor from Aristotle to Gaston Bachelard have provided the strictly indispensable means by which to define, distinguish, analyze, and explain its various modes of operation. Along with this has very often gone the attempt – albeit (so he argues) one that must inevitably fail at a certain point – to hold metaphor accountable to certain well-defined, philosophically articulated terms and conditions and thereby ensure its eventual restoration to the realm of literal meaning, reliable knowledge, and objective truth.

However, that failure is not of the kind that would betoken sloppy thinking or some mere incapacity to think these issues through in a more adequate or clear-sighted way. On the contrary, it emerges to all the more striking and instructive effect in the case of those thinkers to whom Derrida accords close attention in 'White Mythology' since they have approached this topic with the highest degree of analytical acuity and rigour and yet, despite that or for just that reason, run up against an ultimate limit on the extent to which philosophy is able to conceptualize the workings of metaphor or categorize its various types, structures, generic attributes, and so forth. On the one hand this effectively debars philosophy from claiming any privileged adjudicative role vis-à-vis the domain of figural language, not least when it comes to the scrutiny of its own texts which will then – under the kind of exegetical pressure first brought to bear by Nietzsche and carried to a yet higher point of refinement here – always turn out to involve some further range of co-implicated metaphors that eludes any adequate specification in conceptual-philosophical terms. Hence the widespread suspicion of figural language among many canonical thinkers in the mainstream European line of descent whether empiricists like Locke, rationalists like Leibniz, or one such as Kant who sought to achieve a working synthesis between them.[51]

What is evident in the texts of these thinkers – at least once brought to notice through a deconstructive reading – is the extent to which figural language may exceed or outrun their best efforts to contain it within clearly marked conceptual bounds. On the other hand, as Derrida says, there is no question of simply swinging across to the opposite (call it vulgar-deconstructionist) pole and declaring the concept/metaphor distinction to have been just another delusory figment of the long shadow cast by Western 'logocentric' metaphysics. For despite its Nietzschean focus on metaphoricity as the perpetual undoer of claims to conceptual mastery there is also a strongly countervailing line of argument in 'White Mythology' which reminds us that we should have no means of raising these issues – no language, idiom, technical register, or range of analytic resources – were it not for those strictly indispensable concepts and categories that postmodernists (and some deconstructionists) are prone to dismiss while perforce deploying them at every argumentative turn. Indeed there is no caveat that Derrida has entered more often or with greater emphasis when confronted with some instance of this way of thinking and the more-or-less flagrant performative contradiction – or logical conflict between what is claimed and what is presupposed in expressing or articulating that claim – to which it inevitably leads.

V

It is here that I would locate the main difference between deconstruction, properly understood, and postmodernism in so far as the latter is taken to possess a distinctively philosophic aspect as apart from its broader application to diverse, often very loosely related artistic, cultural, and socio-political developments. Briefly put: deconstruction, unlike postmodernism, both acknowledges and – in Derrida's work – exemplifies the claim of logic or reason to constitute a strictly unavoidable commitment or condition of possibility for any critical project that aims to challenge existing or dominant modes of thought. It is on this account firstly that I have sought to undermine the complacent belief among many Anglophone philosophers that theirs is an intellectual culture best defined by its allegiance to just those proprietary 'analytic' virtues, and hence by contrast with the 'continental' habit of plain disregard for logic, reason, and standards of disciplined enquiry. My second objective has been to point out those various developments within the analytic or post-analytic camp that complicate the picture yet further by embracing a range of sceptical, relativist, constructivist, conventionalist, and anti-realist doctrines that take on board all the main components of a fully fledged postmodernist outlook while exhibiting nothing like Derrida's meticulous procedures of argumentation. Nor is that aim in any way compromised – the erroneous assumption commonly made by analytic detractors like Searle – if the upshot of those procedures is most often to reveal complexities of sense that stretch thinking beyond the limits of classical (bivalent) logic.[52]

I should also here emphasize one aspect of Derrida's work that is routinely ignored by admirers and detractors alike but which has a strong bearing on the principal theme of this book, i.e., the need for a naturalized approach to issues in epistemology and philosophy of mind. It has to do with the claim of deconstruction to locate those complexities not in some purely 'textual' domain – some abstract realm of conflicting significations where reality can never obtrude – but rather in the various real-world contexts or fields of application where alone it can find an adequate point of critical-demonstrative purchase. I have argued this case at length elsewhere with particular reference to Derrida's writings about language, music, civil society, politics, aesthetics, and history.[53] More precisely: I have attempted to draw out the various deviant logics that he finds at work wherever it is a question of the will to maintain some established or preconceived order of priority, as for instance between speech and writing, melody and harmony, nature and culture, legitimate (state-authorized) and illegitimate (subversive or counter-state) force, or the artwork itself and whatever belongs – like the frame of a painting and everything that lies beyond or outside that frame – to the order of its merely 'extrinsic' ornaments or outworks. In each case the deviant logic concerned – of 'supplementarity', '*différance*', 'parergonality', and their various kindred yet importantly distinct and context-specific analogues – is such as to highlight those salient passages in Rousseau, Kant, Saussure, Lévi-Strauss and others which reveal the non-coincidence of author's intent, the *vouloir-dire* of express or manifest purport, with what the text actually says or implies as a matter of logical entailment. Yet in no case does this lend credence to the widespread (at least among philosophers) idea of deconstruction as a language-fixated or 'textualist' enterprise bent upon dissolving every aspect of reality into a play of referentially unanchored signifiers or clashing semiotic codes. Rather, it shows to singular effect the way that such textual complications, aporias, or moments of undecidability can be seen to enact in themselves – through a logic none the less rigorous for its deviant (i.e., non-classical or non-bivalent) character – the kinds of obstacle that thought encounters when applied to such inherently dilemma-prone topics as those mentioned above.

Thus, for instance, the upshot of any sustained reflection on the issue of priority between melody and harmony is that it should end by questioning the 'common sense' view championed by Rousseau, i.e., that melody is historically prior and therefore of intrinsically superior value to the extent that it remains more closely in touch with those natural sources of expressive or passionate utterance that had not yet succumbed to the false arts of 'civilized' musical fashion. However that thesis proves to be self-deconstructing since harmony itself turns out to be the precondition for even the most basic, pre-contrapuntal, or (as it might seem) 'purely' melodic of human musical practices and modes of response. There is always some implied harmonic sub-text when we hear a single 'unharmonized' melodic line – say a piece of plainchant or a solo unaccompanied folk-song – since the very act or process of perceiving it *as* a melody requires that we make up that missing dimension (the 'supplement' of harmony) so as to provide the necessary measures of tension, release, suspense, expectation, punctual or delayed cadence, and so forth. Such is the subjective or phenomenological correlate of Derrida's exposition in the logico-semantic mode, that is, his acutely diagnostic reading of the various strictly unavoidable dilemmas that result if one follows out the double and contradictory logic that runs through Rousseau's texts on the origins, history, and nature of music. However there is also – more germane to our interests here – his observation that any single note sung or played (and of course any sequence of notes arranged in some given melodic shape) will always carry along with it a range of harmonic overtones that extend up and down, with diminishing strength, through the resonant circle of fifths. Thus Derrida's practice of deconstruction, so far from leaving us stranded in a realm of free-floating textual signifiers, is in fact anchored firmly in certain properties of the physical or natural world and in certain likewise robust phenomenological truths about the character of human musical response.

One could pursue this case in relation to numerous texts – mostly (though by no means exclusively) from the earlier period of his writing – all of them serving to make the point that deconstruction is not only compatible with a realist ontology and epistemology but actually requires them if its claims are to possess any genuine cogency or force. There are two main aspects to the kind of realism involved here. One has to do with the inbuilt resistance to appropriative treatment that Derrida imputes to the texts in question, that is, with his regarding them (contrary to widespread report) not merely as passive receptacles for whatever zany new interpretation might be dreamed up by the inventive exegete, but rather as holding out strongly against any reading that seeks to impose some preconceived idea of what they should or must signify. The other has to do with the way that those same texts – *qua* objects of a more analytically alert and hence less prejudicial reading – turn out to harbour complexities of logic and sense that resist any straightforward, non-problematical account and thereby signal the existence of kindred complications with respect to those various topics that occupy their focus of interest. In other words, this entails a realist approach both to issues of textual ontology (as against the neo-pragmatists or 'strong'-revisionists for whom interpretation goes all the way down and texts are incapable of putting up any such resistance) and to issues of wider scientific or everyday-practical epistemic consequence (as against the anti-realists, social constructivists, and advocates of cultural-linguistic relativism in its sundry present-day guises).

Thus it is wrong – albeit an error with its source in some of Derrida's less circumspect early pronouncements – to suppose that just because deconstruction is so firmly wedded to the practice of close-reading or so minutely attentive to matters of textual detail therefore it must own allegiance to one or more (maybe all) these camps. Quite simply, such a notion cannot stand up once exposed to the kinds of meticulously reasoned as well as rhetorically sharp-eyed critical exegesis that constitute the bulk of his writing. Again, I must refer the reader to previous publications where I have gone into far more detail as concerns the logico-semantic (especially the modal-logical) dimensions of Derrida's work and also – directly related to that – its realist presuppositions and commitments.[54] Sufficient here to say that these two interlinked aspects are such as to bring that work into a close and potentially fruitful conjunction with the strain of renascent ontological realism that has been a marked feature of analytic philosophy of language over the past three decades.[55] Moreover, as I commented à propos Rousseau on music, this realist component of Derrida's thought sometimes (not atypically) goes along with a distinctive element of philosophic naturalism grounded in the claim that textual complications of the sort discovered – not invented or contrived – by a deconstructive reading will find at least one major source of interpretative warrant in certain salient attributes, properties, or structures of a real-world natural domain.

Needless to say, one would hardly go to Derrida in search of some programmatic statement to that effect, or as a paradigm case of the naturalistic approach to topics in ontology, epistemology, or philosophy of mind and language. Indeed I should remark, in fairness to the prevalent opposing view, that Derrida himself was inclined to impute a certain irony (intended or not) to my characterization of his work – at any rate some key writings of his early period – as implicitly committed to a realist outlook for just the reasons cited above.[56] Still, I should not wish to retract or substantially modify that claim save to press it in the more naturalistic direction that I hope to have defined with sufficient clarity in the course of this book. Thus I would continue to stress the realist presumption bound up with Derrida's meticulous appeals to the textual evidence – no matter how complex or contradictory – and likewise the fact that a deconstructive reading can have no demonstrative or probative force save by taking language to possess both a well-developed referential capacity and a constative dimension that allows adequate room for assessment in terms of propositional content and ascription of classical or bivalent truth/falsehood. Here one can take heart from his vigorous assertions, notably in the second-round rejoinder to Searle, that such values are nowhere rejected in his work but are rather shown – through a mode of argument none the less rigorous for its being conducted by way of textual close-reading – to require a mode of treatment more complex and responsive to anomalous (hence philosophically challenging) cases than anything readily available from the mainstream analytic quarter.

Such, I would contend, is the best way to take his notorious pronouncement in *Of Grammatology* that 'there is nothing outside the text' (more aptly rendered 'there is no "outside" to the text').[57] So far from envisaging a realm of free-floating, unanchored signifiers devoid of any content except that conjured up through the infinitized play of differential relations – the idea of 'unlimited semiosis' that Derrida imports from C. S. Peirce only as a strictly unrealizable limit-point notion – his statement should be read as a claim to just the opposite, i.e., realist, referential, or world-involving effect.[58] That is, it draws attention (granted in a somewhat dramatic and potentially misleading way) to the extent of language's imbrication with reality or the impossibility of any radical disjunction between word and world, as well to the converse point taken up and pressed beyond any reasonable limit by post-structuralists, postmodernists, and 'strong' descriptivists of various hue, namely the extent to which perceptions of reality are influenced (they would say: shaped or determined) by the particular range of concepts and categories that have a place within some particular natural language. In this book I have spent a good deal of time remarking on the prevalence of kindred doctrines among many analytic (or post-analytic) philosophers, and on the easy – or well-nigh inevitable – slide from moderate to other, more extreme and implausible versions of it. Since I have argued here that the only adequate alternative is a thoroughly naturalized conception of mind, world, and linguistic representation and have elsewhere suggested the pertinence of Derrida's thinking in that context I should now say something more about how this on the face of it improbable alignment of interests might work out.[59] For it is vital to the overall integrity of my project that the case be made, and without excessive strain, for an understanding of deconstruction that finds it both accordant with the main precepts of philosophic naturalism and opposed to the various above-mentioned strains of linguistic idealism.

That the case has indeed been made – by some of Derrida's most perceptive commentators across a range of subject-areas – will I think be borne out by a survey of the recent secondary literature, above all those parts of it concerned to explain how deconstruction might productively engage with issues in cognitive science, psychology, and philosophy of mind.[60] A prime exhibit here is his early essay 'Freud and the Scene of Writing' which serves exceptionally well to make the point, i.e., that a 'textualist' emphasis such as Derrida deploys in his reading of Freud's likewise very early (and, until recent years, much neglected) *Project for a Scientific Psychology* is by no means at odds with a naturalized or, indeed, central-state materialist conception of the mental as wholly and exclusively a product of neuro-physical processes and events.[61] What has opened the way to this reassessment is a new approach to the question – as much techno-scientific as cognitive-psychological or philosophic – of just how it is that brains can perform the kinds of immensely complex and, above all, highly context-sensitive operation that enable human beings to respond to their environment with such a singular degree of creative-adaptive intelligence. It was the problem of closing that same explanatory gap – not so much the ultimate challenge of inventing/designing/constructing the hardware and software that might actually exhibit the powers in question but the more immediate conceptual problem of taking this idea on board – which provoked those various a priori arguments against the very possibility of 'strong' AI that have sparked so much philosophical debate over the past four decades. Where the new approach promised to break the impasse (and invalidate that whole greatly overworked line of objection) was by proposing that designers and programmers take their cue from the brain itself, and – more specifically – from its character as an organ specialized for running widely distributed, densely inter-connected, and multi-track parallel processes rather than the linear-sequential procedures favoured by a previous generation of AI theorists.[62] And where this neural-network or connectionist approach links up in a suggestive way with Freud and Derrida is through the latter's treatment of the former's *Project* as a text that presses to the limit with certain distinctly proto-connectionist motifs, albeit couched by Freud in a somewhat archaic idiom – one drawing mainly on nineteenth-century hydraulic and mechanical analogues or metaphors – which lends itself to a deconstructive reading of particular pertinence and force.

However my point, and that of the commentators who have lately picked up on this link, is that Derrida is able to accord the *Project* a more than merely antiquarian interest through his perception that this is a text which itself goes far toward deconstructing its own more crudely reductive modes of psycho-dynamic explanation. The key term here – as in so many of Derrida's early essays – is that of writing (more precisely: *arche-écriture* or 'proto-writing') as the only concept (more precisely: quasi-concept) that comes close to capturing the paradoxical combination of materiality and latency or structural persistence and the endless chain of substitutions, detours, and deferrals that typify the workings of the unconscious as figured in this and later works of Freud. This in turn goes far toward undermining the privileged status that has long been accorded to consciousness – or a certain idea of consciousness – by philosophers and non-philosophers (maybe unwitting heirs of Descartes) alike. As Karen Green puts it, 'Freud's speculations overturn traditional notions of language, in which writing is interpreted by consciousness'. This they do by radically revising or transforming that relationship to the point where 'it now appears that conscious perception itself depends on the prior laying down of traces which form a differential structure and invest that which is available to consciousness with meaning'.[63] Or again, according to F. P. Cilliers: 'Memory is the breaching of pathways, the forming of connections . . . . It is the trace. Just as there is nothing before the trace in language, there is nothing before memory in the brain. Certainly not consciousness.'[64] Hence Derrida's recourse to a whole range of deconstructive key-terms – *différance*, 'trace', 'supplementarity', 'iterability', *retrait*, 'parergonality', and so forth – each of which has its primary source in his reading of some particular text yet all of which can be shown to possess a certain, albeit limited aptitude for transference to other contexts. When this occurs, as when Derrida moves from discussing Rousseau to discussing Saussure or Lévi-Strauss, the effect is once again to focus attention on that intrinsically elusive aspect of language that serves not merely as an analogue to the Freudian unconscious but rather as the very locus of those operations – chief among them metaphor (or substitution) and metonymy (or displacement) – that psychoanalysis reveals as the structural correlates of unconscious thought.[65]

So much is common currency for anyone familiar with his work and, in particular, with the debate around deconstruction and psychoanalysis that developed in the wake of Derrida's critical engagement with Lacan's highly controversial reading of Freud.[66] Less well known – beyond those few who have read the signs of a distinctly naturalistic approach in that early essay on Freud – is his clear affirmation that the *topos* of writing offers our best means of descriptive or explanatory access to the unconscious and its workings. What is thereby revealed is not so much (as assorted anti-realists, 'strong' constructivists, and cultural relativists would have it) the linguistic constitution of reality but rather the reality and indeed, neurologically speaking, the material or physically embodied nature of those various psychic tropes which the Freud of the *Project* tended to express in somewhat mechanistic terms but which he later managed to formulate by way of a primarily linguistic or language-oriented discourse. Still there is a crucial difference between Lacan's wholesale acceptance of the idea that 'the unconscious is structured like a language' – an idea with its source in Saussure, Jakobson, and the project of structuralist linguistics – and Derrida's thesis, in 'Freud and the Scene of Writing', that we yet have something to learn from the early, 'pre-psychoanalytic' Freud and moreover that this has primarily to do with the impossibility of holding language apart from the neuro-physical structures that alone make language possible. Thus (Cilliers again): 'Derrida affirms the materiality of the brain, but insists that matter as such has no meaning, wears no meaning on its face that is not mediated by the discourse in which it appears . . . . With the decentering of the subject, the concept of "mind" as a self-present consciousness is also dislodged, and linked with other examples constrained by the metaphysics of presence'.[67] It is here that Derrida most decisively breaks with the strain of linguistic idealism that has been such a marked and, one would think, embarrassing feature of post-structuralist discourse to some of its politically-minded adherents who could scarcely embark upon their avowed project of changing the world in any substantive or realistic sense if that world were entirely – or 'all the way down' – a product of linguistic interpretation.[68]

VI

This is why his essay on Freud is so crucial for anyone who wishes to save Derrida from those, disciples and detractors alike, for whom deconstruction is indeed just another (albeit, for better or worse, ultra-'radical') variant of the currently widespread linguistic turn. As against this prevailing misconception, it serves to re-emphasize his own emphatically reiterated point: that despite all the complications thrown up by any duly refined or context-sensitive account of language in its referential aspect there is simply no making sense of a doctrine – whether post-structuralist, postmodernist, or cultural-linguistic relativist – which either rejects the very idea of its possessing such an aspect or else treats reference as wholly a product of those diverse signifying codes and conventions that happen to exist within some given language-community at some given time. As I have argued elsewhere, this whole line of thought can be seen to have one of its philosophic sources in the Frege/Russell descriptivist claim that sense 'determines' reference, or that we are able to pick out the sundry items that make up the world (more precisely: the world of our knowledge or acquaintance) through an intermediary 'cluster' of identifying attributes, these latter conceived as belonging to the stock of shared (whether specialist or community-wide) understanding.[69] It is not hard to see how that claim led on, via Quine's and Wittgenstein's very different but between them highly influential forms of radical meaning-holism, to the kinds of extreme anti-realist outlook that might purport to be merely pressing the claim to its logical conclusion.[70]

One response to that outcome – and to other problems with the descriptivist theory, such as its promoting Kuhnian paradigm-relativism or Quinean doubts as to the possibility of inter-lingual (or even intra-lingual) communicative uptake – was the alternative causal theory of naming and necessity developed independently by Kripke and Putnam during the late 1960s.[71] Their new approach promised to resolve those issues by treating the inaugural act of designation as that whereby reference is fixed in the first place and subsequent usage as holding fast referentially throughout and despite whatever shifts might occur in the currency of knowledge, the cluster of related (no longer reference-determining) senses, or the range of identifying properties or attributes assigned to the referent in question. This switch of priorities went along with a markedly realist metaphysical commitment to the existence of natural kinds as paradigm instances of reference-fixing in general, and to the idea of human identity as likewise fixed – held firm across different (e.g., counterfactual) life-histories for this or that individual – at the moment of conception when their genetic-chromosomal make-up is uniquely determined. However, as soon emerged in the course of debate, that alternative account itself faced a number of problems and challenges, whether from continuing adherents to the descriptivist theory or – more to the point here – from naturalists of a different philosophical stripe who thought of the new (Kripke/Putnam) doctrine as a genuine improvement on its 'old' rival metaphysically or ontologically speaking but as insufficiently robust or resourceful in its epistemological aspect.[72] What is further required in order to achieve an adequate level of normative as well as descriptive-explanatory grasp is a naturalized epistemology that makes full allowance for the reference-fixing role of 'external', mind-independent objects and events but which also, quite compatibly with that, finds room for the mind's highly active participant role in the process whereby human beings gain a purchase – a jointly operative power of perceptual-cognitive-linguistic grasp – on the various items that constitute their ambient life-world.

It is in this regard chiefly that recent developments in cognitive science have pointed up the need for a major shift in the way that philosophers conceive the relationship between mind, consciousness, thought, language, and those same worldly realia. Moreover it is in this regard that Derrida's work – especially his early essay on Freud – can be seen to partake of the naturalistic turn that has been such a prominent (and, as I have argued, such a requisite) feature of much recent scientifically-informed philosophical thinking. To some extent this has to do with a familiar aspect of that work, namely its sustained deconstructive critique of all epistemological conceptions – down to and very much including the project of Husserlian transcendental phenomenology – premised on the notion of consciousness as offering privileged first-person epistemic access to its own nature or contents through a mode of punctual, transparent self-knowledge ideally exempt from sceptical doubt.[73]

Of course that doctrine in its pure form is scarcely to be found in any philosopher apart from Descartes, and is subject to considerable qualification even in those texts of Husserl that press furthest with the aim of transcendental or eidetic reduction, as distinct from those others that insist primarily on the origin of all human thought and knowledge in the life-world of shared experience.[74] All the same – a point that has figured centrally throughout this book – it continues to haunt the discourse of philosophy in sundry less extreme yet still conversation-dominating forms, chief among them that whole long-running debate around the 'hard problem' of consciousness. It is in this respect also that Derrida's work has some useful lessons to impart since it brings out most strikingly the extent to which thought can (and must) be capable of carrying on and even – perhaps especially – of achieving its best, most creative or ground-breaking results without the accompaniment of conscious let alone self-conscious awareness. What gives Derrida's case its peculiar demonstrative force is the way that he links the issue of conscious in relation to unconscious/preconscious thought to the issue about speech and writing. Thus he shows with impressively copious documentation how speech stands in – for thinkers from Plato to Husserl and beyond – as an analogue or representative instance of the perfect, unmediated *rapport-à-soi* that is supposed to characterize consciousness just so long as it remains unaffected by the various swerves or detours from origin that are typically ascribed to the malign agency of something that bears either the name or the imputed traits of writing. So it is that Derrida's account of Freud's early *Project* comes to hold out – at least for some of his more cognitive-scientifically informed readers – the prospect of a 'naturalized' deconstruction that would yield no hostages to idealist fortune since the *topoi* of writing and textuality are here very firmly grounded in the kinds of material (no matter how elusive) process and event that have their locus in the human brain.

To be sure, Derrida is at pains to emphasize that same elusive or fugitive property and, through his strategic deployment of terms such as 'trace', 'graft', and 'spacing', to mark the decisive distance between Freud's fundamentally nineteenth-century physics-based conception with its recourse to a range of mechanical, hydraulic, or other such loosely suggestive metaphoric devices and his own (Derrida's) inscriptionalist idiom adopted precisely in order to avoid that liability. Yet he also brings out the very striking points of contact between those two conceptions and the way that Freud's descriptive-explanatory language can be seen as constantly straining toward a more purely differential or relational account that would not be adequately theorized until the belated 1960s rediscovery of Saussurean structural linguistics. Or rather, its articulation at last became possible through the impact of that particular episode in the history of thought along with the range of contemporaneous and (as they now appear) kindred developments – from cybernetics to the first major advances in genetic code-breaking – which Derrida surveys with remarkable prescience in *Of Grammatology*. Thus,

> the entire field covered by the cybernetic *program* [Derrida's italics] will be the field of writing. If the theory of cybernetics is by itself to oust all metaphysical concepts – including the concepts of soul, of life, of value, of choice, of memory – which until recently served to separate the machine from man, it must conserve the notions of writing, trace, grammè, or grapheme, until its own historico-metaphysical character is exposed.[75]

Derrida is able to cite the most forward-looking passages of the *Project* – those that lean most strongly in a 'textualist' or inscriptionalist direction – as evidence not only of Freud's extraordinary acumen but also of the way that a thematics of writing ('proto-writing' in Derrida's extended usage) is itself the clear mark of a thought that has ventured beyond the limits imposed by the idea of self-present conscious thought and speech as its privileged mode of expression. This is yet more impressive, Derrida contends, if one then takes account of the advance in Freud's thinking signalled by his 1925 'Note on the

"Mystic Writing-Pad"' where – 30 years after composing the *Project* – he offers a thoroughly inscriptionalist topology of the unconscious in relation to the workings of perception, memory, desire, repression, and indeed 'consciousness' itself.[76] However the latter is conceived in terms – materialist as well as 'textualist' terms since these are no longer conflicting or mutually exclusive – which effectively remove that aura of impenetrable mystery and depth that nowadays invests the 'hard problem' in cognitive psychology and philosophy of mind.

The opening paragraph of 'Freud and the Scene of Writing' puts this case with exemplary clarity so I shall cite it at some length.

> From the *Project* (1895) to the 'Note on the "Mystic Writing-Pad"' (1925), a strange progression: a problematic of breaching is elaborated only to conform increasingly to a metaphorics of the written trace. From a system of traces functioning according to a model which Freud would have preferred to be a natural one, and from which writing is entirely absent, we proceed toward a configuration of traces which can no longer be represented except by the structure and functioning of writing. At the same time, the structural model of writing, which Freud invokes immediately after the *Project*, will be persistently differentiated and refined in its originality. All the mechanical models will be tested and abandoned, until the discovery of the *Wunderblock*, a writing machine of marvellous complexity into which the whole of the psychical apparatus will be projected.[77]

Derrida's reading has been questioned by some commentators, Elizabeth Wilson among them, who would wish to press further with the claim that Freud's *Project* strikingly prefigures recent advances in what they regard as the immensely promising convergence of interests between neuroscience and cognitive psychology.[78] Thus, for Wilson, it is unfortunate – a failure to follow through on his own best insights – that Derrida should treat it as a sure sign of progress in Freud's thinking that he came replace that 'natural', i.e., neurological mode of descriptive-explanatory theorizing with a 'structural model of writing' supposedly purged of all such (supposedly) reductionist or biologistic residues. As he puts it in a sentence that leads straight on from the passage cited earlier: '[t]he *Wunderblock*, in each of its parts, will realize the apparatus of which Freud said, in the *Project*, "We cannot off-hand imagine an apparatus capable of such complicated functioning", and which he replaces at that time with a neurological fable whose framework and intention, in certain respects, he will never abandon'.[79] For Wilson this can only be seen as falling in – despite the extreme subtlety and insight of Derrida's reading – with that strain of anti-scientific prejudice that has exerted such a strong and (as she believes) damaging influence on the reception of psychoanalysis among those in the humanities and social-science disciplines. It assumes that any physicalist 'reduction' of Freud's discoveries to a naturalistic or neurological register will inherently constitute a backward step or an instance of regression from the epochal advance that Freud achieved when he came to recognize how inadequate were all such inherited nineteenth-century mechanistic models.

On this account – Derrida's account, as Wilson interprets it – one can grasp how decisive was the break that occurred with Freud's subsequent re-envisioning of psychoanalysis as the 'talking cure' only if one accepts that it involved a radical break with every last remnant of 'naïve', 'reductionist', or crudely 'biologistic' thinking. Hence her chief interest in Derrida's essay as yet another symptomatic instance of how 'the division of neurology from psychology is taken so frequently to be the founding moment of psychoanalysis'.[80] And again

> [t]here seems to be a reactive move in Derrida's text against the viability or propriety of neurological models for his own deconstructive project. It appears, at times, that it is Derrida's construction of a certain type of neurology and then its almost immediate exclusion that allows him to construct and enforce his narrative of a metaphorics of writing in Freud's work: neurology becomes the excess in Derrida's reading of Freud.[81]

However it needs pointing out, as against this limiting judgment, that Derrida is very clear as regards the material aspect of writing – especially in the Freudian context – and also as regards the desirable effect of an inscriptionalist or textualist idiom in drawing attention to just that aspect, as opposed to the idealizing tendency promoted by analogies between speech (or the experience of hearing-oneself-speak, *s'entendre-parler*) and the notion of consciousness as involving some mode of uniquely privileged, self-present, first-person access to meaning and truth. Of course that idea has been under attack from various philosophical quarters for a long time now – indeed ever since Descartes first enounced it and encountered a range of highly pertinent objections – so its dislodgement is hardly something that has had to wait either for the ministrations of psychoanalysis or for present-day developments in neuroscience and cognitive psychology. Nor does its recognition in any way depend on Derrida's radically materialist conception of writing as that which alone made it possible for psychoanalysis to transcend both the limits of a purely mechanistic psychodynamics and the opposite temptation of a theory of mind that would succumb once more to the allure of the Cartesian subject-presumed-to-know. Still his reading does very effectively make the point – one that would no doubt meet with stiff resistance among many in the 'cog-sci' camp – that there exists (or should exist) a real convergence of aims between, on the one hand, a psychoanalysis that continues to find room for those elements of the *Project* that are best captured by an inscriptionalist idiom and, on the other, a neuroscience sufficiently attuned to the ubiquity of language as a royal road to the unconscious.

Such, I would argue, is one beneficial consequence of viewing the encounter between psychoanalysis and neuroscience through the lens of Derrida's early essay on Freud. Thus it offers a powerful incentive for approaching those two, at present deeply hostile – or at best mutually suspicious – projects with a view to what they have in common, namely a basic (even discipline-constitutive) interest in pressing right through with the ambition of achieving a naturalized epistemology and philosophy of mind. For, despite all the standard 'textualist' jibes, deconstruction as practiced here turns out to permit – indeed, to mandate – a reading that takes its materialist claims at full philosophical force and which constantly strives to block any recourse to idealist or 'consciousness-first' modes of thought. To this extent it stands in sharp contrast not only to whatever nowadays survives of the purebred Cartesian *cogito* but also to those various compromise positions, such as I have discussed at length in this book, that would seek to discard its worst, philosophically as well as scientifically retrograde features whilst none the less conserving an attenuated version of the anti-naturalist prejudice that has always been its epistemological other half.

Cilliers again brings out what is most radical in Derrida's reading of Freud when he stresses how little room it leaves for consciousness in any of those currently touted guises or disguises. Here we might recall the arguments of Kathleen Wilkes – discussed in Chapter 3 – for taking a similar deflationary line with respect to the 'hard problem' and its tendency to crowd out other, quite likely more productive since less mystification-prone topics of debate.[82] 'What then can be said about consciousness?', Cilliers asks, and offers a distinctly downbeat answer. 'This is a question that should be resisted, because it carries with it all that heavy baggage of "mind", "intention", "feeling", and "reason", fooling us into thinking that these concepts are somehow pure and important. In writing what has been written about memory and perception, everything about consciousness has been said. The question has decomposed.'[83] Indeed I would suggest that Derrida's genealogical account of the Freudian discovery – his patient tracing-out of the process whereby Freud accomplished the passage from a quasi-mechanistic conception of psychic function to one that found its sole distinct model or analogue in writing – is itself the best means of thinking beyond the kind of halfway naturalized approach that has typified so many recent ventures in a broadly kindred direction. Best-known of these is Daniel Dennett's 'multiple drafts' theory of consciousness, that is, his idea of the human mind/brain as a locus of multiple loosely connected, more or less widely dispersed, simultaneously active or operative trains of thought.[84] Among them some may be envisaged as occupying a region of brain/mind space somewhere near the present focus of conscious attention while others occur well out toward the unconscious/preconscious fringes or else in those truly penumbral regions where consciousness seldom if ever intrudes.

The 'hard problem' can therefore be resolved, so Dennett would claim, by addressing it as chiefly a matter of explaining how the mind somehow 'edits' those multiple drafts and decides which of them shall come into conscious focus from one moment to the next. However, as should be evident even from this brief summary, there is more than a suspicion that he is not so much resolving as displacing the problem, or offering an answer that manifestly begs the question since it fails to specify just what (or who) is thereby placed – on no matter how provisional or short-term a contract – in effective editorial control. Besides, there is no very clear statement (despite his book's breezily confident title) as to how philosophy of mind stands to benefit from this idea that consciousness equates for all relevant descriptive or explanatory purposes with the occasional coming-to-light or coming-into-focus of otherwise subliminal thoughts. One then has to ask just what is the nature of that specially attentive mind-state, or just what is meant – in other than metaphoric terms – by all those decidedly Cartesian (ocular, visual or light-based) figures of speech. Nor does it help very much when Dennett argues for a 'heterophenomenological' approach that would accord due weight to the first-person subject and his or her various self-reports but only in balance with a range of other, perhaps conflicting evidence from third-person sources. Indeed, on his account, those sources may sometimes be deemed more expert or authoritative than horse's-mouth witness even if this means setting aside the subject's claim to incorrigible certainty with regard to their present state of thought, feeling, or experience. Yet, despite serving to mark his distance from the brand of 'lone-wolf autophenomenology' that Dennett counts merely a relic of our all-too-recent philosophical past, his alternative proposal still involves the appeal to conscious mind-states (first- or third-person) as sources of the relevant data and also, when it comes to the weighing-up process, as engaged in a kind of communal reflective activity whereby the *cogito* is multiplied, fractured, or distributed as in a hall of mirrors.

His theory thus leaves the way open for anyone who wishes to exploit this curious – although, as we have seen, currently widespread – ambivalence and retain a tactical line of retreat to some other, more overt or unembarrassed version of the old Cartesian view. There is the same distinct impression of having one's cake and eating it about Dennett's idea of the 'intentional stance' – as distinct from the 'physical' or 'design' stances – as an outlook that we can and should adopt when interpreting human actions, motives and beliefs since it offers the best, most reliably efficient (knowledge-conducive or sense-maximizing) way to proceed in such matters. For we are fully entitled to this benefit, he thinks, without needing to go the whole hog and ascribe any kind of first-person verifiable epistemic or psychological reality to the various imputed thoughts, beliefs, desires, or intentions that play a role in our likewise varied interpretative gambits.[85] Even if this strategy might be found to pay off when adopted, say, by a hard-line behaviourist in search of some credibility-enhancing ploy it can scarcely be thought to carry much weight when applied, as here, in the context of an argument that presses much further in a radically materialist direction while on the other hand seeking to conserve far more of what anti-reductionists fondly describe as our full-spectrum 'technicolour phenomenology'. Hence the marked tension, in Dennett's work, between a strong naturalizing impulse that would firmly overrule any concession to the idea of first-person privileged access and the desire – as against flat-out central-state materialists bent upon eliminating every last vestige of folk psychology – to conserve at least some plausible link with the life-world of commonplace inter-subjective experience.[86] This is why his project, unlike Freud's or Derrida's in reading Freud, seems at times to back off from its physicalist credo and opt for something more like the various compromise solutions surveyed in the course of this book where the opt-out clause typically consists of a recourse to this or that qualified version of the consciousness-first or privileged-access idea.

Thus for Derrida, in Richard Harland's words, 'consciousness (at least in the ordinary sense) is an illusion that human beings have invented because they have feared the consequences of a materialist conception of the brain. In this respect, the modern secular notion of mind is really no improvement upon older religious notions of soul or spirit'.[87] To be sure, there is something odd about recruiting Derrida to a cause – that of a materialist or thoroughgoing naturalist philosophy of mind – that can scarcely be said to have occupied a large amount of his time or attention over a long and highly productive philosophical career. However, in response, I would wish to repeat the two chief claims that have formed the basis of my argument throughout this book and, more specifically, my case concerning Derrida's engagement with Freud. First is the thesis that full-strength naturalism signposts the only route that philosophy of mind can take without severe prejudice to its natural-scientific and hence its genuine philosophic credentials. Quite simply we *just are* material creatures in a material world and any attempt to finesse or finagle that fact in however philosophically refined a fashion is sure to run aground on the shoals of a washed-up Cartesian mind-view that dare not quite acknowledge its name. Second is my point that the deconstructive project exemplified in Derrida's early essay on Freud should be thought of not merely as 'compatible' with a radically naturalized (materialist) approach but rather as a singularly powerful means of revealing the problems – the unlooked-for aporias or failures to reckon with its own inherent limitations – that have marked the consciousness-first doctrine from its Cartesian inception down. Again it needs saying that this goes flat against a whole slew of commonplace interpretations (both friendly and hostile) that would place Derrida far out on the textualist-idealist wing. That they are wrong about this and that the naturalizers should be keen to welcome him on board – at least as regards this one highly pertinent aspect of his work – is among the main reasons why philosophers of mind in the 'other' tradition should now take belated (though also, as it happens, very timely) account of Derrida's arguments.

Notes

Introduction

[1] Imre Lakatos, *The Methodology of Scientific Research Programmes* (*Philosophical Papers*, Vol. One) (Cambridge: Cambridge University Press, 1978).

[2] On this topic see Gaston Bachelard, *La formation de l'esprit scientifique* (Paris: Corti, 1938), *Le materialisme rationnel* (Paris: Presses Universitaires de France, 1953), and *The Poetics of Reverie*, trans. Daniel Russell (Boston: Beacon Press, 1971).

[3] For further discussion of these issues concerning the role of unconscious vis-à-vis consciously accessible processes of thought or states of mind, see various notes to Chapters 1, 3 and 6; also – from a wide range of philosophical standpoints – Michel Henry, *The Genealogy of Psychoanalysis*, trans. Douglas Brick (Stanford, CA: Stanford University Press, 1993); Jonathan Lear, *Freud* (London: Routledge, 2005); Michael P. Levine (ed.), *The Analytic Freud: Philosophy and Psychoanalysis* (London: Routledge, 2000); Robert Samuels, *Between Philosophy and Psychoanalysis: Lacan's Reconstruction of Freud* (London: Routledge, 1993).

[4] Jacques Lacan, 'The Insistence of the Letter in the Unconscious, or Reason since Freud', in *Ecrits*, trans. Alan Sheridan (London: Tavistock, 1977), pp. 111–36.

[5] See especially David Chalmers, *The Conscious Mind: In Search of a Fundamental Theory* (Oxford: Oxford University Press, 1996); also – for a range of contrasting views – Jeffrey A. Gray, *Consciousness: Creeping up on the Hard Problem* (Oxford: Oxford University Press, 2004); Joseph Levine, *Purple Haze: The Puzzle of Consciousness* (Oxford: Oxford University Press, 2002); Colin McGinn, *The Mysterious Flame: Conscious Minds in a Material World* (New York: Basic Books, 1999); David M. Rosenthal (ed.), *Materialism and the Mind-Body Problem* (Indianapolis: Hackett, 2000); Sidney Shoemaker, *Physical Realization* (Oxford: Oxford University Press, 2007); Quentin Smith and Aleksandar Jokic (eds), *Consciousness: New Philosophical Perspectives* (Oxford: Oxford University Press, 2003).

[6] On this idea of concept creation as philosophy's chief and proper aim, see Gilles Deleuze and Félix Guattari, *What Is Philosophy?*, trans Hugh Tomlinson and Graham Burchill (London: Verso, 1994).

[7] See Michael Dummett, *Truth and Other Enigmas* (London: Duckworth, 1978) and *The Logical Basis of Metaphysics* (Cambridge, MA: Harvard University Press, 1994).

[8] Philip Kitcher argues a similar case in his (to my mind) highly persuasive book *The Nature of Mathematical Knowledge* (Oxford: Oxford University Press, 1983).

[9] Richard Rorty, *Consequences of Pragmatism* (Brighton: Harvester, 1982) and *Contingency, Irony, and Solidarity* (Cambridge: Cambridge University Press, 1989); also Harold Bloom, *The Anxiety of Influence* (New York: Oxford University Press, 1973) and *A Map of Misreading* (Oxford: Oxford University Press, 1975).

[10] Stephen P. Stich, *From Folk Psychology to Cognitive Science: The Case Against Belief* (Cambridge, MA: MIT Press, 1983) and *Deconstructing the Mind* (Oxford: Oxford University Press, 1999); also Dominic Murphy and Michael Bishop (eds), *Stich and His Critics* (Oxford: Blackwell, 2009).

[11] For some powerful arguments to this effect, see Kathleen V. Wilkes, 'Is Consciousness Important?' *British Journal for the Philosophy of Science*, Vol. 34 (1984), 223–43.

[12] Stephen P. Stich, *The Fragmentation of Reason: Preface to a Pragmatic Theory of Cognitive Evaluation* (Cambridge, MA: MIT Press, 1990); also Joshua Kobe and Shaun Nichols (eds), *Experimental Philosophy* (Oxford: Oxford University Press, 2008).

[13] For a recent outspoken defence of such claims, see Timothy Williamson, *The Philosophy of Philosophy* (Oxford: Blackwell, 2008).

[14] Kitcher, *Nature of Mathematical Knowledge*, see previous note 8.

[15] For primary and secondary sources (Spinoza and Badiou), see notes to Chapters 4 and 5.

[16] For further discussion of these hermeneutic issues from a range of more-or-less 'conservative' or 'liberal' standpoints, see Harold Bloom, *The Anxiety of Influence* and *A Map of Misreading* (see previous note 9); Jacques Derrida, *Of Grammatology*, trans. G. C. Spivak (Baltimore, MD: Johns Hopkins University Press, 1976); Umberto Eco with Christine Brooke-Rose, Jonathan Culler and Richard Rorty, ed. Stefan Collini, *Interpretation and Over-Interpretation* (Cambridge: Cambridge University Press, 1992); E. D. Hirsch, *Validity in Interpretation* (New Haven, CT: Yale University Press, 1967); Christopher Norris, *For Truth in Criticism* (Manchester: Manchester University Press, 1994).

[17] Bertrand Russell, 'On Denoting', *Mind*, New Series, Vol. 14 (1905), 479–93.

[18] See especially P. F. Strawson, *The Bounds of Sense: An Essay on Kant's* Critique of Pure Reason (London: Methuen, 1966).

[19] See for instance Jonathan Bennett, *Learning from Six Philosophers, Vol. One: Descartes, Spinoza, Leibniz; Vol. Two: Locke, Berkeley, Hume* (Oxford: Clarendon Press, 2001).

[20] Ludwig Wittgenstein, *Philosophical Investigations*, trans. G. E. M. Anscombe (Oxford: Blackwell, 1951); J. L. Austin, *Selected Papers*, ed. J. O. Urmson and G. J. Warnock (Oxford: Oxford University Press, 1961).

[21] Deleuze and Guattari, *What Is Philosophy?* See previous note 6.

[22] For further discussion of these developments, see Christopher Norris, *Resources of Realism: Prospects for 'Post-analytic' Philosophy* (London: Macmillan, 1997), *Philosophy of Language and the Challenge to Scientific Realism* (London: Routledge, 2004), and *On Truth and Meaning: Language, Logic and the Grounds of Belief* (London: Continuum, 2006).

[23] For the most influential and thoroughgoing statement of this 'post-analytic' view, see Richard Rorty, *Philosophy and the Mirror of Nature* (Oxford: Blackwell, 1979).

[24] See Chapter 1, note 1 and subsequent entries for an extensive bibliography of recent writings on the topic of naturalism in various philosophical contexts.

[25] Immanuel Kant, *Critique of Pure Reason*, trans. Norman Kemp Smith (London: Macmillan, 1964).

[26] For details of Badiou's major works and commentaries on them, see Chapter 5, notes 1, 5, 13 and 14.

[27] Again the reader is advised to consult my notes to Chapter 4 for extensive reference to Spinoza's works and related scholarly/critical sources.

[28] See especially Lewis Samuel Feuer, *Spinoza and the Rise of Liberalism* (Boston: Beacon Press, 1958); Jonathan Israel, *Radical Enlightenment: Philosophy and the Making of Modernity, 1650–1750* (Oxford: Oxford University Press, 2002); Christopher Norris, *Spinoza and the Origins of Modern Critical Theory* (Oxford: Blackwell, 1991).

[29] Antonio Damasio, *Looking for Spinoza: Joy, Sorrow and the Feeling Brain* (London: Heinemann, 2003).

[30] Donald Davidson, 'Spinoza's Causal Theory of the Affects', in *Truth, Language, and History* (Cambridge: Cambridge University Press, 2005), pp. 295–313.

[31] See for instance Jaegwon Kim, *Supervenience and Mind: Selected Philosophical Essays* (Cambridge: Cambridge University Press, 1993).

[32] Harry Frankfurt, 'Freedom of the Will and the Concept of a Person', in Gary Watson (ed.), *Free Will* (Oxford: Oxford University Press, 2003). See also Frankfurt, *The Importance of What We Care About: Philosophical Essays* (Cambridge: Cambridge University Press, 1988).

[33] John McDowell, *Mind and World* (Cambridge, MA: Harvard University Press, 1994); also Norris, 'McDowell on Kant: Redrawing the Bounds of Sense' and 'The Limits of Naturalism: Further Thoughts on McDowell's *Mind and World*', in *Minding the Gap: Epistemology and Philosophy of Science in the Two Traditions* (Amherst, MA: University of Massachusetts Press, 2000), pp. 172–96 and 197–230.

[34] See Chapter 3, notes 1, 20, 21 and 26 for an extensive listing of works on and around the 'hard problem' of consciousness.

[35] See especially Thomas Nagel, 'What is it Like to be a Bat?' *Philosophical Review*, Vol. 83 (1974), 435–50.

[36] Kant, *Critique of Pure Reason*, see previous note 25. Badiou's dislike (to put it mildly) of Kantian critical philosophy comes across to robust, almost knockabout effect in the following passage, although the reasons for it go deep into his own project and will, I hope, emerge clearly in the course of my Chapter 5.

> Everything in him exasperates me, above all his legalism – always asking *Quid juris?* or 'Haven't you crossed the limit?' – combined, as in today's United States, with a religiosity that is all the more dismal in that it is both omnipresent and vague. The critical machinery he set up has enduringly poisoned philosophy, while giving great succour to the academy, which loves nothing better than to rap the knuckles of the overambitious – something for which the injunction 'You do not have the right!' is a constant boon. Kant is the inventor of the disastrous theme of our finitude. The solemn and sanctimonious declaration that we can have no knowledge of this or that always foreshadows some obscure devotion to the Master of the unknowable, the God of the religions or his placeholders: Being, Meaning, Life . . . To render impracticable all of Plato's shining promises – this was the task of the obsessive from Königsberg, our first *professor.*
>
> Nevertheless, once he broaches some particular question, you are unfailingly obliged, if this question preoccupies you, to pass through him. His relentlessness – that of a spider of the categories – is so great, his delimitation of notions so consistent, his conviction, albeit mediocre, so violent, that, whether you like it or not, you will have to run his gauntlet.
>
> (Alain Badiou, *Logics of Worlds*, trans. Albert Toscano [London: Continuum, 2008], p. 535)

What most irks Badiou about Kant's self-avowed 'revolution' in thought – as much for its ethico-political as its epistemological implications – is the claim to have achieved not only a radical but also, in some substantive sense, a *progressive* break with earlier (whether empiricist or rationalist) philosophical outlooks. On the contrary, according to Badiou it is merest of illusions for Kant to regard his project as doing for philosophy what Copernicus did for astronomy and the other physical sciences when in fact the programmatic Kantian turn from ontology to epistemology results in a placement of the human subject squarely at the centre of all our scientific, ethical, and even – *vide* the intuitionists – mathematical concerns. [This point has been made with particular force by Julian Ryall, currently a PhD candidate at Cardiff University.] In so doing, Badiou asserts, Kant opened the way – no doubt very much against his own express intent – to all sorts of relativist, anti-realist, or framework-internalist thinking, and also to those current, e.g., Wittgensteinian variations on ancient sophistical themes that Badiou subjects to some excoriating commentary in *Manifesto for Philosophy* and elsewhere.

37 See for instance Paul Benacerraf and Hilary Putnam (eds), *The Philosophy of Mathematics: Selected Essays*, 2nd edn. (Cambridge: Cambridge University Press, 1983), pp. 272–94; W. D. Hart (ed.), *The Philosophy of Mathematics* (Oxford: Oxford University Press, 1996); Hilary Putnam, *Mathematics, Matter and Method* (Cambridge: Cambridge University Press, 1975).

38 Donald Davidson, *Essays on Actions and Events* (Oxford: Clarendon Press, 1980).

39 Wittgenstein, *Philosophical Investigations*, previous note 20.

40 Kant, *Critique of Pure Reason*, previous note 25.

41 For further discussion see Richmond Campbell and Bruce Hunter (eds), *Moral Epistemology Naturalized* (Calgary: University of Calgary Press, 2000); Jonathan Knowles, *Norms, Naturalism and Epistemology* (Basingstoke: Palgrave-Macmillan, 2003); Hilary Kornblith, *Knowledge and Its Place in Nature* (Oxford: Clarendon Press, 2002); David Papineau, *Philosophical Naturalism* (Oxford: Blackwell, 1993); Peter Railton, *Facts, Values, and Norms: Essays toward a Morality of Consequence* (Cambridge: Cambridge University Press, 2003); Ralph Wedgwood, *The Nature of Normativity* (Oxford: Clarendon Press, 2007).

42 See Norris, 'Ethics, Autonomy, and the Grounds of Belief', in *On Truth and Meaning*, previous note 22 , pp. 130–154; also 'Choice and Belief', *The Philosophers' Magazine*, Issue 29 (2005), 17–23.

43 For a range of views, see Daniel Dennett, *Elbow Room: The Varieties of Free Will Worth Wanting* (Oxford: Oxford University Press, 1984); Peter van Inwagen, *An Essay on Free Will* (Oxford: Clarendon Press, 1983); Robert Kane, *The Significance of Free Will* (Oxford: Oxford University Press, 2004); Timothy O'Connor, *Persons and Causes: The Metaphysics of Free Will* (Oxford: Oxford: Oxford University Press, 2004); Galen Strawson, *Freedom and Belief* (Oxford: Oxford University Press, 1986); Watson (ed.), *Free Will* (Oxford: Oxford University Press, 2003).

44 Peter T. Geach, 'On Worshipping the Right God', in *God and the Soul* (London: Routledge & Kegan Paul, 1969), pp. 100–16.

45 Gottlob Frege, 'On Sense and Reference', in Max Black and P.T. Geach (eds), *Translations from the Philosophical Writings of Gottlob Frege* (Oxford: Blackwell, 1952), pp. 56–78.

46 Ted Honderich, *A Theory of Determinism*, Vol. One: *The Mind, Neuroscience and Life-Hopes* (Oxford: Clarendon Press, 1988) and Vol. Two: *The Consequences of*

Determinism (Oxford: Clarendon Press, 1988); also *How Free Are You? The Determinism Problem* (Oxford: Oxford University Press, 1993).

47 For detailed reference to primary and secondary texts by and about Spinoza, see notes to Chapter 4.

48 On this topic see especially Jean-Paul Sartre, *Critique of Dialectical Reason*, Vol. 1 *Theory of Practical Ensembles*, trans. A. Sheridan-Smith (London: New Left Books, 1976); Vol. 2 *The Intelligibility of History*, trans. Quintin Hoare (London: Verso, 1991); also – for a sharply opposing view – Maurice Merleau-Ponty, *Adventures of the Dialectic*, trans. Joseph Bien (Evanston, IL: Northwestern University Press, 1973) and *The Prose of the World*, trans. John O'Neill (Evanston, IL: Northwestern University Press, 1974).

49 See for instance the useful compendium of Shelley's atheist (or anti-Christian) pronouncements at http://www.positiveatheism.org/hist/quotes/shelley.htm (accessed on 10 September 2009).

50 J. S. Mill, *An Examination of Sir William Hamilton's Philosophy, and of the Principal Philosophical Questions Raised in his Writings* (London: Longmans, Green and Dyer, 1878), p. 103.

51 Ibid. p. 103.

52 For a full-scale critique of this and other facile anti-Enlightenment jibes, see Norris, *What's Wrong with Postmodernism?* (Hemel Hempstead: Harvester, 1990) and *The Truth About Postmodernism* (Oxford: Blackwell, 1994).

53 See previous note 5; also John R. Searle, *Intentionality: An Essay in the Philosophy of Mind* (Cambridge: Cambridge University Press, 1983) and *The Rediscovery of the Mind* (Cambridge, MA: MIT Press, 1992).

54 See for instance G. E. M. Anscombe, 'Causality and Determinism', in Ernest Sosa (ed.), *Causation and Conditionals* (Oxford: Oxford University Press, 1975), pp. 63–81; also – for an argument against such abusive extrapolations from the quantum-physical domain – Norris, *Quantum Theory and the Flight from Realism: Philosophical Responses to Quantum Mechanics* (London: Routledge, 2000), esp. pp. 134–64.

55 Wittgenstein, *Philosophical Investigations*, see previous note 20 ; also – for one of the most influential statements of this case – Peter Winch, *The Idea of a Social Science and Its Relation to Philosophy* (London: Routledge & Kegan Paul, 1958).

56 See previous notes 42, 43 and 46

57 Kant, *Critique of Practical Reason and Other Writings in Moral Philosophy*, trans. Lewis W. Beck (New York: Garland, 1976).

58 W. V. Quine, *Ontological Relativity and Other Essays* (New York: Columbia University Press, 1969), p. 72.

59 Jonathan Glover, *Humanity: A Moral History of the Twentieth Century* (London: Pimlico, 2001).

60 For further discussion of these issues see Norris, *Truth Matters: Realism, Anti-realism and Response-Dependence* (Edinburgh: Edinburgh University Press, 2002).

61 Gilbert Ryle, *On Thinking* (Oxford: Blackwell, 1979).

62 See especially Paul M. Churchland and Patricia S. Churchland, *On the Contrary: Critical Essays, 1987–1997* (Cambridge, MA: MIT Press, 1998).

63 See especially Paul M. Churchland, *Scientific Realism and the Plasticity of Mind* (Cambridge: Cambridge University Press, 1979) and *Matter and Consciousness* (Cambridge, MA: MIT Press, 1984).

[64] Louis Althusser, *Philosophy of the Encounter: Later Writings, 1978–1987*, trans. G. M. Goshgarian (London: Verso, 2006).

Chapter 1

[1] For further discussion from a wide range of disciplinary and philosophic viewpoints, see Richmond Campbell and Bruce Hunter (eds), *Moral Epistemology Naturalized* (Calgary: University of Calgary Press, 2000); Peter Carruthers, *Phenomenal Consciousness: A Naturalistic Theory* (Cambridge: Cambridge University Press, 1993); Fred Dretske, *Naturalizing the Mind* (Cambridge, MA: MIT Press, 1997); Alvin Goldman, *Epistemology and Cognition* (Cambridge, MA: Harvard University Press, 1986); Jonathan Knowles, *Norms, Naturalism and Epistemology* (Basingstoke: Palgrave-Macmillan, 2003); Hilary Kornblith, *Knowledge and Its Place in Nature* (Oxford: Clarendon Press, 2002); Kornblith (ed.), *Naturalizing Epistemology*, 2nd edn. (Cambridge, MA: MIT Press, 1994); Penelope Maddy, *Second Philosophy: A Naturalistic Method* (Oxford: Oxford University Press, 2007); David Papineau, *Philosophical Naturalism* (Oxford: Blackwell, 1993); Peter Railton, *Facts, Values, and Norms: Essays Toward a Morality of Consequence* (Cambridge: Cambridge University Press, 2003); Jack Ritchie, *Understanding Naturalism* (Stocksfield: Acumen, 2008); Michael Ruse, *Evolutionary Naturalism: Selected Essays* (London: Routledge, 1995); Ralph Wedgwood, *The Nature of Normativity* (Oxford: Clarendon Press, 2007).

[2] See Otto Neurath, Rudolf Carnap and Charles Morris (eds), *Foundations of the Unity of Science: Towards an International Encyclopedia of Unified Science* (Chicago: University of Chicago Press, 1955–1970); also Carnap, *The Unity of Science*, Max Black (ed.) (Bristol: Thoemmes Press, 1995) and Robert L. Causley, *Unity of Science* (Dordrecht: D. Reidel, 1977). For some more recent and strongly dissenting views, see John Dupré, *The Disorder of Things: Metaphysical Foundations of the Disunity of Science* (Cambridge, MA: Harvard University Press, 1993); Peter Gallison and David J. Stump (eds), *The Disunity of Science: Boundaries, Contexts, and Power* (Stanford, CA: Stanford University Press, 1996); Joseph Margolis, *Science without Unity: Reconciling the Human and Natural Sciences* (Oxford: Blackwell, 1987).

[3] W. V. Quine, 'Two Dogmas of Empiricism', in *From a Logical Point of View*, 2nd edn. (Cambridge, MA: Harvard University Press, 1961), pp. 20–46.

[4] For further discussion, see Norris, *Resources of Realism: Prospects for 'Post-analytic' Philosophy* (London: Macmillan, 1997) and *New Idols of the Cave: On the Limits of Anti-realism* (Manchester: Manchester University Press, 1997).

[5] See note 4; also Norris, *Philosophy of Language and the Challenge to Scientific Realism* (London: Routledge, 2004) and *On Truth and Meaning: Language, Logic and the Grounds of Belief* (London: Continuum, 2006).

[6] See especially Jaegwon Kim, *Supervenience and Mind: Selected Philosophical Essays* (Cambridge: Cambridge University Press, 1993).

[7] For a representative sampling of views, see Richard Rorty (ed.), *The Linguistic Turn: Recent Essays in Philosophical Method* (Chicago, IL: University of Chicago Press, 1967) and Kurt Mueller-Vollmer (ed.), *The Hermeneutics Reader: Texts of the German Tradition from the Enlightenment to the Present* (New York: Continuum, 1985).

[8] See for instance David Chalmers, *The Conscious Mind: In Search of a Fundamental Theory* (Oxford: Oxford University Press, 1996); Paul M. Churchland, *Scientific Realism and the Plasticity of Mind* (Cambridge: Cambridge University Press, 1979) and *Matter and Consciousness* (Cambridge, MA: MIT Press, 1984); Jeffrey A. Gray, *Consciousness: Creeping up on the Hard Problem* (Oxford: Oxford University Press, 2004); Joseph Levine, *Purple Haze: The Puzzle of Consciousness* (Oxford: Oxford University Press, 2002); Colin McGinn, *The Mysterious Flame: Conscious Minds in a Material World* (New York: Basic Books, 1999); William S. Robinson, *Understanding Phenomenal Consciousness* (Cambridge: Cambridge University Press, 2004); Quentin Smith and Aleksandar Jokic (eds), *Consciousness: New Philosophical Perspectives* (Oxford: Oxford University Press, 2003).

[9] Donald Davidson, *Essays on Actions and Events* (Oxford: Clarendon Press, 1980).

[10] See especially Kim, *Supervenience and Mind*, previous note 6; also Mark Rowlands, *Supervenience and Materialism* (Aldershot: Avebury, 1995).

[11] For a range of views see Peter van Inwagen, *An Essay on Free Will* (Oxford: Clarendon Press, 1983); Robert Kane, *The Significance of Free Will* (Oxford: Oxford University Press, 2004); Timothy O'Connor, *Persons and Causes: the metaphysics of free will* (Oxford: Oxford University Press, 2004); Galen Strawson, *Freedom and Belief* (Oxford: Oxford University Press, 1986); Gary Watson (ed.), *Free Will* (Oxford: Oxford University Press, 1982).

[12] See previous notes 4 and 5 .

[13] Davidson, *Essays on Actions and Events*, see previous note 9 .

[14] See previous note 8.

[15] See especially F. R. Leavis, *Two Cultures? The Significance of C. P. Snow* (London: Chatto & Windus, 1962); C. P. Snow, *The Two Cultures, and a Second Look* (Cambridge: Cambridge University Press, 1963); also Norris, *Fiction, Philosophy, and Literary Theory: Will the Real Saul Kripke Please Stand Up?* (London: Continuum, 2007) and 'Sexed Equations and Vexed Physicists: The "Two Cultures" Revisited', in *Deconstruction and the Unfinished Project of Modernity* (London: Athlone Press, 2000), pp. 175–201.

[16] For some illuminating discussion, see Paul Johnston, *The Contradictions of Modern Moral Philosophy: Ethics after Wittgenstein* (London: Routledge, 1999).

[17] See for instance John Martin Fischer, Robert Kane, Derk Pereboom, and Manuel Vargas. *Four Views on Free Will* (Oxford: Blackwell, 2007); Robert Kane, *A Contemporary Introduction to Free Will* (Oxford: Oxford University Press, 2005) and Kane (ed.), *Free Will* (Oxford: Blackwell, 2002); also – for a strongly contrasting view – Derk Pereboom, *Living Without Free Will* (Cambridge: Cambridge University Press, 2001).

[18] I offer some arguments to similar effect concerning a related field of philosophical debate in Norris, *Truth Matters: Realism, Anti-realism and Response-Dependence* (Edinburgh: Edinburgh University Press, 2002).

[19] Immanuel Kant, *Critique of Pure Reason*, trans. N. Kemp Smith (London: Macmillan, 1964).

[20] John McDowell, *Mind and World* (Cambridge, MA: Harvard University Press, 1994); Robert C. Brandom, *Making It Explicit: reasoning, representing and discursive commitment* (Cambridge, MA: Harvard University Press, 1994). For critical discussion see Norris, 'McDowell on Kant: Redrawing the Bounds of Sense' and 'The Limits of Naturalism: Further Thoughts on McDowell's *Mind and World*', in

Minding the Gap: Epistemology and Philosophy of Science in the Two Traditions (Amherst, MA: University of Massachusetts Press, 2000), pp. 172–96 and 197–230.

21 See various entries under previous note 8.

22 For a classic treatment of this issue see Frank Jackson, 'What Mary Didn't Know', *The Journal of Philosophy*, Vol. 83, No. 5 (May 1986), 291–5; also previous note 8.

23 See the previous note 17; also Thomas Nagel, *Mortal Questions* (Cambridge: Cambridge University Press, 1979).

24 See especially Mark Addis, *Wittgenstein: Making Sense of Other Minds* (Aldershot: Ashgate, 1999); Severin Schroeder (ed.), *Wittgenstein and Contemporary Philosophy of Mind* (Basingstoke: Palgrave, 2001); David G. Stern, *Wittgenstein on Mind and Language* (Oxford: Oxford University Press, 1995).

25 See especially Daniel C. Dennett, *Consciousness Explained* (Harmondsworth: Penguin, 1993); also Dennett, *Elbow Room: The Varieties of Free Will Worth Wanting* (Oxford: Oxford University Press, 1984).

26 See especially Ian Hacking, *Why Does Language Matter to Philosophy?* (Cambridge: Cambridge University Press, 1975); also Jonathan Bennett, *Learning from Six Philosophers, Vol. One: Descartes, Spinoza, Leibniz* (Oxford: Clarendon Press, 2001).

27 See previous note 3.

28 For further discussion see Norris, *On Truth and Meaning*, previous note 5 .

29 See especially Quine, *Ontological Relativity and Other Essays* (New York: Columbia University Press, 1969).

30 See previous note 1; also Wesley C. Salmon, *Scientific Explanation and the Causal Structure of the World* (Princeton, NJ: Princeton University Press, 1984).

31 Saul A. Kripke, *Naming and Necessity* (Oxford: Blackwell, 1980) and Hilary Putnam, *Mind, Language and Reality* (Cambridge: Cambridge University Press, 1975); also Stephen P. Schwartz (ed.), *Naming, Necessity and Natural Kinds* (Ithaca, NY: Cornell University Press, 1975).

32 Kant, *Critique of Practical Reason and Other Writings in Moral Philosophy*, trans. L. W. Beck (Chicago, IL: University of Chicago Press, 1976).

33 See for instance Rorty, *Consequences of Pragmatism* (Brighton: Harvester, 1982) and *Objectivity, Relativism, and Truth* (Cambridge: Cambridge University Press, 1991).

34 Rorty, 'Texts and Lumps', in *Objectivity, Relativism, and Truth*, see previous note 33, pp. 78–92.

35 William James, *Pragmatism: A New Name for Some Old Ways of Thinking* (New York: Longmans, 1907) and *The Meaning of Truth* (New York: Longmans, 1909).

36 Bertrand Russell, 'William James's Conception of Truth', in Simon Blackburn and Keith Simmons (eds.), *Truth* (Oxford: Oxford University Press, 1999), pp. 69–82; also – for another vigorous rebuttal – W. K. Clifford, 'The Ethics of Belief', in *The Ethics of Belief and Other Essays* (New York: Prometheus Books, 1999).

37 See especially Jonathan Bennett, *A Study of Spinoza's* Ethics (Cambridge: Cambridge University Press, 1984) and *Learning from Six Philosophers, Vol. One*, previous note 26; Alan Donagan, *Spinoza* (Chicago, IL: University of Chicago Press, 1988); Stuart Hampshire, *Spinoza* (Harmondsworth: Penguin, 1951). Two very useful anthologies of commentary and criticism are Genevieve Lloyd (ed.), *Spinoza* (4 vols, London: Routledge, 2001) and Don Garrett (ed.), *The Cambridge Companion to Spinoza* (Cambridge: Cambridge University Press, 1996).

[38] Donald Davidson, 'Spinoza's Causal Theory of the Affects', in *Truth, Language, and History* (Cambridge: Cambridge University Press, 2005), pp. 295–313.

[39] See previous note 24.

[40] See especially Ted Honderich, *A Theory of Determinism*, Vol. One: *The Mind, Neuroscience and Life-Hopes* (Oxford: Clarendon Press, 1988) and Vol. Two: *The Consequences of Determinism* (Oxford: Clarendon Press, 1988); also *How Free Are You? The Determinism Problem* (Oxford: Oxford University Press, 1993).

[41] G. E. M. Anscombe, 'Causality and Determinism', in Ernest Sosa (ed.), *Causation and Conditionals* (Oxford: Oxford University Press, 1975), pp. 63–81; also – for further discussion of Anscombe's claim – Norris, *Quantum Theory and the Flight from Realism: Philosophical Responses to Quantum Mechanics* (London: Routledge, 2000), esp. pp. 134–64.

[42] Gilbert Ryle, *The Concept of Mind* (Harmondsworth: Penguin, 1963).

[43] Ryle, *On Thinking* (Oxford: Blackwell, 1979).

[44] See previous note 24.

[45] Saul Kripke, *Wittgenstein on Rules and Private Language: An Elementary Exposition* (Oxford: Blackwell, 1982); also Alexander Miller and Crispin Wright (eds.), *Rule-Following and Meaning* (Chesham: Acumen, 2002).

[46] Ludwig Wittgenstein, *Philosophical Investigations*, trans. G. E. M. Anscombe (Oxford: Blackwell, 1958), Section 265.

[47] See especially Peter Winch, *The Idea of a Social Science and Its Relation to Philosophy* (London: Routledge & Kegan Paul, 1958) and *Trying to Make Sense* (Oxford: Blackwell, 1987).

[48] For more detailed discussion see Norris, *Philosophy of Language and the Challenge to Scientific Realism*, previous note 5.

[49] See previous note 17.

[50] Rorty, 'Pragmatism, Davidson and Truth', in *Objectivity, Relativism, and Truth*, previous note 33, pp. 126–50 and 'Is Truth a Goal of Inquiry? Donald Davidson versus Crispin Wright', in *Truth and Progress* (Cambridge: Cambridge University Press, 1998), pp. 19–42; also Davidson, 'Afterthoughts, 1987', in Alan R. Malachowski (ed.), *Reading Rorty: Critical Responses to Philosophy and the Mirror of Nature, and Beyond* (Oxford: Blackwell, 1990), pp. 134–8.

[51] McDowell, *Mind and World*, previous note 20 ; also P. F. Strawson, *The Bounds of Sense: An Essay on Kant's* Critique of Pure Reason (London: Methuen, 1966).

[52] McDowell, *Mind and World*, previous note 20.

[53] See especially Frederick C. Beiser, *The Fate of Reason: German Philosophy from Kant to Fichte* (Cambridge, MA: Harvard University Press, 1987).

[54] On this aspect of Spinoza's thought, see Jonathan Israel, *Radical Enlightenment: Philosophy and the Making of Modernity* (Oxford: Oxford University Press, 2002); Christopher Norris, *Spinoza and the Origins of Modern Critical Theory* (Oxford: Blackwell, 1991); Paul Wienpahl, *The Radical Spinoza* (New York: New York University Press, 1979); Yirmiyahu Yovel, *Spinoza and Other Heretics*, Vol. One*: The Marrano of Reason*, and Vol. Two, *The Adventures of Immanence* (Princeton, NY: Princeton University Press, 1989).

[55] See previous notes 1, 4, 5, 30 and 31.

[56] See previous note 37.

[57] For further discussion see D. M. Armstrong and Norman Malcolm, *Consciousness and Causality: A Debate on the Nature of Mind* (Oxford: Blackwell, 1984); Tim Crane

and Sarah Patterson (eds), *A History of the Mind-Body Problem* (London: Routledge, 2000); Margaret Donaldson, *Human Minds: An Exploration* (Harmondsworth, Penguin, 1992); Cynthia Macdonald, *Mind-Body Identity Theories* (London: Routledge, 1989); Stephen Priest, *Theories of the Mind* (Harmondsworth: Penguin, 1991); David M. Rosenthal (ed.), *The Nature of Mind* (New York: Oxford University Press, 1991); G. N. A. Vesey (ed.), *Body and Mind: Readings in Philosophy* (London: Allen & Unwin, 1964).

Chapter 2

1 Harry Frankfurt, 'Freedom of the Will and the Concept of a Person', in Gary Watson (ed.), *Free Will* (Oxford: Oxford University Press, 2003). See also Frankfurt, *The Importance of What We Care About: Philosophical Essays* (Cambridge: Cambridge University Press, 1988).

2 For further discussion from a range of viewpoints, see Peter van Inwagen, *An Essay on Free Will* (Oxford: Clarendon Press, 1983); Robert Kane, *The Significance of Free Will* (Oxford: Oxford University Press, 2004); Timothy O'Connor, *Persons and Causes: The Metaphysics of Free Will* (Oxford: Oxford University Press, 2004); Galen Strawson, *Freedom and Belief* (Oxford: Oxford University Press, 1986); Watson (ed.), *Free Will*, previous note 1.

3 See for instance Alfred Tarski 'The Concept of Truth in Formalised Languages', in *Logic, Semantics and Metamathematics*, trans. J. H. Woodger (Oxford: Oxford University Press, 1956), pp. 152–278; also Rudolf Carnap, *The Logical Structure of the World, and Pseudoproblems in Philosophy*, trans. R. George (Berkeley and Los Angeles: University of California Press, 1969).

4 For a good sampling of views, see A. J. Ayer (ed.), *Logical Positivism* (New York: Free Press, 1959).

5 W. V. Quine, 'Two Dogmas of Empiricism', in *From a Logical Point of View*, 2nd ed. (Cambridge, MA: Harvard University Press, 1961), pp. 20–46; also – for a more detailed account of these developments – Christopher Norris, *Philosophy of Language and the Challenge to Scientific Realism* (London: Routledge, 2004) and *On Truth and Meaning: Language, Logic and the Grounds of Belief* (London: Continuum, 2006).

6 See for instance Bertrand Russell, *Introduction to Mathematical Philosophy* (London: Allen & Unwin, 1930); also Russell, *Foundations of Logic, 1903–1905*, ed. Alasdair Urquhart (London: Routledge, 1994).

7 On this chapter in German post-Kantian philosophy, see David Simpson (ed.), *The Origins of Modern Critical Thought: German Aesthetic and Literary Criticism from Lessing to Hegel* (Cambridge: Cambridge University Press, 1988).

8 Immanuel Kant, *Critique of Practical Reason, and Other Writings in Moral Philosophy*, trans. and ed. L. W. Beck (New York: Garland, 1976).

9 Immanuel Kant, *Critique of Pure Reason*, trans. and ed. Paul Guyer and Allen W. Wood (Cambridge: Cambridge University Press, 1998).

10 For further discussion and extended commentary on various recent approaches to these problems with Kant's doctrine of the faculties, see Norris, 'Kant Disfigured: Ethics, Deconstruction, and the Textual Sublime', in *The Truth About Postmodernism* (Oxford: Blackwell, 1993), pp. 182–256 and *Minding the Gap: Epistemology and*

Philosophy of Science in the Two Traditions (Amherst, MA: University of Massachusetts Press, 2000).

11 See Norris, 'Ethics, Autonomy, and the Grounds of Belief', in *On Truth and Meaning*, previous note 5, pp. 130–54 and 'Choice and Belief', *The Philosophers' Magazine*, Issue 29 (2005), 17–23.

12 William James, *Pragmatism: A New Name for Some Old Ways of Thinking* (New York: Longmans, 1907) and *The Meaning of Truth* (Longmans, 1909); Bertrand Russell, 'William James's Conception of Truth', in Simon Blackburn and Keith Simmons (eds), *Truth* (Oxford: Oxford University Press, 1999), pp. 69–82; also – for another vigorous counter-statement to the Jamesian pragmatist approach – W. K. Clifford, 'The Ethics of Belief', in *The Ethics of Belief and Other Essays* (New York: Prometheus Books, 1999).

13 William Shakespeare, *Hamlet*, Act 3, Scene 1, line 85.

14 See P. F. Strawson, *The Bounds of Sense: An Essay on Kant's* Critique of Pure Reason (London: Methuen, 1966).

15 Strawson, *The Bounds of Sense*, previous note 14; also *Individuals: An Essay in Descriptive Metaphysics* (London: Methuen, 1959).

16 Strawson, *Individuals*, previous note 15.

17 Peter Singer, *Animal Liberation: A New Ethics for Our Treatment of Animals*, 2nd edn. (London: Cape, 1990); also Singer (ed.), *In Defence of Animals* (Oxford: Blackwell, 1985); Tom Regan and Peter Singer (eds), *Animal Rights and Human Obligations* (Englewood Cliffs: Prentice-Hall, 1976).

18 See especially Giorgio Agamben, *The Open: Man and Animal*, trans. Kevin Attell (Stanford, CA: Stanford University Press, 2004); also William Empson, *The Structure of Complex Words* (London: Chatto & Windus, 1951).

19 Agamben, *The Open*, previous note 18.

20 For a classic study of this doctrine in its various historical forms, see Arthur O. Lovejoy, *The Great Chain of Being: A Study in the History of an Idea* (Cambridge, MA: Harvard University Press, 1936).

21 See for instance Roland Barthes, *Sade, Fourier, Loyola*, trans. Richard Miller (London: Cape, 1977) and Jacques Lacan, 'Kant with Sade', trans. James Swenson, *October*, No. 15 (Winter 1989), 55–104 and Slavoj Zizek, 'Kant With (or Against) Sade?', *New Formations*, No. 35 (1998), 93–107.

22 See previous note 8.

23 St Augustine, *De Bono Coniugali, De Sancta Virginitate*, trans. P. G. Walsh (Oxford: Oxford University Press, 2001).

24 Plato, *Phaedrus*, trans. Robin Waterfield (Oxford: Oxford University Press, 2002), sections 246a–254e.

25 See previous notes 8 and 9; also Gilles Deleuze, *Kant's Critical Philosophy: The Doctrine of the Faculties*, trans. Hugh Tomlinson and Barbara Habberjam (London: Athlone Press, 1984).

26 For a scholarly, perceptive and unusually angled approach to these themes, see Geoffrey Galt Harpham, *The Ascetic Imperative in Culture and Criticism* (Chicago: University of Chicago Press, 1987); also Norris, 'Kant Disfigured', previous note 10.

27 See previous notes 21 and 26.

28 For a classic treatment of this episode in literary history, see Ian Watt, *The Rise of the Novel: Studies in Defoe, Richardson and Fielding* (London: Chatto & Windus, 1957).

[29] See for instance Lorraine Code, *Epistemic Responsibility* (Hanover, NH: University Press of New England, 1987); M. DePaul and L. Zagzebski (eds), *Intellectual Virtue: Perspectives from Ethics and Epistemology* (Oxford: Oxford University Press, 2002); A. Fairweather and L. Zagzebski (eds), *Virtue Epistemology: Essays on Epistemic Virtue and Responsibility* (Oxford: Oxford University Press, 2001); L. Zagzebski, *Virtues of the Mind: An Inquiry into the Nature of Virtue and the Ethical Foundations of Knowledge* (Cambridge: Cambridge University Press, 1996).

[30] Daniel C. Dennett, *Elbow Room: The Varieties of Free Will Worth Wanting* (Oxford: Oxford University Press, 1984).

[31] See for instance John Foster, *The Immaterial Self: A Defence of the Cartesian Dualist Conception of the Mind* (London: Routledge, 1991); also – from a different, more qualified dualist standpoint – Karl R. Popper and John C. Eccles, *The Self and Its Brain* (London: Routledge, 1998).

[32] See previous note 18.

[33] John McDowell, *Mind and World* (Cambridge, MA: Harvard University Press, 1994); also Christopher Norris, 'McDowell on Kant: Redrawing the Bounds of Sense' and 'The Limits of Naturalism: Further Thoughts on McDowell's *Mind and World*', in *Minding the Gap*, see previous note 10 , pp. 172–96 and 197–230.

[34] See previous notes 5 and 10.

[35] I put this case more fully in Norris, *Truth Matters: Realism, Anti-Realism and Response-Dependence* (Edinburgh: Edinburgh University Press, 2002).

[36] See especially Roderick M. Chisholm, 'The Agent as Cause', in M. Brand and D. Walton (eds), *Action Theory* (Dordrecht: D. Reidel, 1977), pp. 199–211; also Chisholm, *Person and Object* (La Salle, IL: Open Court, 1976).

[37] See previous notes 15, 16 and 33.

[38] On the waning of this aprioristic way of thinking about issues in epistemology and philosophy of mind, see J. Alberto Coffa, *The Semantic Tradition from Kant to Carnap: To the Vienna Station* (Cambridge: Cambridge University Press, 1991).

[39] See for instance David M. Rosenthal (ed.), *Materialism and the Mind-Body Problem* (Indianapolis: Hackett, 2000).

[40] For a representative range of views, see Peter Carruthers, *Phenomenal Consciousness: A Naturalistic Theory* (Cambridge: Cambridge University Press, 2003); Jeffrey A. Gray, *Consciousness: Creeping up on the Hard Problem* (Oxford: Oxford University Press, 2004); Joseph Levine, *Purple Haze: The Puzzle of Consciousness* (Oxford: Oxford University Press, 2002); Colin McGinn, *The Mysterious Flame: Conscious Minds in a Material World* (New York: Basic Books, 1999); William S. Robinson, *Understanding Phenomenal Consciousness* (Cambridge: Cambridge University Press, 2004); John R. Searle, *Intentionality: An Essay in the Philosophy of Mind* (Cambridge: Cambridge University Press,. 1983), *The Rediscovery of the Mind* (Cambridge, MA: MIT Press, 1992), *Consciousness and Language* (Cambridge: Cambridge University Press, 2002); Sidney Shoemaker, *Physical Realization* (Oxford: Oxford University Press, 2007); Quentin Smith and Aleksandar Jokic (eds.), *Consciousness: New Philosophical Perspectives* (Oxford: Oxford University Press, 2003).

[41] See especially entries for Searle, previous note 40.

[42] See McGinn, *The Mysterious Flame*, previous note 40.

[43] Kant, *Critique of Pure Reason*, see previous note 9.

[44] See especially Jaegwon Kim, *Supervenience and Mind: Selected Philosophical Essays* (Cambridge: Cambridge University Press, 1993); also Kim (ed.), *Supervenience*

(Dartmouth: Ashgate, 2002); Mark Rowlands, *Supervenience and Materialism* (Aldershot: Avebury, 1995).

45 See previous notes 3 and 4.

46 Bertrand Russell, 'Mathematical Logic as Based on the Theory of Types', *American Journal of Mathematics*, Vol. 30 (1908), 222–62; also Russell, *My Philosophical Development* (London: Routledge & Kegan Paul, 1959).

47 See previous note 8.

48 See various entries under previous note 40; also Quentin Smith and Aleksandar Jokic (eds.), *Consciousness: New Philosophical Perspectives* (Oxford: Oxford University Press, 2003).

49 See especially Donald Davidson, *Essays on Actions and Events* (Oxford: Clarendon Press, 1980).

50 Peter Carruthers, *The Animal Issue: Moral Theory in Practice* (Cambridge: Cambridge University Press, 1992); also Carruthers, *The Nature of the Mind: An Introduction* (London: Routledge, 2004) and *Consciousness: Essays from a Higher-Order Perspective* (Oxford: Oxford University Press, 2005).

51 See previous note 17.

52 Jeremy Bentham, *Introduction to the Principles of Morals and Legislation*, eds. J. H. Burns and H. L. A. Hart (London: Athlone Press, 1970), Chapter 17, para 4, note.

53 See previous notes 31, 40, 44 and 50.

Chapter 3

1 For further discussion from a range of viewpoints, see David J. Chalmers, *The Conscious Mind: In Search of a Fundamental Theory* (Oxford: Oxford University Press, 1996); Tim Crane, *The Mechanical Mind: A Philosophical Introduction to Minds, Machines and Mental Representation* (London: Routledge, 2003); Margaret Donaldson, *Human Minds: An Exploration* (Harmondsworth: Penguin, 1993); William G. Lycan, *Consciousness and Experience* (Cambridge, MA: MIT Press, 1996); Joseph Levine, *Purple Haze: The Puzzle of Consciousness* (Oxford: Oxford University Press, 2002); Cynthia Macdonald, *Mind-Body Identity Theories* (London: Routledge, 1989); Stephen Priest, *Theories of the Mind* (Harmondsworth: Penguin, 1991); David M. Rosenthal (ed.), *The Nature of Mind* (New York: Oxford University Press, 1991); Quentin Smith and Aleksandar Jokic (eds.), *Consciousness: New Philosophical Perspectives* (Oxford: Oxford University Press, 2003); Lawrence Weiskrantz and Martin Davies (eds.), *Frontiers of Consciousness* (Oxford: Oxford University Press, 2008).

2 See especially Ludwig Wittgenstein, *Philosophical Investigations*, trans. and ed. G. E. M. Anscombe (Oxford: Blackwell, 1958) and J. L. Austin, *Philosophical Papers* (Oxford: Clarendon Press, 1961).

3 Martin Heidegger, *Poetry, Language, Thought*, trans. Albert Hofstadter (New York: Harper & Row, 2001).

4 For a critical overview of these and related developments, see Christopher Norris, *What's Wrong with Postmodernism?* (Hemel Hempstead: Harvester-Wheatsheaf, 1990), *The Truth about Postmodernism* (Oxford: Blackwell, 1993), *New Idols of the*

Cave: On the Limits of Anti-realism (Manchester: Manchester University Press, 1997); *Resources of Realism: Prospects for 'Post-analytic' Philosophy* (London: Macmillan, 1997).

[5] William Empson, *The Structure of Complex Words*, 2nd edn. (London: Chatto & Windus, 1965).

[6] See especially Noam Chomsky, *Language and Mind* (New York: Harcourt, Brace, Jovanovich, 1972), *Reflections on Language* (New York: Pantheon, 1975), *Knowledge of Language: Its Nature, Origin, and Use* (New York: Praeger, 1986), *Language and Problems of Knowledge: The Managua Lectures* (Cambridge, MA: MIT Press, 1988).

[7] Chomsky, *Cartesian Linguistics* (New York: Harper & Row, 1966).

[8] Norris, 'Modularity, Nativism, and Reference-Fixing: On Chomsky's Internalist Assumptions', in Norris, *Language, Logic and Epistemology: A Modal-realist Approach* (London: Macmillan, 2004), pp. 111–49.

[9] See for instance Alexander George (ed.), *Reflections on Chomsky* (Oxford: Blackwell, 1989); Gilbert Harman (ed.), *On Noam Chomsky: Critical Essays* (Garden City, NY: Anchor Press, 1974); Asa Kasher (ed.), *The Chomskian Turn* (Oxford: Blackwell, 1991); James McGilvray (ed.), *The Cambridge Companion to Chomsky* (Cambridge: Cambridge University Press, 2005).

[10] Chomsky, *Lectures on Government and Binding* (Berlin: Mouton de Gruyter, 1993); also Gert Webelhuth (ed.), *Government and Binding and the Minimalist Programme: Principles and Parameters in Syntactic Theory* (Oxford: Blackwell, 1995).

[11] Immanuel Kant, *Critique of Pure Reason*, trans. N. Kemp Smith (London: Macmillan, 1964).

[12] René Descartes, *Discourse on Method and Other Writings*, trans. Arthur Wollaston (Harmondsworth: Penguin, 1960).

[13] See for instance Chomsky, *American Power and the New Mandarins* (New York: Pantheon, 1967), *Deterring Democracy* (London: Verso, 1991), *World Orders Old and New* (New York: Columbia University Press, 1994).

[14] Kant 'Transcendental Aesthetic', in *Critique of Pure Reason*, pp. 65–82, previous note 11.

[15] For some highly illuminating commentary, see J. Alberto Coffa, *The Semantic Tradition from Kant to Carnap: To the Vienna Station* (Cambridge: Cambridge University Press, 1991).

[16] See especially Michael Dummett, *Truth and Other Enigmas* (London: Duckworth, 1978) and *The Logical Basis of Metaphysics* (Duckworth, 1991); also Neil Tennant, *Anti-realism and Logic* (Oxford: Clarendon Press, 1987) and *The Taming of the True* (Oxford: Oxford University Press, 1997).

[17] Norris, *New Idols of the Cave*, previous note 4 , *Resources of Realism*, previous note 4, *Truth Matters: Realism, Anti-realism and Response-dependence* (Edinburgh: Edinburgh University Press, 1992).

[18] See previous note 16.

[19] Edmund Husserl, *Formal and Transcendental Logic*, trans. Dorion Cairns (The Hague: Martinus Nijhoff, 1969) and *Experience and Judgement: Investigations in a Genealogy of Logic*, trans. James S. Churchill and Karl Ameriks (Evanston, IL: Northwestern University Press, 1973).

[20] See for instance Jeffrey Alan Gray, *Consciousness: Creeping up on the Hard Problem* (Oxford: Oxford University Press, 2004); Jonathan Shear (ed.), *Explaining*

Consciousness: The 'Hard Problem' (Cambridge, MA: MIT Press, 1997); Evan Thompson (ed.), *The Problem of Consciousness: New Essays in Phenomenological philosophy of mind* (Calgary, Alta: University of Calgary Press, 2003).

21 The *locus classicus* here is Thomas Nagel's essay 'What is it Like to be a Bat?', *Philosophical Review*, Vol. 83 (1974), 435–50. See also Nagel, 'Subjective and Objective', in his *Mortal Questions* (Cambridge: Cambridge University Press, 1979), pp. 196–213 and William S. Robinson, *Understanding Phenomenal Consciousness* (Cambridge University Press, 2004).

22 See previous notes 1, 20 and 21.

23 Gilbert Ryle, *The Concept of Mind* (London: Hutchinson, 1949).

24 See C. H. Langford, 'The Notion of Analysis in Moore's Philosophy', in P. A. Schlep (ed.), *The Philosophy of G. E. Moore* (La Salle: Open Court, 1968), pp. 321–41 and G. E. Moore, 'A Reply to My Critics', ibid, pp. 535–687.

25 Michael Devitt. *Ignorance of Language* (Oxford: Oxford University Press, 2006).

26 Owen Flanagan, *Consciousness Reconsidered* (Cambridge, MA: MIT Press, 1992); Colin McGinn, *The Character of Mind: An Introduction to the Philosophy of Mind*, 2nd edn. (Oxford: Oxford University Press, 1991) and *The Mysterious Flame: Conscious Minds in a Material World* (New York: Basic Books, 1999).

27 See especially Gilles Deleuze and Félix Guattari, *What Is Philosophy?*, trans. Hugh Tomlinson and Graham Burchill (London: Verso, 1994).

28 See previous note 2.

29 For a useful conspectus of these two branches of the analytic enterprise, see Richard Rorty (ed.), *The Linguistic Turn: Recent Essays in Philosophical Method* (Chicago, IL: University of Chicago Press, 1967).

30 Baruch Spinoza, *Ethics*, in *The Collected Writings of Spinoza*, trans. Edwin Curley (Princeton, NJ: Princeton University Press, 1985); also Jonathan Bennett, *A Study of Spinoza's Ethics* (Cambridge: Cambridge University Press, 1984).

31 For useful commentary see Alan Donagan, *Spinoza* (Chicago, IL: University of Chicago Press, 1988); Stuart Hampshire, *Spinoza* (Harmondsworth: Penguin, 1951); G. H. R. Parkinson, *Spinoza's Theory of Knowledge* (Oxford: Clarendon Press, 1953).

32 Donald Davidson, 'Spinoza's Causal Theory of the Affects', in *Truth, Language, and History* (Cambridge: Cambridge University Press, 2005), pp. 295–313.

33 See previous notes 20 and 21.

34 W. V. Quine. 'Two Dogmas of Empiricism', in *From a Logical Point of View*, 2nd ed. (Cambridge, MA: Harvard University Press, 1961), pp. 20–46; p. 38.

35 Quine, 'Two Dogmas of Empiricism', previous note 34 and *Ontological Relativity and Other Essays* (New York: Columbia University Press, 1961); also Thomas S. Kuhn, *The Structure of Scientific Revolutions*, 2nd edn. (Chicago: University of Chicago Press, 1961).

36 For further argument to this effect, see Norris, 'The Limits of *Whose* Language?: Wittgenstein on Logic, Mathematics, and Science', in *Language, Logic and Epistemology*, previous note 8, pp. 66–110.

37 Kant, *Critique of Pure Reason*, previous note 11.

38 See previous note 36.

39 See Wittgenstein, *Philosophical Investigations*, previous note 2, Sections 201–92; Saul Kripke, *Wittgenstein on Rules and Private Language: An Elementary Exposition*

(Oxford: Blackwell, 1982); also Alexander Miller and Crispin Wright (eds), *Rule-Following and Meaning* (Chesham: Acumen, 2002).

40 See previous note 36.

41 Wittgenstein, *Philosophical Investigations*, previous note 2 and *On Certainty*, trans. and ed. G. E. M. Anscombe and G. H. von Wright (Oxford: Blackwell, 1969).

42 I present further arguments to this effect in Norris, 'The Limits of *Whose* Language?', previous note 36, and 'Kripkenstein's Monsters: Anti-realism, Scepticism, and the Rule-following Debate', in *On Truth and Meaning: Language, Logic and the Grounds of Belief* (London: Continuum, 2006), pp. 155–202.

43 See previous note 39.

44 See Norris, *Truth Matters*, previous note 17.

45 See previous note 16.

46 These issues are posed most sharply with respect to the debate between realists (or Platonists) and anti-realists (or intuitionists) in philosophy of mathematics. See especially Paul Benacerraf, 'What Numbers Could Not Be', in Benacerraf and Hilary Putnam (eds), *The Philosophy of Mathematics: Selected Essays*, 2nd edn. (Cambridge: Cambridge University Press, 1983), pp. 272–94; also W. D. Hart (ed.), *The Philosophy of Mathematics* (Oxford: Oxford University Press, 1996); Philip Kitcher, *The Nature of Mathematical Knowledge* (Oxford: Oxford University Press, 1983); Hilary Putnam, *Mathematics, Matter and Method* (Cambridge: Cambridge University Press, 1975).

47 Philip Pettit, *The Common Mind: An Essay on Psychology, Society, and Politics* (Oxford: Oxford University Press, 1992).

48 Wittgenstein, *Philosophical Investigations*, previous note 2, Section 265.

49 See previous note 34.

50 See previous note 35.

51 For a classic and widely influential presentation of this aspect-relativist case, see Norwood Russell Hanson, *Patterns of Discovery: An Enquiry into the Perceptual Foundations of Science* (Cambridge: Cambridge University Press, 1958); also – for a detailed critique of Hanson's and related lines of thought – Norris, *Philosophy of Language and the Challenge to Scientific Realism* (London: Routledge, 2004).

52 See especially Plato, *Ion*, in *Early Socratic Dialogues*, ed. Trevor J. Saunders (Harmondsworth: Penguin, 1987) and *The Republic*, trans. Desmond Lee (Penguin, 1974), Books II, III and X.

53 Norris, *Platonism, Music and the Listener's Share* (London: Continuum, 2006).

54 William Empson, *Seven Types of Ambiguity*, 2nd edn. (London: Chatto & Windus, 1953 [1930]).

55 Kathleen V. Wilkes, 'Is Consciousness Important?', *British Journal for the Philosophy of Science*, Vol. 34 (1984), 223–43; p. 229.

56 For further discussion of this presently widespread way of thinking, see Norris, *Minding the Gap: Epistemology and Philosophy of Science in the Two Traditions* (Amherst, MA: University of Massachusetts Press, 1997).

57 See Gilbert Ryle, *The Concept of Mind*, previous note 23 and *On Thinking* (Oxford: Blackwell, 1979); also Ira Altman, *The Concept of Intelligence: A Philosophical Analysis* (Lanham, MD: University Press of America, 1997).

58 Wilkes, 'Is Consciousness Important?' previous note 55, p. 238.

59 Daniel C. Dennett, *Consciousness Explained* (Harmondsworth: Penguin, 1991); also *Kinds of Minds: Toward an Understanding of Consciousness* (London: Weidenfeld & Nicolson, 1996) and *Brainstorms: Philosophical Essays on Mind and Psychology* (Penguin, 1997).

60 Edmund Husserl, *Cartesian Meditations: An Introduction to Phenomenology*, trans. Dorion Cairns (Dordrecht: Kluwer, 1995).

61 William James, *The Principles of Psychology* (Cambridge, MA: Harvard University Press, [1890]1981).

62 Wilkes, 'Is Consciousness Important?', previous note 55, p. 241.

63 See especially Paul M. Churchland, *Scientific Realism and the Plasticity of Mind* (Cambridge: Cambridge University Press, 1979) and *Matter and Consciousness* (Cambridge, MA: MIT Press, 1984).

64 Wilkes, 'Is Consciousness Important?' previous note 55, p. 241.

65 See notes to Chapter 4 for relevant sources.

66 Timothy Williamson, *The Philosophy of Philosophy* (Oxford: Blackwell, 2009).

67 Ibid. p. 21.

68 On the 'object' side of this metaphysically-based naturalizing argument, see especially Saul Kripke, *Naming and Necessity* (Oxford: Blackwell, 1980); Hilary Putnam, *Mind, Language and* Reality (Cambridge: Cambridge University Press, 1975); Stephen P. Schwartz (ed.), *Naming, Necessity and Natural Kinds* (Ithaca, NY: Cornell University Press, 1977).

Chapter 4

1 See for instance, from various philosophic and political perspectives, Etienne Balibar, *Spinoza and Politics* (London: Verso, 1998); Lewis Samuel Feuer, *Spinoza and the Rise of Liberalism* (Beacon Press: Boston, 1958); Jonathan Israel, *Radical Enlightenment: Philosophy and the Making of Modernity, 1650–1750* (Oxford: Oxford University Press, 2002); Christopher Norris, *Spinoza and the Origins of Modern Critical Theory* (Oxford: Blackwell, 1991); Gideon Segal and Yirmiyahu Yovel (eds), *Spinoza* (Aldershot: Ashgate, 2002); Paul Wienpahl, *The Radical Spinoza* (New York: New York University Press, 1979); Yovel, *Spinoza and Other Heretics*, Vol. One: *The Marrano of Reason,* and Vol. Two: *The Adventures of Immanence* (Princeton, NJ: Princeton University Press, 1989).

2 See especially Stephen Nadler, *Spinoza's Heresy: Immortality and the Jewish Mind* (Oxford: Oxford University Press, 2002); Samuel J. Preuss, *Spinoza and the Irrelevance of Biblical Authority* (Cambridge: Cambridge University Press, 2001); Steven B. Smith, *Spinoza, Liberalism and the Question of Jewish Identity* (New Haven, CT: Yale University Press, 1997); Theo Verbeek, *Spinoza's Theologico-Political Treatise: Exploring 'the Will of God'* (London: Ashgate, 2003); Yovel, *Spinoza and Other Heretics*, previous note 1.

3 See various entries under previous notes 1 and 2; also Frederick C. Beiser, *The Fate of Reason: German Philosophy from Kant to Fichte* (Cambridge, MA: Harvard University Press, 1978); Moira Gatens and Genevieve Lloyd, *Collective Imaginings: Spinoza Past and Present* (London: Routledge, 1999); Leszek Kolakowski, 'The Two Eyes of Spinoza', in Marjorie Grene (ed.), *Spinoza: A Collection of Critical Essays* (Garden City, NY: Anchor Books, 1973), pp. 279–94.

[4] Louis Althusser, *For Marx*, trans. Ben Brewster (London: New Left Books, 1969) and 'Elements of Self-Criticism', in *Essays in Self-Criticism* (New Left Books, 1976), pp. 101–61; Althusser and Etienne Balibar, *Reading Capital*, trans. Brewster (New Left Books, 1970); Balibar, *Spinoza and Politics*, see previous note 1 ; Pierre Macherey, *Hegel ou Spinoza?* (Paris: Maspero, 1979) and *In a Materialist Way: Selected Essays*, ed. Warren Montag, trans. Ted Stolze (London: Verso, 1998).

[5] See especially Gilles Deleuze, *Spinoza: Practical Philosophy*, trans. Robert Hurley (San Francisco: City Lights Books, 1988) and *Expressionism in Philosophy: Spinoza*, trans. Martin Joughin (New York: Zone Books, 1992); also the various references to Spinoza in Deleuze and Félix Guattari, *A Thousand Plateaus* (*Capitalism and Schizophrenia*, Vol. Two), trans. Brian Massumi (Minneapolis: University of Minnesota Press, 1987).

[6] Spinoza, *Theologico-Political Treatise*, trans. Samuel Shirley (Indianapolis: Hackett Publishing, 2001). See entries under previous note 2; also Israel, *Radical Enlightenment*, previous note 1; Robert J. McShea, *The Political Philosophy of Spinoza* (New York: Columbia University Press, 1968); Richard H. Popkin, *The History of Scepticism from Erasmus to Spinoza* (Berkeley and Los Angeles: University of California Press, 1979); Leo Strauss, *Spinoza's Critique of Religion*, trans. E. M. Sinclair (New York: Schocken Books, 1965); Silvain Zac, *Spinoza et l'interpretation de l'écriture* (Paris: Presses Universitaires de France, 1965) and *Philosophie, Théologie, Politique dans l'oeuvre de Spinoza* (Paris: Vrin, 1979).

[7] Norris, *Spinoza and the Origins of Modern Critical Theory*, previous note 1.

[8] Donald Davidson, 'Spinoza's Causal Theory of the Affects', in *Truth, Language, and History* (Cambridge: Cambridge University Press, 2005), pp. 295–313; Alan Donagan, *Spinoza* (Chicago, IL: University of Chicago Press, 1988); Grene (ed.), *Spinoza: A Collection of Critical Essays*, see previous note 3; Stuart Hampshire, *Spinoza* (Harmondsworth: Penguin, 1951); G. H. R. Parkinson, *Spinoza's Theory of Knowledge* (Oxford: Clarendon Press, 1953). Two anthologies that offer a useful conspectus of past and present scholarship, commentary and criticism are Genevieve Lloyd (ed.), *Spinoza* (4 vols., London: Routledge, 2001) and Don Garrett (ed.), *The Cambridge Companion to Spinoza* (Cambridge: Cambridge University Press, 1996).

[9] See previous notes 4 and 5.

[10] See previous notes 4 and 8.

[11] For further discussion see Norris, *Spinoza and the Origins of Modern Critical Theory*, see previous note 1.

[12] Spinoza, *Ethics*, in *The Collected Writings of Spinoza*, trans. Edwin Curley (Princeton, NJ: Princeton University Press, 1985).

[13] Jonathan Bennett, *A Study of Spinoza's* Ethics, (Cambridge: Cambridge University Press, 1984), p. 373.

[14] Ibid. p. 374.

[15] Jacques Derrida, '*Différance*', in *Margins of Philosophy*, trans. Alan Bass (Chicago, IL: University of Chicago Press, 1982), pp. 3–27.

[16] See entries under previous note 8.

[17] Antonio Damasio, *Looking for Spinoza: Joy, Sorrow and the Feeling Brain* (London: Heinemann, 2003). See also Jerome Neu, *Emotion, Thought and Therapy: A Study of Hume and Spinoza, and the Relationship of Philosophical Theories of the Emotions to Psychological Theories of Therapy* (London: Routledge & Kegan Paul, 1978).

[18] Immanuel Kant, *Critique of Pure Reason*, trans. N. Kemp Smith (London: Macmillan, 1964).

[19] For further argument to this effect, see Christopher Norris, *Minding the Gap: Epistemology and Philosophy of Science in the Two Traditions* (Amherst, MA: University of Massachusetts Press, 2000); *Truth Matters: Realism, Anti-realism and Response-dependence* (Edinburgh: Edinburgh University Press, 2002); *Philosophy of Language and the Challenge to Scientific Realism* (London: Routledge, 2004); *On Truth and Meaning: Language, Logic and the Grounds of Belief* (London: Continuum, 2006).

[20] See previous notes 4 and 5.

[21] See previous note 6.

[22] See for instance Edwin M. Curley, *Behind the Geometrical Method: A Reading of Spinoza's* Ethics (Princeton, NJ: Princeton University Press, 1988).

[23] See previous note 6.

[24] Antonio Negri, *The Savage Anomaly: The Power of Spinoza's Metaphysics and Politics,* trans. Michael Hardt (Minneapolis: University of Minnesota Press, 1991).

[25] See especially Hans Reichenbach, *The Rise of Scientific Philosophy* (Chicago, IL: University of Chicago Press, 1938).

[26] Bennett, *A Study of Spinoza's* Ethics, previous note 13.

[27] Donagan, *Spinoza,* see previous note 8.

[28] See especially Israel, *Radical Enlightenment,* previous note 1; also Margaret Gullan-Whur, *Within Reason: A Life of Spinoza* (London: Jonathan Cape, 1998) and Steven Nadler, *Spinoza: A Life* (Cambridge: Cambridge University Press, 1999).

[29] See especially Balibar, *Spinoza and Politics* and Negri, *The Savage Anomaly* (previous notes 1 and 24).

[30] See previous notes 1, 2, 3 and 6.

[31] Donagan, *Spinoza,* previous note 8, pp. 10–11.

[32] See previous note 4.

[33] Althusser and Balibar, *Reading Capital,* previous note 4, pp. 188–9.

[34] Ibid. p. 16.

[35] Deleuze, *Spinoza: Practical Philosophy,* previous note 5, p. 129.

[36] Deleuze and Guattari, *A Thousand Plateaus,* see previous note 5 and *The Anti-Oedipus* (*Capitalism and Schizophrenia,* Vol. One, trans. Robert Hurley, Mark Seem, and Helen R. Lane (Minneapolis: University of Minnesota Press, 1977).

[37] See previous notes 1, 3, 5 and 22.

[38] Derrida, *The Ear of the Other: Texts and Discussions,* trans and ed. Christie V. McDonald, Claude Lévesque and Peggy Kamuf (New York: Schocken Books, 1985).

[39] Derrida, *Of Grammatology,* trans. Gayatri C. Spivak (Baltimore: Johns Hopkins University Press, 1976), p. 158.

[40] Ibid. p. 158.

[41] Bennett, *A Study of Spinoza's* Ethics, see previous note 13.

[42] Davidson, 'Spinoza's Causal Theory of the Affects' (see previous note 8).

[43] For further elucidation of these arguments, see Davidson, *Essays on Actions and Events* (Oxford: Clarendon Press, 1980).

[44] Spinoza, *Ethics,* previous note 12, Book III, Proposition 1.

[45] Spinoza, *Ethics,* previous note 12, Book II, Prop. 2, Note.

[46] Davidson, 'Spinoza's Causal Theory of the Affects', see previous note 8, p. 310.

[47] Ibid.

[48] Ibid. p. 312.

[49] Althusser and Balibar, *Reading Capital,* previous note 4, p. 102.

50 See especially Tim Crane and Sarah Patterson (eds), *A History of the Mind-Body Problem* (London: Routledge, 2000); also D. M. Armstrong and Norman Malcolm, *Consciousness and Causality: A Debate on the Nature of Mind* (Oxford: Blackwell, 1984); Cynthia Macdonald, *Mind-Body Identity Theories* (London: Routledge, 1989).

51 For further discussion see entries under previous note 50; also Margaret Donaldson, *Human Minds: An Exploration* (Harmondsworth: Penguin, 1992); Stephen Priest, *Theories of the Mind* (Harmondsworth: Penguin, 1991); David M. Rosenthal (ed.), *The Nature of Mind* (New York: Oxford University Press, 1991).

52 See especially Ludwig Wittgenstein, *Philosophical Investigations*, trans. G. E. M. Anscombe (Oxford: Blackwell, 1959); also Richard Rorty (ed.), *The Linguistic Turn: Recent Essays in Philosophical Method* (Chicago, IL: University of Chicago Press, 1967) and – from a dissenting standpoint – C. W. K. Mundle, *A Critique of Linguistic Philosophy* (Oxford: Clarendon Press, 1970) and Norris, *Philosophy of Language and the Challenge to Scientific Realism* (London: Routledge, 2004).

53 Spinoza, *On the Improvement of the Understanding*, in R. H. M. Elwes (ed. and trans.), *The Chief Works of Benedict de Spinoza* Vol. Two, (New York: Dover, 1951), pp. 3–41.

54 For some highly relevant commentary, see Ian Hacking, *Why Does Language Matter to Philosophy?* (Cambridge: Cambridge University Press, 1975); also Norris, *Spinoza and the Origins of Modern Critical Theory*, previous note 1, pp. 103–42; G. H. R. Parkinson, 'Language and Knowledge in Spinoza', in Grene (ed.), *Spinoza: A Collection of Critical Essays*, previous note 3, pp. 73–100; David Savan, 'Spinoza and Language', in S. Paul Kashap (ed.), *Studies in Spinoza: Critical and Interpretative Essays* (Berkeley and Los Angeles: University of California Press, 1972), pp. 236–48.

55 Spinoza, *On the Improvement of the Understanding*, previous note 53, p. 33.

56 Savan, 'Spinoza and Language', previous note 54, p. 239.

57 See the essays collected in Rorty, *The Linguistic Turn*, previous note 52, for a representative sampling of work in both these traditions, i.e., the Frege-Russell or *echt*-analytic and the Wittgenstein-Austin or 'ordinary-language' lines of descent.

58 See previous note 43.

59 See previous notes 1, 2 and 6.

60 For the most recent and controversial statement of this position, see David J. Chalmers, *The Conscious Mind: In Search of a Fundamental Theory* (Oxford: Oxford University Press, 1996).

61 See previous notes 6, 24 and 28.

62 See especially Gaston Bachelard, *The Philosophy of No*, trans. G. C. Waterston (New York: Orion Press, 1969) and *The New Scientific Spirit*, trans. Arthur Goldhammer (Boston: Beacon Press, 1984); also Mary Tiles, *Bachelard: Science and Objectivity* (Cambridge: Cambridge University Press, 1984).

63 See previous note 4; also Althusser, 'Ideology and Ideological State Apparatuses', in *'Lenin and Philosophy' and Other Essays*, trans. Ben Brewster (London: New Left Books, 1977), pp. 121–73.

64 See especially Ted Benton, *The Rise and Fall of Structural Marxism* (London: New Left Books, 1984) and Gregory Elliott, *Althusser: The Detour of Theory* (London: Verso, 1987).

65 Israel, *Radical Enlightenment*, previous note 1.

[66] See Feuer, *Spinoza and the Rise of Liberalism*, and other entries under previous notes 1, 2, 6 and 24.
[67] Damasio, *Looking for Spinoza*, previous note 17.
[68] Bennett, *A Study of Spinoza's* Ethics, previous note 13.
[69] Hampshire, *Spinoza*, see previous note 8.
[70] For classic examples of this approach, see P. F. Strawson, *Individuals: An Essay in Descriptive Metaphysics* (London: Methuen, 1959) and *The Bounds of Sense: An Essay on Kant's* 'Critique of Pure Reason' (London: Methuen, 1966).
[71] For representative contemporary surveys, see Graham Bird (ed.), *A Companion to Kant* (Oxford: Blackwell, 2006); Paul Guyer (ed.), *The Cambridge Companion to Kant* (Cambridge: Cambridge University Press, 1992).
[72] See especially Robert Hanna, *Kant and the Foundations of Analytic Philosophy* (Oxford: Oxford University Press, 2001).
[73] See Norris, *Minding the Gap: Philosophy of Science and Epistemology in the Two Traditions*, previous note 19; also J. Alberto Coffa, *The Semantic Tradition from Kant to Carnap: To the Vienna Station* (Cambridge: Cambridge University Press, 1991).
[74] John McDowell, *Mind and World* (Cambridge, MA: Harvard University Press, 1994); see also Norris, 'McDowell on Kant: Redrawing the Bounds of Sense' and 'The Limits of Naturalism: Further Thoughts on McDowell's *Mind and World*', in *Minding the Gap*, previous note 19, pp. 172–96 and 197–230.
[75] Norris, *Minding the Gap*, previous note 19.
[76] For further discussion see Coffa, *The Semantic Tradition from Kant to Carnap*.
[77] In this mode of 'rational reconstruction', see especially Bertrand Russell, *A History of Western Philosophy*, 2nd edn. (London: Unwin, 1961) and Jonathan Bennett, *Learning from Six Philosophers, Vol One: Descartes, Spinoza, Leibniz; Vol. Two: Locke, Berkeley, Hume* (Oxford: Clarendon Press, 2001).
[78] Badiou's writings are extensively referenced in my notes to Chapter 5 so I have not thought it necessary to duplicate the details here.
[79] See Norris, *Minding the Gap*, previous note 19.
[80] Timothy Williamson, *The Philosophy of Philosophy* (Oxford: Blackwell, 2007).
[81] Ibid. p. 46.
[82] Ibid. p. 285.
[83] See entries under previous note 19.
[84] Again, these works of Badiou are documented fully in my notes to Chapter 5 – along with various sources and references for the reader unacquainted with these issues in the philosophy of mathematics – so I shall not annotate them here.

Chapter 5

[1] See especially Alain Badiou, *Being and Event*, trans. Oliver Feltham (London: Continuum, 2005) and *Theoretical Writings*, ed. and trans. Ray Brassier and Alberto Toscano (London: Continuum, 2004).
[2] For a representative sampling, see Richard Rorty (ed.), *The Linguistic Turn: Recent Essays in Philosophical Method* (Chicago, IL: University of Chicago Press, 1967).
[3] See especially Michael Dummett, *Truth and Other Enigmas* (London: Duckworth, 1978); also Crispin Wright, *Truth and Objectivity* (Cambridge, MA: Harvard University

Press, 1992) and Christopher Norris, *Truth Matters: Realism, Anti-realism and Response-dependence* (Edinburgh: Edinburgh University Press, 2002).

4 See Ludwig Wittgenstein, *Philosophical Investigations*, trans. G. E. M. Anscombe (Oxford: Blackwell, 1951), sections 201–92 ; Saul Kripke, *Wittgenstein on Rules and Private Language: An Elementary Exposition* (Oxford: Blackwell, 1982); Alexander Miller and Crispin Wright (eds), *Rule-Following and Meaning* (Chesham: Acumen, 2002).

5 See especially Badiou, *Manifesto for Philosophy*, trans. Norman Madarasz (Albany, NY: State University of New York Press, 1999) and *Infinite Thought: Truth and the Return to Philosophy*, trans. Oliver Feltham and Justin Clemens (London: Continuum, 2003).

6 See for instance Martin Heidegger, *Poetry, Language, Thought*, trans. Albert Hofstadter (New York: Harper & Row, 1972) and *Early Greek Thinking*, trans. David F. Krell and Frank Capuzzi (New York: Harper & Row, 1975).

7 For some interesting commentary in this regard see David Macey, *Lacan in Contexts* (London: Verso, 1988); also Robert Samuels, *Between Philosophy and Psychoanalysis: Lacan's Reconstruction of Freud* (New York: Routledge, 1993).

8 See Abraham A. Fraenkel, *Elements of Set Theory*, rev. edn. (Amsterdam: North-Holland, 1973).

9 Paul J. Cohen, *Set Theory and the Continuum Hypothesis* (New York: W. Benjamin, 1966).

10 See especially Paul Benacerraf, 'What Numbers Could Not Be', in Benacerraf and Hilary Putnam (eds), *The Philosophy of Mathematics: Selected Essays*, 2nd edn. (Cambridge: Cambridge University Press, 1983), pp. 272–94; W. D. Hart (ed.), *The Philosophy of Mathematics* (Oxford: Oxford University Press, 1996); Hilary Putnam, *Mathematics, Matter and Method* (Cambridge: Cambridge University Press, 1975).

11 See for instance Neil Tennant, *The Taming of the True* (Oxford: Clarendon Press, 2002); also Norris, *Philosophy of Language and the Challenge to Scientific Realism* (London: Routledge, 2004).

12 E.P. Wigner, 'The Unreasonable Effectiveness of Mathematics in the Physical Sciences', in *Symmetries and Reflections* (Bloomington, IN: Indiana University Press, 1967), pp. 228–38.

13 For further discussion from a range of viewpoints, see J. Aronson, R. Harré and E. Way, *Realism Rescued: How Scientific Progress is Possible* (London: Duckworth, 1994); Michael Devitt, *Realism and Truth*, 2nd edn. (Oxford: Blackwell, 1986); Jarrett Leplin (ed.), *Scientific Realism* (California: University California Press, 1984); Peter Lipton, *Inference to the Best Explanation* (London: Routledge, 1993); Norris, *Resources of Realism: Prospects for 'Post-analytic' Philosophy* (London: Macmillan, 1997); Stathis Psillos, *Scientific Realism: How Science Tracks Truth* (London: Routledge, 1999).

14 Dummett, *Truth and Other Enigmas*, previous note 3.

15 Gottlob Frege, *The Foundations of Arithmetic: A Logico-mathematical Enquiry into the Concept of Number*, trans. J. L. Austin (Oxford: Blackwell, 1950).

16 Philip Kitcher, *The Nature of Mathematical Knowledge* (London: Oxford University Press, 1984).

17 See Mark Johnston, 'How to Speak of the Colours', *Philosophical Studies*, Vol. 68 (1992), 221–63, and 'Objectivity Refigured', in J. Haldane and C. Wright (eds),

Realism, Representation and Projection (Oxford: Oxford University Press, 1993), pp. 85–130; Philip Pettit, 'Realism and Response Dependence', *Mind*, Vol. 100 (1991), 597–626; Crispin Wright, 'Moral Values, Projection, and Secondary Qualities', *Proceedings of the Aristotelian Society*, Supplementary Vol. 62 (1988), 1–26.

18 Gilles Deleuze and Félix Guattari, *What Is Philosophy?*, trans. Hugh Tomlinson and Graham Burchill (London: Verso, 1991).

19 See for instance A. J. Ayer, 'Novelist-Philosophers V: Jean-Paul Sartre', *Horizon*, Vol. 12 (July 1945), 12–26 and Vol. 12 (August 1945), 101–10.

20 See John R. Searle, 'Reiterating the Differences: A Reply to Derrida', *Glyph*, Vol. 1 (Baltimore: Johns Hopkins University Press, 1977), pp. 198–208; also Derrida, 'Afterword: Toward an Ethic of Conversation', in Gerald Graff (ed.), *Limited Inc* (Evanston, IL: Northwestern University Press, 1989), pp. 111–54.

21 For an informative general account, see Michael Potter, *Set Theory and its Philosophy: A Critical Introduction* (Oxford: Oxford University Press, 2004).

22 Kurt Gödel, *On Formally Undecidable Propositions of* Principia Mathematica *and Related Systems*, trans. B. Meltzer (New York: Basic Books, 1962).

23 This marked shift of emphasis in the reception-history of Badiou's work – from a predominantly cultural-theoretical to a more philosophically oriented mode of engagement – can be seen very clearly in the relevant sections of Paul Ashton's very useful 'Bibliography of Work On and By Alain Badiou in English', *Cosmos and History: The Journal of Natural and Social Philosophy*, Vol. 2, Nos. 1 and 2 (2006), 313–26, available online at http://www.cosmosandhistory.org/index.php/journal/article/viewFile/124/73 (accessed on 22 September 2009).

24 These readings are contested and Badiou's thinking given its philosophic due in more recent studies such as Peter Hallward, *Badiou: A Subject to Truth* (Minneapolis: University of Minnesota Press, 2003) and Hallward (ed.), *Think Again: Alain Badiou and the Future of Philosophy* (London: Continuum, 2004). For further remarks on the reception-history to broadly similar effect, see Justin Clemens, 'Doubles of Nothing: The Problem of Binding Truth to Being in the Work of Alain Badiou', *Filozofski vestnik*, Vol. 16, No. 2 (2005), 97–112. The best short guide is Jason Barker, *Alain Badiou: A Critical Introduction* (London: Pluto, 2002).

25 For more on the relationship between these two lines of thought, see Christopher Norris, *Fiction, Philosophy and Literary Theory: Will the Real Saul Kripke Please Stand Up?* (London: Continuum, 2007).

26 See especially Roger Penrose, *The Emperor's New Mind: Concerning Computers, Minds, and the Laws of Physics* (London: Oxford University Press, 1999) and Jerrold J. Katz, *Realistic Rationalism* (Cambridge, MA: M.I.T. Press, 1998).

27 Kurt Gődel, 'What Is Cantor's Continuum Problem?', in Benacerraf and Putnam (eds), *Philosophy of Mathematics*, previous note 10, pp. 470–85.

28 Immanuel Kant, *Critique of Pure Reason*, trans. N. Kemp Smith (London: Macmillan, 1964).

29 Norris, *Truth Matters: Realism, Anti-realism and Response-dependence*, previous note 3.

30 Wittgenstein, *Philosophical Investigations*, previous note 4.

31 See Gottlob Frege, 'On Sense and Reference', in Max Black and P. T. Geach (eds), *Translations from the Philosophical Writings of Gottlob Frege* (Oxford: Blackwell,

1952), pp. 56–78 and Bertrand Russell, 'On Denoting', *Mind*, Vol. 14 (1905), 479–93.

32 Cora Diamond, *The Realistic Spirit: Wittgenstein, Philosophy and the Mind* (Cambridge, MA: M.I.T. Press, 1991).

33 Dummett, *Truth and Other Enigmas*, previous note 3; also *The Logical Basis of Metaphysics* (London: Duckworth, 1991) and *The Seas of Language* (Oxford: Clarendon Press, 1993).

34 Badiou, *Being and Event*, trans. Oliver Feltham (London: Continuum, 2005), p. 342.

35 Ibid. p. 342.

36 Ibid. p. 88.

37 Ibid. p. 87.

38 Ibid. p. 337.

39 Ibid. p. 337.

40 Roland Barthes, *Mythologies*, trans. Annette Lavers (London: Jonathan Cape, 1972), 112.

41 Badiou, *Being and Event*, previous note 34, p. 340.

42 Ibid. p. 429.

43 See especially L. E. J. Brouwer, *Brouwer's Cambridge Lectures on Intuitionism*, ed. D. van Dalen (Cambridge: Cambridge University Press, 1981).

44 For a representative sampling see Michael Lane (ed.), *Structuralism: A Reader* (London: Cape, 1970).

45 Jean-Paul Sartre, *Critique of Dialectical Reason*, Vol. One, *Theory of Practical Ensembles*, trans. Alan Sheridan-Smith (London: New Left Books, 1976) and Vol. Two, *The Intelligibility of History*, trans Quintin Hoare (London: Verso, 1994).

46 Karl Marx, *The Eighteenth Brumaire of Louis Napoleon* (London: Lawrence & Wishart, 1954), p. 12.

47 Badiou, *Being and Event*, previous note 34, p. 406.

48 See especially Hans Reichenbach, *Science and Prediction* (Chicago, IL: University of Chicago Press, 1953).

49 Badiou, *Being and Event*, previous note 34, p. 403.

50 Ibid. p. 303.

51 Ibid. pp. 403 and 406.

52 See for instance – among the most resourceful and ingenious defences – Joseph Margolis, *The Truth About Relativism* (Oxford: Blackwell, 1991) and – with its own very definite range of axes to grind – Barbara Herrnstein Smith, *Belief and Resistance: The Dynamics of Contemporary Intellectual Controversy* (Cambridge, MA: Harvard University Press, 1997).

53 On this distinction, see also Bernard Williams, *Truth and Truthfulness: An Essay in Genealogy* (Princeton, NJ: Princeton University Press, 2002).

54 See Otto Neurath, Rudolf Carnap and Charles Morris (eds), *Foundations of the Unity of Science: Towards an International Encyclopedia of Unified Science* (Chicago, IL: University of Chicago Press, 1955–70); also Carnap, *The Unity of Science*, ed. Max Black (Bristol: Thoemmes Press, 1995) and Robert L. Causley, *Unity of Science* (Dordrecht: D. Reidel, 1977).

55 W. V. Quine, 'Two Dogmas of Empiricism', in *From a Logical Point of View*, 2nd edn. (Cambridge, MA: Harvard University Press, 1961), pp. 20–46.

[56] See for instance John McDowell, *Mind and World* (Cambridge, MA: Harvard University Press, 1994); also Christopher Norris, 'McDowell on Kant: Redrawing the Bounds of Sense' and 'The Limits of Naturalism: Further Thoughts on McDowell's *Mind and World*', in *Minding the Gap: Epistemology and Philosophy of Science in the Two Traditions* (Amherst, MA: University of Massachusetts Press, 2000), pp. 172–96 and 197–230.

[57] See especially Badiou, *Theoretical Writings*, trans. Ray Brassier (London: Continuum, 2004); also – on this distinctively 'French' strain of critical-rationalist thought – Norris, *Minding the Gap*, previous note 56.

[58] See Norris, *Resources of Realism*, previous note 13; also *New Idols of the Cave: On the Limits of Anti-realism* (Manchester: Manchester University Press, 1997) and *Hilary Putnam: Realism, Reason and the Uses of Uncertainty* (Manchester: Manchester University Press , 2002).

[59] See especially Frege, 'The Thought: A Logical Inquiry', in Robert M. Harnish (ed.), *Basic Topics in the Philosophy of Language* (Hemel Hempstead: Harvester-Wheatsheaf, 1994), pp. 517–35.

[60] Badiou, *Metapolitics*, trans. Jason Barker (London: Verso, 2005); *Polemics*, trans. Steve Corcoran (London: Verso, 2006); *Century*, trans. Alberto Toscano (Cambridge: Polity Press, 2007).

[61] See especially Badiou, *Ethics: An Essay on the Understanding of Evil*, trans. Peter Hallward (London: Verso, 2001).

[62] Kitcher, *The Nature of Mathematical Knowledge* (see previous note 16).

[63] Badiou, *Theoretical Writings*, previous note 1; Plato, *Meno*, ed. E. Seymer Thompson (London: Macmillan, 1901); also Wright, *Truth and Objectivity* see previous note 3.

[64] Wittgenstein, *Philosophical Investigations*, previous note 4; also Peter Winch, *The Idea of a Social Science and Its Relation to Philosophy* (London: Routledge & Kegan Paul, 1958) and *Trying to Make Sense* (Oxford: Blackwell, 1987).

[65] See especially Badiou, *Logics of Worlds*, trans. Alberto Toscano (London: Continuum, 2008).

[66] For further discussion see Norris, *On Truth and Meaning: Language, Logic and the Grounds of Belief* (London: Continuum, 2006).

[67] See Wigner, 'The Unreasonable Effectiveness of Mathematics in the Physical Sciences' (see previous note 12).

[68] See especially Richard Boyd, 'The Current Status of Scientific Realism', in Leplin (ed.), *Scientific Realism*, previous note 13, pp. 41–82 and Hilary Putnam, *Mind. Language and Reality* (Cambridge: Cambridge University Press, 1975); also Gilbert Harman, 'Inference to the Best Explanation', *Philosophical Review*, Vol. 74 (1965), pp. 88–95; Peter Lipton, *Inference to the Best Explanation* (see previous note 13). For a sceptical view of such arguments, see Larry Laudan, 'A Confutation of Convergent Realism', *Philosophy of Science*, Vol. 48 (1981), 19–49 and – in response – J. L. Aronson, 'Testing for Convergent Realism', *British Journal for the Philosophy of Science*, Vol. 40 (1989), 255–60.

[69] For a recent stage in his long series of attempts to overcome this and other vexatious dualisms, see Hilary Putnam, *The Threefold Cord: Mind, Body, and World* (New York: Columbia University Press, 1999); also Norris, *Hilary Putnam: Realism, Reason and the Uses of Uncertainty* (Manchester: Manchester University Press, 2002).

[70] See previous note 56.

[71] For further discussion from a range of viewpoints, see Ansgar Beckerman, Hans Flohr and Jaegwon Kim (eds.), *Emergence or Reduction?: Essays on the Prospects of Nonreductive Physicalism* (New York & Berlin: de Gruyter, 1992); David Charles and Kathleen Lennon (eds.), *Reduction, Explanation and Realism* (Oxford: Clarendon Press, 1992); Steven W. Horst, *Beyond Reduction: Philosophy of Mind and Post-reductionist Philosophy of Science* (Oxford: Oxford University Press, 2008).

[72] See Norris, *What's Wrong with Postmodernism* (Hemel-Hempstead: Harvester-Wheatsheaf, 1970) and *The Truth about Postmodernism* (Oxford: Blackwell, 1993).

[73] For a wide range of views on this development in recent (loosely so-called) analytic philosophy, see see previous notes 3, 11, 33 and 58.

Chapter 6

[1] See following notes 7, 9 and 13.

[2] Christopher Norris, *Minding the Gap: Epistemology and Philosophy of Science in the Two Traditions* (Amherst, MA: University of Massachusetts Press, 2000),

[3] For some highly illuminating discussion, see Timothy Williamson, *The Philosophy of Philosophy* (Oxford: Blackwell, 2007).

[4] See following notes 11, 12 and 52.

[5] See for instance Richmond Campbell and Bruce Hunter (eds), *Moral Epistemology Naturalized* (Calgary: University of Calgary Press, 2000); Jonathan Knowles, *Norms, Naturalism and Epistemology* (Basingstoke: Palgrave-Macmillan, 2003); Hilary Kornblith, *Knowledge and Its Place in Nature* (Oxford: Clarendon Press, 2002); David Papineau, *Philosophical Naturalism* (Oxford: Blackwell, 1993); Ralph Wedgwood, *The Nature of Normativity* (Oxford: Clarendon Press, 2007).

[6] Cf. Stephen P. Stich, *From Folk Psychology to Cognitive Science: The Case Against Belief* (Cambridge, MA: MIT Press, 1983) and *Deconstructing the Mind* (Oxford: Oxford University Press, 1999).

[7] See following notes 47 and 52.

[8] See following notes 60 and 61.

[9] For a classic instance, see Theodor W. Adorno and Max Horkheimer, *Dialectic of Enlightenment*, trans. John Cumming (New York: Seabury Press, 1972); also Adorno, *Negative Dialectics*, trans. E. B. Ashton (London: Routledge & Kegan Paul, 1973).

[10] For further discussion of these and related developments see Christopher Norris, *What's Wrong with Postmodernism?* (Hemel Hempstead: Harvester-Wheatsheaf, 1990) and *The Truth about Postmodernism* (Oxford: Blackwell, 1994).

[11] This distinction is well made by Joseph Margolis in *The Truth about Relativism* (Oxford: Blackwell, 1991).

[12] I discuss these developments at greater length in Norris, *Deconstruction and the Unfinished Project of Modernity* (London: Athlone, 2000).

[13] See for instance Simon Blackburn, *Truth: A Guide* (Oxford: Oxford University Press, 2005) and Paul Boghossian, *Fear of Knowledge: Against Relativism and Constructivism* (Oxford: Oxford University Press, 2006).

[14] Ludwig Wittgenstein, *Philosophical Investigations*, trans. G. E. M. Anscombe (Oxford: Blackwell, 1958); Richard Rorty, *Consequences of Pragmatism* (Brighton: Harvester, 1982).

[15] For a representative sampling, see Rorty (ed.), *The Linguistic Turn: Essays in Philosophical Method* (Chicago, IL: University of Chicago Press, 1967).

[16] I take a critical view of these ideas in Norris, *Resources of Realism: Prospects for 'Post-analytic' Philosophy* (London: Macmillan, 1997) and *New Idols of the Cave: On the Limits of Anti-realism* (Manchester: Manchester University Press, 1997).

[17] See previous note 14; also Rorty, *Objectivity, Relativism, and Truth* (Cambridge: Cambridge University Press, 1991) and *Truth and Progress* (Cambridge University Press, 1998).

[18] For further discussion see Norris, *Truth Matters: Realism, Anti-realism and Response-dependence* (Edinburgh: Edinburgh University Press, 2005).

[19] Wittgenstein, *Philosophical Investigations*; also Donald Davidson, 'On the Very Idea of a Conceptual Scheme', in *Inquiries into Truth and Interpretation* (Oxford: Oxford University Press, 1984), pp. 183–98.

[20] See especially W. V. Quine, 'Two Dogmas of Empiricism', in *From a Logical Point of View*, 2nd edn. (Cambridge, MA: Harvard University Press, 1961), pp. 20–46.

[21] Robert C. Brandom, *Making It Explicit: Reasoning, Representing, and Discursive Commitment* (Cambridge, MA: Harvard University Press, 1994); also Brandom, *Articulating Reasons: An Introduction to Inferentialism* (Cambridge, MA: Harvard University Press, 2000).

[22] See especially Jerry Fodor and Ernest LePore, 'Why Meaning (Probably) Isn't Conceptual Role', *Mind and Language*, Vol. 6, No. 4 (1991), 328–43.

[23] For further discussion of the 'Euthyphro Contrast', see Norris, *Truth Matters*, previous note 18, and Crispin Wright, *Truth and Objectivity* (Cambridge, MA: Harvard University Press, 1991).

[24] Wright, *Truth and Objectivity*, previous note 23 .

[25] See for instance Jürgen Habermas, *Theory of Communicative Action*, 2 vols., trans. Thomas McCarthy (London: Heinemann, 1984 and 1987).

[26] Michael Dummett, *Truth and Other Enigmas* (London: Duckworth, 1978) and *The Logical Basis of Metaphysics* (Cambridge, MA: Harvard University Press, 1994).

[27] Jürgen Habermas, 'From Kant to Hegel: On Robert Brandom's Pragmatic Philosophy of Language', *European Journal of Philosophy*, Vol 8 (2000), 322–55; Robert Brandom, 'Facts, Norms, and Normative Facts: A Reply to Habermas', *European Journal of Philosophy*, Vol. 8 (2000), 356–74.

[28] Habermas, *The Unfinished Project of Modernity: Twelve Lectures*, trans Frederick Lawrence (Cambridge: Polity Press, 1987).

[29] Dummett, *Truth and Other Enigmas*, previous note 26; also *Truth and the Past* (New York: Columbia University Press, 2004).

[30] Dummett, *Elements of Intuitionism* (Oxford: Oxford University Press, 1977).

[31] Nelson Goodman, *Ways of Worldmaking* (Indianapolis: Bobbs-Merrill, 1978).

[32] See Rorty (ed.), *The Linguistic Turn*, previous note 15.

[33] Dummett, *The Logical Basis of Metaphysics*, previous note 26, p. 6.

[34] Ibid. p. 7.

[35] For further reflections on the Rumsfeld pronouncement in the context of this realism/anti-realism debate, see Norris, *Epistemology: Key Concepts in Philosophy* (London: Continuum, 2006), pp. 18–24.

[36] See especially David Detmer, *Challenging Postmodernism: Philosophy and the Politics of Truth* (New York: Humanity Books, 2003); also Norris, *Uncritical Theory: Postmodernism, Intellectuals and the Gulf War* (Amherst, MA: University of Massachusetts Press, 1992).

[37] See Dummett, *Truth and the Past*, previous note 29. Among his writings on more explicitly social and political themes, see Dummett, *Voting Procedures* (Oxford: Oxford University Press, 1985); *Principles of Electoral Reform* (Oxford: Oxford University Press, 1997; *On Immigration and Refugees* (London: Routledge, 2001).

[38] Dummett, *Voting Procedures*, previous note 37, p. 6.

[39] Anat Matar, *Modernism and the Language of Philosophy* (London: Routledge, 2006).

[40] Ibid. p. 37.

[41] Dummett, *The Logical Basis of Metaphysics*, previous note 33, p. 7.

[42] See entries under previous notes 16, 18 and 35.

[43] I develop these comparisons more fully in Norris, 'Saussure, Linguistic Theory and Philosophy of Science', in Carol Sanders (ed.), *The Cambridge Companion to Saussure* (Cambridge: Cambridge University Press, 2004), pp. 219–39.

[44] On this chapter of intellectual history, see especially Frederick C. Beiser, *The Fate of Reason: German Philosophy from Kant to Fichte* (Cambridge, MA: Harvard University Press, 1987).

[45] See previous notes 21 and 27; also John McDowell, *Mind and World* (Cambridge, MA: Harvard University Press, 1994); Christopher Norris, 'McDowell on Kant: Redrawing the Bounds of Sense' and 'The Limits of Naturalism: Further Thoughts on McDowell's *Mind and World*', in *Minding the Gap*, previous note 2, pp. 172–96 and 197–230.

[46] McDowell, *Mind and World*; Immanuel Kant, *Critique of Pure Reason*, trans. Norman Kemp Smith (London: Macmillan, 1964).

[47] Jacques Derrida, 'The Supplement of Copula: Philosophy *before* Linguistics', in *Margins of Philosophy*, trans. Alan Bass (Chicago, IL: University of Chicago Press), pp. 175–205; also Emile Benveniste, *Problems in General Linguistics*, trans. Mary Meek (Coral Gables: University of Miami Press, 1971).

[48] See previous note 19.

[49] Derrida, 'Cogito and the History of Madness', in *Writing and Difference*, trans Alan Bass (London: Routledge & Kegan Paul, 1978), pp. 31–63; also Michel Foucault, *Folie et deraison: histoire de la folie à l'age classique* (Paris: Plon, 1961).

[50] Derrida, 'White Mythology: Metaphor in the Text of Philosophy', in *Margins of Philosophy*, previous note 47, pp. 207–71.

[51] See also Paul de Man, 'The Epistemology of Metaphor', *Critical Inquiry*, Vol. 5, No. 1 (Fall 1978), 13–30.

[52] John R. Searle, 'Reiterating the Differences: A Reply to Derrida', *Glyph*, Vol. 1 (1977), 198–208. For more encouraging signs of an 'analytic' readiness to engage thoughtfully with Derrida's work, see among others Simon Glendinning (ed.), *Arguing with Derrida* (Oxford: Blackwell, 2001); Reed Way Dasenbrock (ed.), *Re-Drawing the Lines: Analytic Philosophy, Deconstruction, and Literary Theory* (Minneapolis: University of Minnesota Press, 1989); Newton Garver and Seung-Chong Lee, *Derrida and Wittgenstein* (Philadelphia, PA: Temple University Press, 1994); Christopher Norris and David Roden (eds.), *Jacques Derrida*, 4 vols. (London: Sage, 2003); Norris, 'Derrida on Rousseau: deconstruction as philosophy of logic', ibid. Vol. 2, pp. 70–124; Graham Priest, 'Derrida and Self-Reference', *Australasian Journal of Philosophy*, Vol. 72 (1994), 103–111; Samuel C. Wheeler, *Deconstruction as Analytic Philosophy* (Stanford, CA.: Stanford University Press, 2000).

[53] See for instance Norris, 'Derrida on Rousseau', previous note 52.

[54] For further argument to this effect, see entries under previous notes 12 and 16.

[55] I discuss this emergent realist strain in recent (well, post-1975!) Anglo-American philosophic thought in Norris, *Philosophy of Language and the Challenge to Scientific Realism* (London: Routledge, 2004).

[56] See his slightly puzzled but broadly approving remarks in Derrida, 'As if it were Possible, "within such limits" . . .', in Michel Meyer (ed.), *Questioning Derrida* (Aldershot: Ashgate, 2001), pp. 96–119; pp. 112–13.

[57] Derrida, *Of Grammatology*, trans. G. C. Spivak (Baltimore, MD: Johns Hopkins University Press, 1976), p. 158.

[58] See C. S. Peirce, *Selected Writings of Peirce*, ed. Justus Buchler (London: Kegan Paul, Trench, Trubner, 1940).

[59] See previous notes 12, 16 and 52.

[60] See especially Karen Green, 'Brain Writing and Derrida', in Norris and Roden (eds), *Jacques Derrida*, previous note 52, Vol. Two, pp. 362–82; also F. P. Cilliers, 'The Brain, the Mental Apparatus and the Text: A Post-structural Neuropsychology', ibid. Vol. Two, pp. 383–98; Elizabeth A. Wilson, 'Projects for a Scientific Psychology: Freud, Derrida, and Connectionist Theories of Cognition', ibid. Vol. Three, pp. 309–336; Gordon Globus, 'Derrida and Connectionism: *Différance* in Neural Nets', *Philosophical Psychology*, Vol. Five (1992), 183–97.

[61] Derrida, 'Freud and the Scene of Writing', in *Writing and Difference*, previous note 49, pp. 196–231; Sigmund Freud, *Project for a Scientific Psychology, The Standard Edition*, Vol. One, pp. 295–397; also – for the text that seems to have sparked much of this currently renewed interest – Karl H. Pribram and Merton M. Gill, *Freud's 'Project' Re-Assessed* (London: Hutchinson, 1976).

[62] See especially William Bechtel and Adele Abrahamsen, *Connectionism and the Mind: An Introduction to Parallel Processing in Networks* (Oxford: Blackwell, 1991).

[63] Green, 'Brain Writing and Derrida', previous note 60, pp. 370–1.

[64] Cilliers, 'The Brain, the Mental Apparatus and the Text', previous note 60, p. 392.

[65] Derrida, *Of Grammatology*, previous note 57.

[66] See Shoshana Felman (ed.), *Psychoanalysis and Literature – the Question of Reading: Otherwise* (Baltimore: Johns Hopkins University Press, 1982).

[67] Cilliers, 'The Brain, the Mental Apparatus and the Text', previous note 60, p. 391.

[68] For further argument to this effect, see Norris, *Reclaiming Truth: Contribution to a Critique of Cultural Relativism* (London: Lawrence & Wishart, 1996).

[69] Norris, *On Truth and Meaning: Language, Logic and the Grounds of Belief* (London: Continuum, 2006).

[70] Gottlob Frege, 'On Sense and Reference', in Peter Geach and Max Black (eds), *Translations from the Philosophical Writings of Gottlob Frege* (Oxford: Blackwell, 1952), pp. 56–78 and Bertrand Russell, 'On Denoting', *Mind*, Vol. 14 (1905), 479–93; also previous notes 6 and 12.

[71] See especially Saul A. Kripke, *Naming and Necessity* (Oxford: Blackwell, 1980) and Hilary Putnam, 'Is Semantics Possible?', 'The Meaning of "Meaning"', and 'Language and Reality', in *Mind, Language and Reality* (Cambridge: Cambridge University Press, 1975), pp. 139–52, 215–71, and 272–90; also Leonard Linsky (ed.), *Reference and Modality* (Oxford: Oxford University Press, 1971); Stephen Schwartz (ed.), *Naming, Necessity, and Natural Kinds* (Ithaca, NY: Cornell University Press, 1977); David Wiggins, *Sameness and Substance* (Oxford: Blackwell, 1980).

72 See Schwartz (ed.), *Naming, Necessity, and Natural Kinds,* previous note 71; also A. W. Moore (ed.), *Meaning and Reference* (Oxford: Oxford University Press, 1993).

73 See especially Derrida, *Edmund Husserl's 'The Origin of Geometry': An Introduction,* trans. John P. Leavey (Pittsburgh, PA: Duquesne University Press, 1973) and *'Speech and Phenomena' and Other Essays on Husserl's Theory of Signs,* trans. David B. Allison (Evanston, IL: Northwestern University Press, 1973); also previous notes 36, 38 and 39.

74 See for instance Edmund Husserl, *Logical Investigations,* 2 vols., trans. J. N. Findlay (New York: Humanities Press, 1970) and *Ideas: General Introduction to Pure Phenomenology,* trans. W. R. Boyce Gibson (London: Collier Macmillan, 1975).

75 Derrida, *Of Grammatology,* previous note 65, p. 9.

76 Freud, 'A Note upon the "Mystic Writing-Pad"', *The Standard Edition,* Vol. 19, pp. 227–32.

77 Derrida, 'Freud and the Scene of Writing', previous note 61, p. 200.

78 Wilson, 'Projects for a Scientific Psychology', previous note 60, pp. 309–336.

79 Derrida, 'Freud and the Scene of Writing', previous note 61, p. 200.

80 Wilson, 'Projects for a Scientific Psychology', previous note 60, p. 310.

81 Ibid. p. 322.

82 Kathleen V. Wilkes, 'Is Consciousness Important?', *British Journal for the Philosophy of Science,* Vol. 34 (1984), 223–43; p. 229.

83 Cilliers, 'The Brain, the Mental Apparatus and the Text', previous note 60, p. 393.

84 Daniel C. Dennett, *Consciousness Explained* (Harmondsworth: Penguin, 1991); also *Kinds of Minds: Toward an Understanding of Consciousness* (London: Weidenfeld & Nicolson, 1996) and *Brainstorms: Philosophical Essays on Mind and Psychology* (Harmondsworth: Penguin, 1997).

85 Dennett, *The Intentional Stance* (Bradford, MA: MIT Press, 1987).

86 See especially Paul M. Churchland, *Scientific Realism and the Plasticity of Mind* (Cambridge: Cambridge University Press, 1979) and *Matter and Consciousness* (Cambridge, MA: MIT Press, 1984).

87 Richard Harland, *Superstructuralism: The Philosophy of Structuralism and Poststructuralism* (London: Methuen, 1987).

Index of Names